Magneto-Inductive Communication and Localization

Fundamental Limits with Arbitrary Node Arrangements

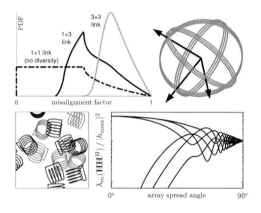

Gregor Dumphart

λογος

Series in Wireless Communications
edited by:
Prof. Dr. Armin Wittneben
Eidgenössische Technische Hochschule
Institut für Kommunikationstechnik
Sternwartstr. 7
CH-8092 Zürich

E-Mail: wittneben@nari.ee.ethz.ch
Url: http://www.nari.ee.ethz.ch/

Bibliographic information published by the Deutsche Nationalbibliothek

The Deutsche Nationalbibliothek lists this publication in the
Deutsche Nationalbibliografie; detailed bibliographic data are
available in the Internet at http://dnb.d-nb.de .

ISBN 978-3-8325-5483-5
ISSN 1611-2970

Logos Verlag Berlin GmbH
Georg-Knorr-Str. 4, Geb. 10, D-12681 Berlin
Tel.: +49 030 42 85 10 90
Fax: +49 030 42 85 10 92
INTERNET: http://www.logos-verlag.de

Diss. ETH No. 26890

Magneto-Inductive Communication and Localization: Fundamental Limits with Arbitrary Node Arrangements

A thesis submitted to attain the degree of

DOCTOR OF SCIENCES of ETH ZURICH

(Dr. sc. ETH Zurich)

presented by

GREGOR DUMPHART

Dipl.-Ing., Graz University of Technology

born on August 27, 1987

citizen of Austria

accepted on the recommendation of

Prof. Dr.-Ing. Armin Wittneben, examiner

Prof. Dr.-Ing. Robert Schober, co-examiner

2020

Day of Complete Draft: November 15th, 2019
Day of Examination: July 23rd, 2020
Last Format Adaptation: March 14th, 2022

Abstract

Wireless sensors are a key technology for many current or envisioned applications in industry and sectors such as biomedical engineering. In this context, magnetic induction has been proposed as a suitable propagation mechanism for wireless communication, power transfer and localization in applications that demand a small node size or operation in challenging media such as body tissue, fluids or soil. Magnetic induction furthermore allows for load modulation at passive tags as well as improving a link by placing passive resonant relay coils between transmitter and receiver. The existing research literature on these topics mostly addresses static links in well-defined arrangement, i.e. coaxial or coplanar coils. Likewise, most studies on passive relaying consider coil arrangements with equidistant spacing on a line or grid. These assumptions are incompatible with the reality of many sensor applications where the position and orientation of sensor nodes is determined by their movement or deployment.

This thesis addresses these shortcomings by studying the effects and opportunities in wireless magnetic induction systems with arbitrary coil positions and orientations. As prerequisite, we introduce appropriate models for near- and far-field coupling between electrically small coils. Based thereon we present a general system model for magneto-inductive networks, applicable to both power transfer and communication with an arbitrary arrangement of transmitters, receivers and passive relays. The model accounts for strong coupling, noise correlation, matching circuits, frequency selectivity, and relevant communication-theoretic nuances.

The next major part studies magnetic induction links between nodes with random coil orientations (uniform distribution in 3D). The resulting random coil coupling gives rise to a fading-type channel; the statistics are derived analytically and the communication-theoretic implications are investigated in detail. The study concerns near- and far-field propagation modes. We show that links between single-coil nodes exhibit catastrophic reliability: the asymptotic outage probability $\epsilon \propto \mathrm{SNR}^{-1/2}$ for pure near-field or pure far-field propagation, i.e. the diversity order is 1/2 (even 1/4 for load modulation). The diversity order increases to 1 in the transition between near and far field. We furthermore study the channel statistics and implications for randomly oriented coil arrays with various spatial diversity schemes.

A subsequent study of magneto-inductive passive relaying reveals that arbitrarily deployed passive relays give rise to another fading-type channel: the channel coefficient is governed by a non-coherent sum of phasors, resulting in frequency-selective fluctuations similar to multipath radio channels. We demonstrate reliable performance gains when these fluctuations are utilized with spectrally aware signaling (e.g. waterfilling) and that optimization of the relay loads offers further and significant gains.

We proceed with an investigation of the performance limits of wireless-powered medical in-body sensors in terms of their magneto-inductive data transmission capabilities, either with a transmit amplifier or load modulation, in free space or conductive medium (muscle tissue). A large coil array is thereby assumed as power source and data sink. We employ previous insights to derive design criteria and study the interplay of high node density, passive relaying, channel knowledge and transmit cooperation in detail. A particular focus is put on the minimum sensor-side coil size that allows for reliable uplink transmission.

The developed models are then used in a study of the fundamental limits of node localization based on observations of magneto-inductive channels to fixed anchor coils. In particular, we focus on the joint estimation of position and orientation of a single-coil node and derive the Cramér-Rao lower bound on the estimation error for the case of complex Gaussian observation errors. For the five-dimensional non-convex estimation problem we propose an alternating least-squares algorithm with adaptive weighting that beats the state of the art in terms of robustness and runtime. We then present a calibrated system implementation of this paradigm, operating at 500 kHz and comprising eight flat anchor coils around a 3 m × 3 m area. The agent is mounted on a positioner device to establish a reliable ground truth for calibration and evaluation; the system achieves a median position error of 3 cm. We investigate the practical performance limits and dominant error source, which are not covered by existing literature.

The thesis is complemented by a novel scheme for distance estimation between two wireless nodes based on knowledge of their wideband radio channels to one or multiple auxiliary observer nodes. By exploiting mathematical synergies with our theory of randomly oriented coils we utilize the random directions of multipath components for distance estimation in rich multipath propagation. In particular we derive closed-form distance estimation rules based on the differences of path delays of the extractable multipath components for various important cases. The scheme does not require precise clock synchronization, line of sight, or knowledge of the observer positions.

Kurzfassung

Drahtlose Sensoren gelten als Schlüsseltechnologie für viele aktuelle und künftige Anwendungen, etwa industrieller Art oder in der Medizintechnik. Magnetische Induktion gilt als geeigneter Ausbreitungsmechanismus für drahtlose Kommunikation, Energieübertragung und Positionsbestimmung in Sensoranwendungen, die nur sehr kleine Geräte erlauben oder in schwierigen Umgebungen operieren, z.b. in Gewebe, Flüssigkeiten oder unterirdisch. Magnetische Induktion ermöglicht darüber hinaus Lastmodulation an passiven Sensoren sowie Übertragungsverbesserungen durch das Platzieren von passiven resonanten Relayspulen zwischen Sender und Empfänger. Die dazugehörige Fachliteratur befasst sich hauptsächlich mit wohldefinierten Anordnungen von koaxialen oder koplanaren Spulen, welche äquidistant auf einer Linie oder Gitter platziert sind. Diese Annahmen sind jedoch unvereinbar mit der Realität vieler Sensoranwendungen, in denen Knotenpositionen und -ausrichtungen meist durch Mobilität oder Einsatzzweck bestimmt sind.

Diese Arbeit reagiert auf diese Mängel, indem die Auswirkungen und Möglichkeiten von beliebigen Spulenpositionen und -ausrichtungen in magnetisch-induktiven Übertragungssystemen untersucht werden. Vorbereitend führen wir adäquate Modelle für Nah- und Fernfeldkopplung zwischen elektrisch kleinen Spulen ein. Darauf aufbauend präsentieren wir ein allgemeines Systemmodell, das Energieübertragung und Kommunikation in einer beliebigen Anordnung von Sendespulen, Empfangsspulen und passiven Relays beschreibt. Das Modell berücksichtigt starke Kopplung, Rauschkorrelation, Anpassung, Frequenzabhängigkeit und relevante kommunikationstheoretische Nuancen.

Der nächste grosse Abschnitt befasst sich mit der Übertragung zwischen Spulen mit zufälliger Ausrichtung (Gleichverteilung in 3D), wobei die resultierende zufällige Spulenkopplung zu einem Fadingkanal führt. Wir leiten dessen statistische Verteilung her und untersuchen die kommunikationstheoretischen Eigenschaften sowohl für Nah- als auch Fernfeldausbreitung. Wir zeigen, dass die Übertragung zwischen einzelner solcher Spulen katastrophale Zuverlässigkeit aufweist: Die asymptotische Ausfallwahrscheinlichkeit erfüllt $\epsilon \propto \mathrm{SNR}^{-1/2}$ für reine Nah- oder Fernfeldausbreitung, d.h. der Diversitätsexponent beträgt $1/2$ (sogar $1/4$ für Lastmodulation). Wir zeigen, dass der Wert im Nah-Fern-Übergang auf 1 steigt. Des Weiteren studieren wir räumliche Diversitätsverfahren für zufällig gedrehte Spulenarrays und die resultierende Kanalstatistik.

Ein Abschnitt über passive Relays zeigt zunächst, dass diese in zufälliger Anordnung ebenfalls Fading hervorrufen: Eine nicht kohärenten Summe von komplexwertigen Zeigern bestimmt den Kanalkoeffizienten, was (ähnlich der Mehrwegeausbreitung) frequenzabhängige Schwankungen zur Folge hat. Wir demonstrieren, dass die Nutzung dieser Schwankungen mittels sendeseitiger Kanalkenntnis (z.b. Waterfilling) und vor allem die Optimierung der Relaylasten verlässlich für erhebliche Verbesserungen sorgen.

Ein weiterführender Teil untersucht die Performancegrenzen sehr kleiner medizinischer in-vivo Sensoren in puncto induktiver Datenübertragung, entweder mittels Sendeverstärker oder Lastmodulation, für Freiraumausbreitung oder in einem leitenden Medium (Muskelgewebe). Ein grosses Spulenarray ausserhalb des Körpers wird als Leistungsquelle und Datensenke betrachtet. Wir leiten Designkriterien aus früheren Erkenntnisse ab und untersuchen die Auswirkungen hoher Knotendichte, passivem Relaying, Kanalkenntnis und kooperativer Übertragung im Detail. Ein besonderer Fokus liegt auf der minimalen sensorseitigen Spulengrösse für zuverlässige Übertragung.

Die entwickelten Modelle werden dann in einer Untersuchung der magnetischinduktiven Knotenlokalisierung, basierend auf Kanalmessungen zu Ankerspulen, verwendet. Der Fokus liegt auf den grundlegenden Grenzen der gemeinsamen Schätzung von Position und Ausrichtung eines Knotens; hierfür wird die Cramér-Rao-Schranke für den Fall komplexer Gaussscher Messfehler hergeleitet. Für dieses fünfdimensionale nichtkonvexe Schätzproblem schlagen wir eine alternierende und adaptiv gewichtete Methode der kleinsten Quadrate vor, die hinsichtlich Robustheit und Laufzeit den Stand der Technik schlägt. Anschliessend stellen wir eine kalibrierte Systemimplementierung dieses Paradigmas vor, die bei 500 kHz arbeitet und acht flache Ankerspulen um eine 3 m × 3 m Fläche verwendet. Um eine zuverlässige Ground Truth für Kalibrierung und Auswertung sicherzustellen ist der mobile Knoten auf einer Positioniervorrichtung montiert. Das System erreicht einen Medianpositionsfehler von 3 cm. Wir untersuchen die praktischen Leistungsgrenzen und die (im momentanen Wissensstand unbekannte) dominante Fehlerquelle solcher Systeme.

Ein ergänzendes Kapitel widmet sich einem neuen Ansatz zur Abstandsschätzung zwischen zwei drahtlosen Knoten, deren breitbandige Funkkanäle zu einem oder mehreren Beobachtern bekannt sind. Dabei nutzen wir die zufälligen Richtungen von Mehrwegekomponenten bei ausgeprägter Mehrwegeausbreitung aus. Basierend auf den Verzögerungsunterschieden der extrahierbaren Mehrwegekomponenten leiten wir Abstandsschätzungsformeln für mehrere wichtige Fälle her. Der Ansatz erfordert weder genaue Synchronisation, Sichtverbindung noch Kenntnis der Beobachterpositionen.

Contents

Notation

Scalars x are written lowercase italic, vectors $\mathbf{x}$ lowercase boldface and matrices $\mathbf{X}$ uppercase boldface. An exception are established physical field vectors like $\vec{E}$ and $\vec{B}$, but these occur only briefly in Sec. 2.1 and are otherwise represented by their complex phasors $\mathbf{e}$ and $\mathbf{b}$. All vectors are column vectors unless transposed explicitly. $\hat{\mathbf{x}}$ indicates an estimate of $\mathbf{x}$ (*not* a unit vector). $\|\mathbf{x}\|$ is the Euclidean norm of vector $\mathbf{x}$. We denote the $N \times N$ unit matrix by $\mathbf{I}_N$, the $M \times N$ all-ones matrix by $\mathbf{1}_{M \times N}$, and the imaginary unit by j (fulfilling $j^2 = -1$). We will use the trace $\mathrm{tr}(\mathbf{X})$, determinant $\det(\mathbf{X})$, and the m-th eigenvalue $\lambda_m(\mathbf{X})$ of a matrix $\mathbf{X}$. For a random variable x we denote the probability density function (PDF) as $f_x(x)$ and the cumulative distribution function (CDF) as $F_x(x)$, i.e. we use the same symbol for the random variable and the realization to avoid an unnecessarily bloated notation. We will often just write $f(x)$ for the PDF of x when there is no risk of confusion. The indicator function $\mathbb{1}_S(x)$ for some set S is characterized by $\mathbb{1}_S(x) = 1$ if $x \in S$ and $\mathbb{1}_S(x) = 0$ otherwise.

List of Common Symbols

Symbol	Set or Value	Description
Communication and Power Transfer:		
N_T	$\mathbb{N}$	number of transmitters
N_R	$\mathbb{N}$	number of receivers
N_Y	$\mathbb{N}$	number of passive relays
P_T	$\mathbb{R}_+ \cdot \mathrm{W}$ (watt)	transmit power (active power)
$\mathbf{x}$	$\mathbb{C}^{N_\mathrm{T}} \cdot \sqrt{\mathrm{W}}$	transmit signal vector
$\mathbf{y}$	$\mathbb{C}^{N_\mathrm{R}} \cdot \sqrt{\mathrm{W}}$	received signal vector
$\mathbf{G}$	$\mathbb{C}^{N_\mathrm{R} \times N_\mathrm{T}}$	current gain matrix
$\mathbf{H}$	$\mathbb{C}^{N_\mathrm{R} \times N_\mathrm{T}}$	channel matrix
$\mathbf{h}$	$\mathbb{C}^{N_\mathrm{R}}$	channel vector (SIMO)
h	$\mathbb{C}$	channel coefficient (SISO)
η	$[0,1]$	power transfer efficiency
Δ_f	$\mathbb{R}_+ \cdot \mathrm{Hz}$ (hertz)	narrow bandwidth
$\mathbf{K}$	$\mathbb{C}^{N_\mathrm{R} \times N_\mathrm{R}} \cdot \mathrm{W}$	noise covariance matrix
$\bar{\mathbf{H}}$	$\mathbb{C}^{N_\mathrm{R} \times N_\mathrm{T}} \cdot \mathrm{W}^{-\frac{1}{2}}$	noise-whitened channel matrix
SNR	$\mathbb{R}_+$	signal-to-noise ratio
D	$\mathbb{R}_+ \cdot \mathrm{bit/s}$	data rate
C	$\mathbb{R}_+ \cdot \mathrm{bit/s}$	channel capacity
Link Geometry:		
$\mathbf{p}_\mathrm{T}$	$\mathbb{R}^3 \cdot \mathrm{m}$	transmitter position
$\mathbf{p}_\mathrm{R}$	$\mathbb{R}^3 \cdot \mathrm{m}$	receiver position
r	$\mathbb{R}_+ \cdot \mathrm{m}$	Euclidean distance $r = \|\mathbf{p}_\mathrm{R} - \mathbf{p}_\mathrm{T}\|$
$\mathbf{o}_\mathrm{T}$	$\mathbb{R}^3, \|\mathbf{o}_\mathrm{T}\| = 1$	orientation of transmit coil axis
$\mathbf{o}_\mathrm{R}$	$\mathbb{R}^3, \|\mathbf{o}_\mathrm{R}\| = 1$	orientation of receive coil axis
$\mathbf{u}$	$\mathbb{R}^3, \|\mathbf{u}\| = 1$	transmitter-to-receiver direction $\mathbf{u} = (\mathbf{p}_\mathrm{R} - \mathbf{p}_\mathrm{T})/r$
$\boldsymbol{\beta}_\mathrm{NF}$	$\mathbb{R}^3, \beta_\mathrm{NF} \in [\frac{1}{2}, 1]$	scaled near field $\boldsymbol{\beta}_\mathrm{NF} = \frac{1}{2}(3\mathbf{u}\mathbf{u}^\mathrm{T} - \mathbf{I}_3)\mathbf{o}_\mathrm{T}$
$\boldsymbol{\beta}_\mathrm{FF}$	$\mathbb{R}^3, \beta_\mathrm{FF} \in [0, 1]$	scaled far field $\boldsymbol{\beta}_\mathrm{FF} = (\mathbf{I}_3 - \mathbf{u}\mathbf{u}^\mathrm{T})\mathbf{o}_\mathrm{T}$
J_NF	$J_\mathrm{NF} \in [-1, 1]$	near-field alignment factor $J_\mathrm{NF} = \mathbf{o}_\mathrm{R}^\mathrm{T} \boldsymbol{\beta}_\mathrm{NF}$
J_FF	$J_\mathrm{FF} \in [-1, 1]$	far-field alignment factor $J_\mathrm{FF} = \mathbf{o}_\mathrm{R}^\mathrm{T} \boldsymbol{\beta}_\mathrm{FF}$
$\mathbf{O}_\mathrm{T}$	$\mathbb{R}^{3 \times N_\mathrm{T}}$	transmit array orientation $\mathbf{O}_\mathrm{T} = [\mathbf{o}_{\mathrm{T},1} \dots \mathbf{o}_{\mathrm{T},N_\mathrm{T}}]$
$\mathbf{O}_\mathrm{R}$	$\mathbb{R}^{3 \times N_\mathrm{R}}$	receive array orientation $\mathbf{O}_\mathrm{R} = [\mathbf{o}_{\mathrm{R},1} \dots \mathbf{o}_{\mathrm{R},N_\mathrm{R}}]$
$\mathbf{Q}_\mathbf{u}$	$\mathbb{R}^{3 \times 3}, \mathbf{Q}_\mathbf{u}^\mathrm{T}\mathbf{Q}_\mathbf{u} = \mathbf{I}_3$	orthogonal matrix fulfilling $\mathbf{u}^\mathrm{T}\mathbf{Q}_\mathbf{u} = [1\ 0\ 0]$

Symbol	Set or Value	Description

Physics and Circuits:

Symbol	Set or Value	Description
f	$\mathbb{R} \cdot \mathrm{Hz}$	temporal frequency
ω	$\mathbb{R}_+ \cdot \frac{1}{\mathrm{s}}$	radial frequency $\omega = 2\pi f$
λ	$\mathbb{R}_+ \cdot \mathrm{m}$	wavelength $\lambda = \frac{c}{f}$
k	$\mathbb{R}_+ \cdot \frac{1}{\mathrm{m}}$	wave number (spatial frequency) $k = \frac{2\pi}{\lambda} = \frac{\omega}{c}$
R	$\mathbb{R}_+ \cdot \Omega$ (ohm)	resistance
X	$\mathbb{R} \cdot \Omega$	reactance
Z	$\mathbb{C} \cdot \Omega$	impedance $Z = R + jX$
L	$\mathbb{R}_+ \cdot \mathrm{H}$ (henry)	self-inductance
M	$\mathbb{R} \cdot \mathrm{H}$	mutual inductance
R_{ref}	$50\,\Omega$	reference impedance
μ	$\mathbb{R}_+ \cdot \frac{\mathrm{T} \cdot \mathrm{m}}{\mathrm{A}}$	permeability
$\mathbf{Z}_{\mathrm{C:RT}}$	$\mathbb{C}^{N_\mathrm{R} \times N_\mathrm{T}} \cdot \Omega$	transmitter-to-receiver mutual impedances
$\mathbf{Z}_{\mathrm{C:T}}$	$\mathbb{C}^{N_\mathrm{T} \times N_\mathrm{T}} \cdot \Omega$	transmit array mutual impedances
$\mathbf{Z}_{\mathrm{C:R}}$	$\mathbb{C}^{N_\mathrm{R} \times N_\mathrm{R}} \cdot \Omega$	receive array mutual impedances
$\mathbf{Z}_{\mathrm{3DoF}}$	$\mathbb{C}^{3 \times 3} \cdot \Omega$	mut. imp. $\mathbf{Z}_{\mathrm{3DoF}} = \mathbf{Q_u}\,\mathrm{diag}(Z_{\mathrm{RT}}^{\mathrm{coax}}, Z_{\mathrm{RT}}^{\mathrm{copl}}, Z_{\mathrm{RT}}^{\mathrm{copl}})\,\mathbf{Q_u^T}$
$\mathbf{H}_{\mathrm{3DoF}}$	$\mathbb{C}^{3 \times 3}$	description of spatial channels $\mathbf{H}_{\mathrm{3DoF}} = \frac{\mathbf{Z}_{\mathrm{3DoF}}}{\sqrt{4 R_\mathrm{T} R_\mathrm{R}}}$

Localization:

Symbol	Set or Value	Description
N	$\mathbb{N}$	number of anchor coils
$\mathbf{h}_{\mathrm{meas}}$	$\mathbb{C}^N$	channel vector by measurement
$\mathbf{h}_{\mathrm{model}}$	$\mathbb{C}^N$	channel vector by model
$\boldsymbol{\epsilon}$	$\mathbb{C}^N$	model error $\boldsymbol{\epsilon} = \mathbf{h}_{\mathrm{meas}} - \mathbf{h}_{\mathrm{model}}$
$\mathbf{K}_{\boldsymbol{\epsilon}}$	$\mathbb{C}^{N \times N}$	model error covariance matrix $\mathbf{K}_{\boldsymbol{\epsilon}} = \mathbb{E}[\boldsymbol{\epsilon}\boldsymbol{\epsilon}^\mathrm{H}]$
$\hat{\mathbf{p}}_{\mathrm{ag}}$	$\mathbb{R}^3 \cdot \mathrm{m}$	estimate of agent position
$\hat{\mathbf{o}}_{\mathrm{ag}}$	$\mathbb{R}^3, \|\hat{\mathbf{o}}_{\mathrm{ag}}\| = 1$	estimate of agent orientation
ϕ_{ag}	$(-\pi, \pi]$	agent orientation azimuth angle
θ_{ag}	$[0, \pi]$	agent orientation polar angle
$\boldsymbol{\psi}$	$\mathbb{R}^5$	estimation parameter vector $\boldsymbol{\psi} = [\mathbf{p}_{\mathrm{ag}}^\mathrm{T}, \phi_{\mathrm{ag}}, \theta_{\mathrm{ag}}]^\mathrm{T}$
$\boldsymbol{\mathcal{I}}_{\boldsymbol{\psi}}$	$\mathbb{R}^{5 \times 5}$	Fisher information matrix of $\boldsymbol{\psi}$

List of Acronyms

This work avoids acronyms for the most part but they occur in various variables, subscripts and plots.

Acronym	Meaning
CDF	cumulative distribution function
CRLB	Cramér-Rao lower bound
LNA	low-noise amplifier
MIMO	multiple-input and multiple-output
MISO	multiple-input and single-output
MLE	maximum-likelihood estimate
MQS	magnetoquasistatic
MRC	maximum-ratio combining
PDF	probability density function
PEB	position error bound
PSD	power spectral density
PTE	power transfer efficiency
RFID	radio-frequency identification
RX	receiver
SC	selection combining
SIMO	single-input and multiple-output
SISO	single-input and single-output
SNR	signal-to-noise ratio
TX	transmitter
UMVUE	uniform minimum-variance unbiased estimate
UWB	ultra-wideband

Chapter 1

Motivation and Contributions

This chapter describes contemporary research goals regarding wireless sensors and the associated need for wireless communication, powering, and localization. We discuss the potential benefits of magnetic induction and our goals in this context, associated open research problems, the corresponding state of the art and its shortcomings as well as the structure and contributions of this dissertation.

1.1 Wireless Sensors: Technological Situation

Information and communication technology has revolutionized most processes in industry, health care, business administration and daily life. In particular, remarkable advances in integrated circuits, computing, sensors, displays and battery technology gave rise to powerful wireless communication technology. Prominent examples are tablet computers and smart phones equipped with antennas and chipsets for local area networking via the IEEE 802.11ac standard [1], cellular networking via LTE-Advanced [2], and reception of navigation satellite signals. They are capable of *reliable digital communication over wireless channels* with high data rate and can *determine their location* within a few meters of accuracy [3]. These devices are rather large and expensive [4] and have considerable energy consumption [5,6]. Apart from such consumer electronics, modern wireless communication technology also finds important uses in devices for sensing and actuation (henceforth referred to as *wireless sensors*). The topic has received a lot of attention by the wireless industry and research community, mostly under the umbrella of *wireless sensor networks* (WSN) [7–11] and the *Internet of Things* (IoT) [12–15]. Wireless sensors are used for all kinds of sensing and monitoring tasks in the military [9,16], power grid [17], large machines [18] and a multitude of industrial processes [10]. Envisioned environmental applications comprise the detection of hazardous materials and contamination cleanup [19]. Medical in-body applications of wireless sensors could disrupt the field of health care: future *medical microrobots* are expected to provide untethered diagnostic sensing, targeted drug delivery and treatment (e.g. removing a kidney stone or tumor) [20–25].

1

Wireless sensors rely on wireless technology to transmit acquired sensor data, receive commands, coordinate actions, and to determine their location [8, 26]. Their *technical requirements and limitations* are however stricter than those of consumer electronics. First, many target applications require wireless sensors to be deployed in vast numbers, which constrains the unit cost and thus also the hardware complexity. Secondly, wireless sensors are usually battery-powered but required to stay operational for a long period of time [7]. The use of low-power hardware and transmission schemes can remedy the problem [11, 27, 28], but even despite these measures a wireless sensor may be energy-limited to an extend where the fulfillment of its basic tasks is in jeopardy. This holds especially for the task of transmitting vast data to a remote data sink [28, 29]. The problem is even more pronounced when the maximum device size is constrained by the application [23, 26, 30, 31]: with current technology, a severely size-constrained wireless sensor can not be equipped with a battery of any useful capacity [26] (although some progress is made in that regard [32]). A prime example are medical microrobots which must be sufficiently small to fit in cavities of the human body in a minimally invasive way. Their application-specific maximum device size ranges from ≈ 3 cm for gastrointestinal cameras down to a few µm for maneuvering the finest capillary vessels [22, 23, 31]. As an alternative to a battery, energy can be supplied via the electromagnetic field (wireless power transfer) [7, 33, 34] or gathered from environmental processes (energy harvesting) [14, 35].

Most contemporary wireless technology relies on conventional radio with antennas whose size is matched to the employed wavelength λ for efficient radiation and reception of electromagnetic waves. It is the technology of choice for long-range communication because the link amplitude gain h (a.k.a. channel coefficient) of a free-space radio link decays with only $h \propto r^{-1}$ versus the link distance r. *Conventional radio is however inadequate for certain wireless sensor applications.* The link gain is usually way below -50 dB and thus insufficient for wireless power transfer [36, Fig. 5.1]. Further significant attenuation occurs when an antenna shall fit into a small device because then the realizable aperture is limited [37, Sec. 8.4]. Likewise, for conventional radio, a maximum antenna size implies a maximum wavelength λ, i.e. a minimum carrier frequency f_c. For example, a dipole antenna whose $\lambda/2$ length is set to just 0.5 mm (e.g. because it is integrated into a medical microrobot) radiates efficiently at $f_c = 300$ GHz. Fields of such large frequency may however be subject to severe medium attenuation [38], e.g. caused by conducting body fluids and tissue [39, 40]. Other challenging propagation environments for wireless sensor applications are the underground [41–43], underwater [44], oil reservoirs [45–48], engines [18], hydraulic systems [48] and battlefields [9].

High-frequency radio waves interact with the environment: they are reflected, scattered and diffracted by objects. They are thus hard to predict in dense environments [49], which constitutes a huge problem for accurate radio localization. In particular, multipath propagation and non-line-of-sight situations deteriorate time-of-arrival localization schemes [50] and, likewise, the associated multipath fading and shadowing cause fluctuations that deteriorate received-signal-strength schemes heavily [51].

1.2 Magnetic Induction for Wireless Sensors

Low-frequency magnetic induction is an *alternative propagation mechanism* to conventional radio. It uses antenna coils whose dimensions are significantly smaller than the employed wavelength. Such electrically small coils feature a very small radiation resistance and thus usually a small overall coil resistance (determined by the ohmic resistance of the coil wire). This allows to drive a strong current through a resonant[1] transmit coil with a given available transmit power, resulting in a strong generated magnetic field and strong induced currents at a resonant receive coil.

The chosen wavelength will often be larger than the intended link distance. In this case the receiver is in the near field of the transmit coil and the link amplitude gain h effectively scales like $h \propto r^{-3}$. This limits the usable range of low-frequency magnetic induction. Another disadvantage is that a low carrier frequency naturally limits the communication bandwidth and thus the achievable data rate.

Yet, in comparison to the described problems of conventional radio, low-frequency magnetic induction offers *various advantages* to wireless sensor technology:

1. Low-frequency magnetic fields *penetrate* various relevant *materials* (e.g. tissue, soil, water) with little attenuation [41, 54, 55] due to the large wavelength and favorable material permeability. Water for example hardly affects the magnetic field ($\mu_\mathrm{r} \approx 1$) but attenuates the electric field amplitude by a factor of $\epsilon_\mathrm{r} \approx 80$.

2. Low-frequency magnetic fields hardly interact with the environment and can thus be predicted by a free-space model [56–61]. Also, the amplitude gain of a magneto-inductive link is very sensitive to position and orientation of the transmit and receive coils (cf. $h \propto r^{-3}$) and thus bears rich geometric information. Magnetic induction is thus *suitable for accurate wireless localization*.

---

[1]Many texts present resonance as a distinctive aspect of magnetic induction. However, we note that usually any radio antenna is operated at resonance in the sense that its electrical reactance is compensated by reactive matching circuits in order to maximize the antenna current for a given available transmit power [52, 53].

3. Increasing the *number of coil turns* is a very effective means of increasing the link gain in order to realize *strong mid-range links*. To some extend the number of turns can be increased while maintaining a coil geometry that is integrable into a device of *limited volume* (e.g. a cylindrical casing). No equivalent mechanism is available for electric antennas [62, Sec. 5.2.3].

4. At very low frequencies, the use of high-permeability *magnetic cores* can vastly increase the link gain.

5. With a low carrier frequency (i.e. a large carrier period time), *phase synchronization between distributed nodes* can feasibly be established. Cooperating sensors can then form a distributed antenna array for beamforming to achieve an array gain, a diversity gain, and possibly even a spatial multiplexing gain.

6. The severe path loss of near-field systems allows for vast *spatial reuse* and *security against remote eavesdropping* (e.g. for contactless payment with NFC).

The mere presence of a passive resonant coil can cause a significant alteration of the local magnetic field. This can be utilized to a technological advantage in various ways:

7. Inductive radio-frequency identification (RFID) tags use the effect for data transmission via *load modulation*. Thereby a tag modulates information bits by switching between two different termination loads for its coil. The receiver (an RFID reader) detects the field changes to decode the transmitted bits. [63]

8. One can place passive resonant coils between a transmitter-receiver pair in order to act as *passive relays*; a technique also known as magneto-inductive waveguide. The primary magnetic field generated by the transmit coil induces currents in the passive relay coils, giving rise to a secondary magnetic field which propagates to the receiver. This can improve the link. [64–68]

9. Significant link gain improvements can be achieved by putting a resonant passive relay coil right next to the transmit coil (coaxial and as a part of the transmitter device) and/or next to the receive coil (coaxial and as a part of the receiver device). Such *multi-coil designs*, which utilize the effect of strongly coupled magnetic resonances, allow for capable wireless power transfer systems. [69–75]

In summary, low-frequency magnetic induction with multi-turn coils is a suitable propagation mechanism for short- and mid-range power transfer, localization, and communication (either from an active transmitter or from a passive tag that uses load modulation). This holds especially for small devices in harsh propagation environments.

4

It furthermore allows for link improvements by placing passive resonant coils. Major drawbacks are the severe path loss and the small bandwidth. The low-frequency aspect is henceforth implied for magnetic induction and will not be pointed out repeatedly.

1.3 State of the Art, Open Issues, Contributions

1.3.1 Opening Remarks: Greater Goal and Focus

This dissertation is motivated by the greater (and currently open) problem of understanding the full capabilities of magnetic induction in the context of wireless sensors when the full technological potential is utilized. This problem context has been formulated in detail in the dissertation of Slottke [26]. A particularly interesting regime are small-scale applications with a potentially high node density such as medical microrobots. We desire a thorough understanding of the interplay of wireless powering, reach, radiation, achievable rates, the impact of coil arrangement and channel knowledge, outage and diversity, node cooperation, array techniques and mutual coupling, spatial degrees of freedom, passive relaying, miniaturization and high node density, as well as load modulation. A good understanding thereof would allow for an educated comparison to competing propagation mechanisms for medical microrobots, namely ultrasonic acoustic waves [76], molecular communication [24], and optical approaches [77].

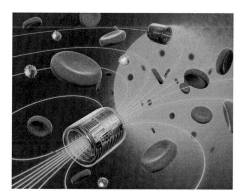

Figure 1.1: Concept art of several medical microrobots operating inside a blood vessel. They are equipped with a single-layer solenoid coil for wireless transmission and reception via magnetic induction. Integrated circuits for sensing and digital logic are indicated.

Clearly this greater problem divides into a multitude of subproblems, a subset of which will be addressed by the dissertation at hand. The remainder of the section describes this problem subset in detail, in relation to the current state of the art and with a focus on the physical layer and signal processing research literature.

1.3.2 Magneto-Inductive Coupling Models

State of the Art: All fundamental aspects of coil coupling are covered by classical electromagnetism; the general approach to coupling problems (e.g. via Maxwell's equations) is however associated with numerical approaches to partial differential equations [78]. Existing formal studies of communication or power transfer via magnetic induction (e.g., [65–68]) thus, to the best of our knowledge, all employ at least two simplifying assumptions: (i) the coils are electrically small and AC circuit theory applies, (ii) the magnetoquasistatic assumption, i.e. no radiation occurs whatsoever.

Identified Shortcomings: The magnetoquasistatic assumption requires that the wavelength exceeds the link distance by orders of magnitude. This limits a model's scope of validity and is particularly problematic because there is engineering incentive for choosing a small wavelength: using a larger frequency results in a larger induced receiver voltage. Furthermore, radiation can be desired to increase the reach (cf. mid-field power transfer [75]) and to obtain an additional phase-shifted field component which could help against receive-coil misalignment. Radiation should thus be considered in the analysis of a magneto-inductive link, even for electrically small coils.

Chapter and Contribution: In Cpt. 2 we work out coupling formulas for electrically-small coils that do include radiation. In particular we present (i) a formula for arbitrary coil geometries and (ii) a dipole-type formula based on linear algebra, which allows for convenient interfacing with communication theory.

Associated Publications: The formulas appeared in our paper [79, Eq. 11 and 12] and the dipole formula was used in our paper on magneto-inductive localization [80].

1.3.3 Modeling and Analysis of Magneto-Inductive Links

State of the Art: The established approach to modeling magneto-inductive links uses an equivalent circuit description. This way the maximum power transfer efficiency between two coils (magnetoquasistatic regime) was stated by Ko [81, Eq. 6]. A rudimentary analysis of the channel capacity in thermal noise was given in [82] for coaxial coils and the assumption of a flat channel over the 3 dB-bandwidth of the system. They observe a rate-optimal coil Q-factor depending on the distance. Sun and Akyildiz studied magnetic induction with passive relaying for underground communication between coaxial coils in [41] and compared the approach to conventional radio for different soil conditions. They study the bit error rate of narrowband BPSK. In [65] they investigate the communication limits of underground networks of coplanar coils for various network topologies while exploiting the spatial reuse advantage. They use the 3 dB bandwidth as communication bandwidth and assume a flat channel thereover. The papers [83, 84] are dedicated to the optimization of technical parameters for capacity maximization of magneto-inductive channels, whereby the evaluation in [83] assumes a frequency-flat channel and noise spectral density over a heuristically chosen communication bandwidth. Kisseleff et al. [67, 85] formulate the channel capacity under due consideration of colored noise (thermal noise shaped by the receiver circuit) and the proper capacity-achieving spectral power allocation via waterfilling. They furthermore study practical digital transmission schemes over the frequency-selective (and thus time-dispersive) magneto-inductive channel in [86] and simultaneous wireless information and power transfer in [87]. The work in [88] investigates user cooperation for magneto-inductive communication.

Antenna arrays offer crucial advantages to wireless systems, namely array gain, diversity, and spatial multiplexing [89]. The use of arrays is thus vastly popular in radio communications [1, 2, 90] and has been proposed for magnetic induction for wireless power beamforming [91–93] and selection combining [94], underwater sensor networks [44], localization [59, 95–98], and beamforming for body-area sensor networks [99].

Identified Shortcomings: Simplifying assumptions are prevalent in the literature, e.g., the exclusive use of a dipole coupling model, narrowband assumptions, weak coupling, coaxial arrangement, thermal noise only, white noise, and heuristic spectral power allocation. The multi-stage transformer model [41, 65, 83] disregards coupling between non-neighboring coils but results in a more complicated formalism than a general approach (e.g. in terms of impedance matrices [26, 87, 100]). Most work assumes just a series capacitance as matching circuit even though it does in general not maximize power delivery from source to load.

A coil array usually exhibits mutual coupling among the associated coils. Such inter-array coupling has significant implications for matching and performance that are well-understood for MIMO radio communications [53, 101–103] but, to the best of our knowledge, are currently not considered by research on magnetic induction.

In conclusion, we identify the lack of a *well-structured general system model for magneto-inductive links* that would remedy the described shortcomings.

A related shortcoming is that load modulation has not received attention from communication theorists despite its use in disruptive RFID technology with great commercial success [63].

Chapter and Contribution: Cpt. 3 presents a concise and general system model for magneto-inductive communication and power transfer in any arrangement of transmitter (or transmit array) and receiver (or receive array). It accounts for array coupling, the statistics of noise signals from various sources, the desired matching strategy and its frequency-dependent effects. We state the channel capacity for narrowband and broadband cases, for a constraint on the available sum power or on the per-node powers. We discuss special cases such as weakly-coupled links, perfectly-matched links, orthogonal coil arrays as well as the associated degrees of freedom in detail. We furthermore present a treatise on cooperative load modulation with a reader array and the associated communication-theoretic performance limits.

Associated Publications: A summary of the MIMO system model appeared in our paper [79, Sec. II and III].

1.3.4 Impact of Arbitrary Coil Arrangement

State of the Art: The location of a wireless sensor is determined by its movement and the deployment strategy (which is either arbitrary or according to some application-specific criterion). In any case, the sensor position and orientation can be considered random by the communications engineer, which amounts to considering a random channel [83]. For magneto-inductive sensors an unfavorable coil orientation may result in severe link attenuation or outage. This trade-off between favorable coil arrangement and mobility has been noted in [33,83] and is the topic of [104] which studies the connectivity of magneto-inductive ad-hoc sensors. In [83] they identified the outage capacity as a meaningful performance measure of randomly arranged magneto-inductive communication links.

An appropriate coil coupling model provides a formal description of coil misalignment (i.e. the link attenuation due to deviations from coaxial arrangement), cf. Sec. 1.3.2. Most studies of coil misalignment are concerned with small lateral or angular deviations in the context of efficient short-distance power transfer [105–109]. The specific coil geometries must be considered in this regime, which complicates a mathematical analysis. At larger distances the much simpler dipole model (e.g. as stated in [67]) is appropriate.

Identified Shortcomings: The referenced studies [105–109] do not address the effect of a fully random coil orientation on the link gain, although this circumstance is to be expected for wireless sensors with high mobility such as medical microrobots. The impact of an arbitrary node orientation on the performance (and performance statistics) has not been studied so far, neither for single-input single-output (SISO) links nor for links with coil arrays. The need for an appropriate statistical channel model is highlighted by [110] who assumed a Rayleigh fading model for the effect of RFID coil misalignment because of the lack of a better model. Similarly, [83] worked with a Gaussian-distributed channel capacity with heuristically chosen variance.

The effect of coil arrays with a diversity combining scheme for misalignment mitigation so far (to the best of our knowledge) has also not been studied formally.

Chapter and Contribution: Cpt. 4 presents an analytic study of the statistics of the random fading-type channel that arises with random coil orientations (with a uniform distribution in 3D). The outage implications on the power transfer efficiency and channel capacity are investigated in detail. The SISO case is shown to exhibit catastrophic outage behavior: the diversity order is $1/2$ for pure near-field or pure far-field propagation (even $1/4$ for load modulation). The diversity order increases to 1 when both modes are present. The results are contrasted with the channel statistics for randomly oriented coil arrays after the application of a spatial diversity scheme.

Associated Publications: The channel statistics results for the pure near-field case with and without diversity combining appeared in our paper [111].

1.3.5 Magneto-Inductive Passive Relaying

State of the Art: Magneto-inductive passive relaying (as described in Item 8 of Sec. 1.2) was first proposed by Shamonina et al. [64] as a novel method of forming a waveguide. Thereby the relay elements were assumed in coaxial and equidistant arrangement between the transmitter and receiver coils. The merits of the concept for wireless powering or communication have been studied by [41, 65–68] for regular arrangements and with simplifying assumptions on the node couplings. For example, in [68] they analyze magneto-inductive communication over a 2-D grid of relays. The authors of [41, 65] consider only couplings between neighboring coils in networks of equidistant coplanar relays. The work contains an analysis of failure or misplacement of a single relay. The effect of coupling between non-adjacent relays in such a setup was studied in [66] for wireless power transfer. In [67] the communication performance of magneto-inductive relaying networks is optimized by adjusting the coil orientations for interference zero-forcing. The notion of random errors in relay deployment locations was introduced by [83] and its effect was investigated as part of [86]. The link SNR statistics for one randomly deployed relay were investigated in [26, Fig. 4.13]. Various researches pointed out the complicated effect of the relay density on the channel frequency response [41, 64, 83, 112, 113] and on the noise spectral density [114].

Identified Shortcomings: The literature on magneto-inductive passive relaying considers very specific regular arrangements or just small deviations thereof while simplifying assumptions on the node couplings are prevalent. We envision a cooperative scheme in (possibly very dense and arbitrarily arranged) magneto-inductive networks by which idle nodes may act as relays to improve the channel between the currently operating transmitter-receiver pair. Thereby we consider the node locations and orientations as completely arbitrary because sensor networks are often mobile or deployed in an ad-hoc fashion. The effects and technical merits of passive relays in such dense and random configurations are currently unknown and not described by any existing model. Dense swarms of nodes are of particular interest because they are an important envisioned use case for medical microrobots [115], where passive relaying might yield significant gains for magneto-inductive power transfer or communication.

Chapter and Contribution: In Cpt. 5 we analyze magneto-inductive passive relaying and its impact on the channel for arbitrary arrangements. Their effect is rigorously integrated into the system model of Cpt. 3 with one simple formula. A numerical evaluation of the channel statistics shows that randomly deployed relays cause frequency-selective fading: they can cause significant channel improvement or attenuation, depending on the density and individual geometric realization of the network. This is primarily caused by a non-coherent superposition of individual relay contributions to the link coefficient. The practical merits of such passive relay swarms are thus limited when a fixed operating frequency is used, but adapting the transmit signal to the frequency-selective channel allows for significant gains. For better utilization of the relays channel we study an optimization scheme based on the deactivation of individual relays by load switching and demonstrate considerable and reliable improvements.

Associated Publications: The content appeared in parts in our paper [116].

1.3.6 Magneto-Inductive Medical In-Body Sensors: Wireless Powering, Capabilities, Feasibility

State of the Art: Magnetic induction with its suitability for miniaturization and the other outlined advantages has been proposed for wireless-powered small-scale sensors with potential medical in-body applications [26, 117]. To this effect, the work of [26] contains an investigation on the miniaturization limits of magneto-inductive sensors. The authors of [118] discuss wireless powering of medical implants under consideration of tissue absorption. Coil designs for 4 mm-sized bio-implants and the resulting power transfer efficiency (PTE) in free space and tissue are presented in [119].

As discussed earlier, antenna arrays play a crucial role for modern wireless technology. The reach of energy-limited wireless nodes can be improved by forming a distributed array through user cooperation (e.g., see [28, 29, 88]), depending on the availability of channel knowledge and distributed phase synchronization.

Identified Shortcomings: The state of the art lacks an understanding of the behavior and performance capabilities of small magneto-inductive wireless nodes under exploitation of all technological aspects (arrays, cooperation, passive relaying, load modulation). This holds especially true for dense swarm networks, a relevant use case in envisioned applications of medical microrobots [22, 115] and an opportunity to the wireless engineer: physical layer cooperation between in-body devices allows for an array gain and spatial diversity in the uplink. Furthermore, dense swarms of strongly-coupled resonant coils can give rise to a passive relaying effect, associated with the complicated frequency-selective channel described in Sec. 1.3.5. These channel fluctuations should be exploited by the signaling scheme.

Chapter and Contribution: Cpt. 6 presents a technical evaluation of magnetic induction for small-scale in-body sensors. The sensors are assumed to receive power wirelessly and transmit data to a massive external coil array, which serves as data sink and power source (1 W). We discuss key aspects of the wireless channel and appropriate link design and compare propagation in muscle tissue to free space. For sensor devices 5 cm deep beneath the skin and an assumed 50 nW required chip activation power we project a minimum coil size of about 0.3 mm. However, a coil larger than 1 mm can be necessary depending on the data rate and reliability requirements as well as the availability of channel knowledge. We compare the cases of full channel knowledge, no knowledge, and sensor location knowledge. We find that an operating frequency of 300 MHz is suitable for this use case, although a much smaller frequency must be chosen if a larger penetration depth is desired. Moreover, we study resonant sensor nodes in dense swarms, a key aspect of envisioned biomedical applications. In particular, we investigate the occurring passive relaying effect and cooperative transmit beamforming. We show that the frequency- and location-dependent signal fluctuations

11

in such swarms allow for significant performance gains when utilized with adaptive matching, spectrally-aware signaling and node cooperation. We show that passive relays are particularly capable in this context when their load capacitance is optimized and, furthermore, that load optimization can compete with active transmission if the receiving external device can measure with high fidelity (e.g., if thermal noise is the only impairment).

Associated Publications: Some of the content appeared in similar form in our paper [79, Sec. IV and V].

1.3.7 Magneto-Inductive Localization

State of the Art: In dense propagation environments, radio localization faces severe challenges from radio channel distortions such as line-of-sight blockage or multipath propagation [51, 120, 121]. Magnetic near-fields, in contrast, are hardly affected by the environment as long as no major conducting objects are nearby [56–61]. In consequence the magnetic near-field at some position relative to the source (a driven coil or a permanent magnet) can be predicted accurately with a free-space model. This enables the localization of an agent coil in relation to stationary coils of known locations (anchors) [58–61, 98]. In particular, position and orientation estimates can be obtained by fitting a channel model to measurements [59–61, 98]. The problem of estimating position and orientation of a single-coil agent or a permanent magnet was tackled least-squares estimation problem by [122–125]. Various system implementations for magneto-inductive localization have been published, e.g. [57–61].

Identified Shortcomings: The fundamental limits of magneto-inductive 3D localization are not addressed by existing research even though a rich set of tools for this purpose has been developed for radio localization [50, 126, 127].

The least-squares approach to position and orientation estimation with a gradient-based solver is slow and unreliable because the cost function is non-convex and has a five-dimensional parameter space. A fast and robust solution remains as an open algorithmic problem.

While magnetic induction presents the prospect of highly accurate localization (cf. Item 2 of Sec. 1.2) only mediocre accuracy is reported for practical systems, with a relative position error of at least 2% [57–61] (details are given in Cpt. 7). Thereby it is unclear which error source causes the accuracy bottleneck. Candidates are noise and interference, quantization, an inadequate signal model, the estimation algorithm, poor calibration, unconsidered radiation and field distortion due to nearby conductors.

Chapter and Contribution: In Cpt. 7 we first derive the Cramér-Rao lower bound (CRLB) on the position error for unknown agent orientation, based on the complex-valued dipole model from Cpt. 2 and a Gaussian error model. Therewith we study the potential localization accuracy on the indoor scale.

For this parametric estimation problem we find that numerical standard approaches are slow and often highly inaccurate due to missing the global cost function minimum. To this effect we design two fast and robust localization algorithms, enabled by a dimensionality reduction from 5D to 3D via eliminating the agent orientation parameters and by means for smoothing the cost function.

Based on these algorithms we present a system implementation with flat spiderweb coils tuned to 500 kHz. We evaluate the achievable accuracy in an office setting after a thorough calibration. During the calibration procedure we attempt to compensate field distortions and multipath propagation. The measurements are acquired with a multiport network analyzer, i.e. the agent is tethered and furthermore mounted on a controlled positioner device. We investigate the different sources of error and conjecture that field distortions due to reinforcement bars cause the accuracy bottleneck. Using the CRLB we project the potential accuracy in more ideal circumstances.

Associated Publications: Our paper [128] contains the proposed WLS3D algorithm and the CRLB result for the magnetoquasistatic case. These were generalized by our paper [80] which also presents the system implementation and evaluation.

1.3.8 Wideband Radio Localization

The work described in the following relates to wideband radio localization of wireless sensors in indoor environments with rich multipath propagation. It was conducted in the context of an industry project.

State of the Art: Most proposals for wireless radio localization rely on distance estimates to fixed infrastructure nodes (anchors) to determine the position of a mobile node [120], e.g. via trilateration. Cooperative network localization furthermore employs the distances between different mobile nodes [120, 121, 129, 130]. Given an exchanged radio signal between two nodes, a distance estimate can be obtained from the received signal strength (RSS) or the time of arrival (TOA).

Identified Shortcomings: RSS-based estimates have poor accuracy due to signal fluctuations [51, 131]. A TOA-based estimate can be very accurate but requires wideband signaling, a round-trip protocol for synchronization [50, 120] and involved hardware. It furthermore suffers from synchronization errors and processing de-

13

lays [50, 132–134]. Yet the main problem is ensuring a sufficient number of anchors in line of sight (LOS) to all relevant mobile positions [135]. TOA thus exhibits a large relative error at short distances and is not well-suited for dense and crowded settings.

Chapter and Contribution: Cpt. 8 presents a novel paradigm for (short) distance estimation between ultra-wideband radio nodes in dense multipath environments, with various significant advantages over state-of-the-art indoor localization schemes. The scheme does not consider the channel between the two nodes whose distance is of interest but instead consider the presence an observer node. Consequently, the distance estimate is obtained by comparing the channels to that observer. We use the assumption of multipath components with random direction with a uniform distribution in 3D and, this way, utilize mathematical synergies with Cpt. 4.

Associated Publications: The content of Cpt. 8 appeared in our paper [136] and the core idea resulted in the patent applications [137, 138].

1.4   Acknowledgments and Joint Work

A number of people supported the creation of this thesis with technical advice and contributions. In particular I would like to express my gratitude to my supervisor Armin Wittneben for his guidance and countless crucial pointers, Robert Schober for acting as referee, Marc Kuhn and Henry Schulten for many important hints and discussions, Tim Rüegg for many things, Eric Slottke for advice on magneto-inductive technology and scientific computing, Yahia Hassan for advice on circuit theory and system modeling, Erwin Riegler and the contributors at stackexchange.com for mathematical consulting, Steven Kisseleff for fruitful discussions, and all collaborators listed in the following. Parts of Cpt. 5 emerged in collaboration with Eric Slottke. An earlier SISO-case variant of the system model in Cpt. 3 was used in [26] based on an earlier unpublished document [139] by the author of the thesis at hand (cf. [26, cited reference 33]). This work [139] comprised passive relaying evaluations of the kind [26, Fig. 4.9, 4.11, 4.12]. The collaboration with Henry Schulten in [140] laid the foundation of Appendix E. Cpt. 6 and Cpt. 7 benefited from simulations by Bertold Bitachon [141] and Bharat Bhatia [142], respectively. The localization system implementation in Cpt. 7 received valuable contributions by Christoph Sulser, Manisha De [143], Bharat Bhatia [144], and Henry Schulten. The idea of alternating position and orientation estimates in Cpt. 7 is from Wolfgang Utschick. The work on Cpt. 8 was supported by the Commission for Technology and Innovation CTI, Switzerland and conducted in cooperation with

Schindler Aufzüge AG. This ultra-wideband approach received inputs by Marc Kuhn, Malte Göller [145] and Robert Heyn and employed the ray tracer used in [127,135,146] which was graciously provided by the group of Klaus Witrisal at Graz University of Technology. Fig. 1.1 was created in collaboration with Philipp Gosch; a snippet was used on the cover of [26] with permission.

On a personal note, I would like to express my gratitude to my family for continuously supporting my endeavours. To my father Franz and sister Helga for introducing me to engineering and higher mathematics. To Rahel for much joy and comfort during the later PhD stages. To Armin, Marc and all colleagues at the Wireless Communications Group for their support, many life lessons, and fostering a constructive and respectful work culture. To my Telematik colleagues from TU Graz for the many spirited exchanges and their invaluable support. To the SPSC researchers at TU Graz for introducing us to the engineering sciences and the beautiful associated theories. To Alfred Strauß for imposing his emphatic reality checks on the Styrian youth and Joachim Maderer for a precious hard-line introduction to computer science. To every visitor and Signalöler for compensating the sometimes stuffy ways of Zurich. To everyone who supported my relocation, despite the turmoil. To the great scientists of ages past. To the taxpayers of Switzerland and Austria for funding my education and research. And finally I want to thank all the keen people at the ITET department and the IEEE for tirelessly advancing our field.

Chapter 2

Essential Physics for Electrically Small Coils

In Sec. 1.3.2 we argued that radiative propagation modes should be included in a magneto-inductive coupling model, even if the involved coils are electrically small (i.e. much smaller than the employed wavelength). To this effect, Sec. 2.1 derives respective formulas for the mutual impedance between coils. We furthermore state necessary coil self-impedance formulas for relevant coil geometries in Sec. 2.2 and a description of coil interaction in terms of impedance matrices in Sec. 2.3. The exposition is preceded by a wrap-up of the essential physics.

2.1 Mutual Impedance Between Wire Loops

When an electric current i_T (complex-valued phasor, unit ampere) is applied at the terminals of a transmit antenna, the resulting induced voltage at a receive antenna is

$$v_R = Z_{RT} \, i_T \tag{2.1}$$

where Z_{RT} is the complex-valued *mutual impedance* Z_{RT} (a.k.a. transimpedance) between the two antennas. It is a key quantity for the description of a wireless link. This section is concerned with mathematical descriptions of Z_{RT} between two coils, given their wire geometries and relative posture. The situation is illustrated in Fig. 2.1. The specific objectives of this section are:

- Providing an insightful derivation of the general formula (2.16) for Z_{RT} between electrically small thin-wire coils, comprising near- and far-field propagation.

- Introduction of the simple linear-algebraic formula (2.23) for Z_{RT}, valid for coils whose turns have consistent surface orientation and for link distances appreciably larger than the coil dimensions.

Before we dive into details about magnetic induction we want to wrap up key principles of electromagnetism in order to recall the mechanisms and establish the

17

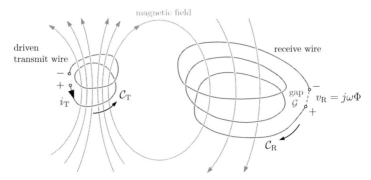

Figure 2.1: A basic magneto-inductive link. A current i_T drives the transmit coil wire whose geometry is described by the one-dimensional smooth curve $\mathcal{C}_T$ (whose direction is illustrated as well). The induced voltage v_R is measured between the terminals of the smooth curve $\mathcal{C}_R$, which describes the geometry of the receive coil wire.

notation. All quantities are in SI units. It is assumed that the reader is familiar with vector fields over space and time and with the basiscs of vector calculus such as the curl and divergence of fields as well as line and surface integrals.

Electromagnetism describes the forces on electrically charged particles due to the presence and movement of other electrical charges (the so-called field sources). Wireless systems use this mechanism in the fashion "move electrons at the transmitter to make electrons move at the receiver". In particular, the force $\vec{F} = q\,(\vec{E} + \vec{v} \times \vec{B})$ applies to a particle with electrical charge q and velocity $\vec{v}$ where $\vec{E}$ and $\vec{B}$ are the electric and magnetic field, respectively, at the particle position [147, Cpt. 18]. These fields arise due to the field sources, which are described by the volume charge density ϱ and the current density $\vec{J}$. Calculating $\vec{E}$ and $\vec{B}$ from given ϱ and $\vec{J}$ over space and time is a difficult problem; a complete framework to do so is given by Maxwell's famous four equations [148, 149] which are well-documented in modern physics literature [147] and wireless engineering literature [52, 62, 150, 151]. To describe Maxwell's equations in a nutshell, charges are sources and sinks of $\vec{E}$ according to the divergence $\nabla \cdot \vec{E} = \varrho/\varepsilon_0$ while $\vec{B}$ has no such sources or sinks, i.e. $\nabla \cdot \vec{B} = 0$. The law $\nabla \times \vec{B} = \mu_0(\vec{J} + \varepsilon_0\,\partial \vec{E}/\partial t)$ states that a solenoidal $\vec{B}$-field arises around a current or around a time-variant electric field (hence called a displacement current). Finally, by the law of induction $\nabla \times \vec{E} = -\partial \vec{B}/\partial t$ from Faraday [152], a solenoidal $\vec{E}$-field arises around a time-variant magnetic field.

For most purposes in wireless engineering the above laws are unnecessarily general and a description for harmonic quantities at some radial frequency $\omega = 2\pi f$ suffices.

Following the proposal of [52, Eq. 1.14] we write Maxwell's equations in phasor notation

$$\nabla \cdot \mathbf{e} = \rho/\varepsilon_0 \ , \tag{2.2}$$

$$\nabla \cdot \mathbf{b} = 0 \ , \tag{2.3}$$

$$\nabla \times \mathbf{e} = -j\omega \, \mathbf{b} \ , \tag{2.4}$$

$$\nabla \times \mathbf{b} = \mu_0 \left(\mathbf{j} + j\omega \, \varepsilon_0 \mathbf{e} \right) \tag{2.5}$$

in terms of complex-valued phasors $\mathbf{e}$, $\mathbf{b}$, ρ, $\mathbf{j}$. The original quantities relate to the phasors via

$$\vec{B} = \sqrt{2} \ \mathrm{Re}\{\mathbf{b}e^{j\omega t}\} \tag{2.6}$$

and so forth. Thereby e is Euler's number and j is the imaginary unit. An overview of the relevant quantities for the following exposition is given in Table 2.1.

Symbol	Set or Value	Description
i_T	$\mathbb{C} \cdot \mathrm{A}$ (ampere)	transmit current phasor
v_R	$\mathbb{C} \cdot \mathrm{V}$ (volt)	receive voltage phasor
ρ	$\mathbb{C} \cdot \frac{\mathrm{C}}{\mathrm{m}^3}$	charge density phasor
$\mathbf{j}$	$\mathbb{C}^3 \cdot \frac{\mathrm{A}}{\mathrm{m}^2}$	current density phasor
$\mathbf{e}$	$\mathbb{C}^3 \cdot \frac{\mathrm{V}}{\mathrm{m}}$	electric field phasor
$\mathbf{b}$	$\mathbb{C}^3 \cdot \frac{\mathrm{V \cdot s}}{\mathrm{m}^2} = \mathbb{C}^3 \cdot \mathrm{T}$ (tesla)	magnetic field (a.k.a. flux density) phasor
Φ	$\mathbb{C} \cdot \mathrm{V} \cdot \mathrm{s}$	magnetic flux phasor
φ	$\mathbb{C} \cdot \mathrm{V}$	electric potential phasor
$\mathbf{a}$	$\mathbb{C}^3 \cdot \mathrm{T} \cdot \mathrm{m}$	magnetic vector potential phasor
$\mathcal{C}_T$	$\subset \mathbb{R}^3 \cdot \mathrm{m}, \dim(\mathcal{C}_T) = 1$	directed curve, describes transmit wire
$\mathcal{C}_R$	$\subset \mathbb{R}^3 \cdot \mathrm{m}, \dim(\mathcal{C}_R) = 1$	directed curve, describes receive wire
r	$\mathbb{R} \cdot \mathrm{m}$	distance between wire points
$d\boldsymbol{\ell}$	$\mathbb{R}^3 \cdot \mathrm{m}$	directed length element
$d\mathbf{s}$	$\mathbb{R}^3 \cdot \mathrm{m}^2$	directed surface element
μ_0	$\approx 4\pi \cdot 10^{-7} \frac{\mathrm{T \cdot m}}{\mathrm{A}}$	vacuum permeability [153, App. 2]
ε_0	$\approx 8.854 \cdot 10^{-12} \frac{\mathrm{C}}{\mathrm{V \cdot m}}$	vacuum permittivity
c	$= \frac{1}{\sqrt{\mu_0 \varepsilon_0}} \approx 3 \cdot 10^8 \frac{\mathrm{m}}{\mathrm{s}}$	vacuum speed of light

Table 2.1: Overview of the relevant physical quantities of Sec. 2.1. Every complex phasor quantity x represents a harmonic signal $X(t) = \sqrt{2} \ \mathrm{Re}\{xe^{j\omega t}\}$ whereby both x and $X(t)$ are position-dependent, which is not denoted explicitly. The factor $\sqrt{2}$ ensures that $|x|^2$ equals the mean square value of $X(t)$. The same conversion rule $\vec{X}(t) = \sqrt{2} \ \mathrm{Re}\{\mathbf{x}e^{j\omega t}\}$ holds between a field vector $\vec{X}$ and its complex phasor representation $\mathbf{x}$.

A more compact description of electromagnetic effects is given by the electric potential φ and the magnetic vector potential $\mathbf{a}$. The relations

$$\mathbf{e} = -\nabla\varphi - j\omega\,\mathbf{a}\;, \tag{2.7}$$

$$\mathbf{b} = \nabla \times \mathbf{a} \tag{2.8}$$

constitute a full description of $\mathbf{e}$ and $\mathbf{b}$ over space. In many circumstances, φ and $\mathbf{a}$ can be calculated more easily than $\mathbf{e}$ and $\mathbf{b}$. In particular, one can calculate ϕ from ρ and $\mathbf{a}$ from $\mathbf{j}$ separately. [147, Sec. 18–6 and 21-3]

2.1.1 Voltage Induced in a Receive Wire

We consider a surface $\mathcal{A}$ (a two-dimensional manifold) with boundary $\mathcal{C}_R = \partial\mathcal{A}$ (a closed curve). On both sides of the vector equation $\nabla \times \mathbf{e} = -j\omega\,\mathbf{b}$ from (2.4) we form the surface integral over $\mathcal{A}$, giving $\iint_{\mathcal{A}}(\nabla \times \mathbf{e})\cdot d\mathbf{s} = -j\omega\iint_{\mathcal{A}}\mathbf{b}\cdot d\mathbf{s}$. By applying Stokes' theorem to the left-hand side we obtain the law of induction in integral form

$$\oint_{\mathcal{C}_R} \mathbf{e} \cdot d\boldsymbol{\ell}_R = -j\omega\,\Phi \tag{2.9}$$

where Φ is the magnetic flux through surface $\mathcal{A}$,

$$\Phi = \iint_{\mathcal{A}} \mathbf{b} \cdot d\mathbf{s}_R = \oint_{\mathcal{C}_R} \mathbf{a} \cdot d\boldsymbol{\ell}_R\;. \tag{2.10}$$

The latter formulation in terms of the line integral of $\mathbf{a}$ is a welcome mathematical simplification. It follows from $\mathbf{b} = \nabla \times \mathbf{a}$ in conjunction with Stokes' theorem.

Now consider a thin receive wire along a non-closed curve $\mathcal{W} \subset \mathcal{C}_R$, i.e. with two terminals and a gap between (see Fig. 2.1). The voltage across the wire terminals is

$$v_R = -\oint_{\mathcal{C}_R} \mathbf{e} \cdot d\boldsymbol{\ell}_R = j\omega\,\Phi \tag{2.11}$$

by the following argument (which is analogous to [147, Sec. 22-1]). Let $\mathcal{G}$ denote the straight line between the terminals, running from the end terminal to the start terminal of $\mathcal{W}$ such that the union $\mathcal{C}_R = \mathcal{W} \cup \mathcal{G}$ forms a closed curve. Inside the wire, $\mathbf{e}$ must be zero if the material has high conductivity. Then $\int_{\mathcal{W}} \mathbf{e}\cdot d\boldsymbol{\ell}_R = 0$ and the law of induction (2.9) dictates $\oint_{\mathcal{C}_R} \mathbf{e}\cdot d\boldsymbol{\ell}_R = \int_{\mathcal{G}} \mathbf{e}\cdot d\boldsymbol{\ell}_R = -j\omega\,\Phi$. Let v_R denote the voltage across $\mathcal{W}$, i.e. between start terminal (considered as plus pole) and end terminal (minus pole). This path is along the gap but in opposite direction as $\mathcal{G}$, hence $V_{\mathrm{R}} = -\int_{\mathcal{G}} \mathbf{e} \cdot d\boldsymbol{\ell}_{\mathrm{R}} = j\omega\,\Phi$, which proves (2.11).[1]

2.1.2 Fields Generated by a Driven Coil

The magnetic field $\mathbf{b}$ at some reference position $\mathbf{p}_{\mathrm{R}}$ due to a current density $\mathbf{j}$ in an infinitesimal volume element dV_{T} at position $\mathbf{p}_{\mathrm{T}}$ is given by

$$\mathbf{b}(\mathbf{p}_{\mathrm{R}}) = \frac{\mu_0}{4\pi} \frac{\mathbf{j} \times (\mathbf{p}_{\mathrm{R}} - \mathbf{p}_{\mathrm{T}})}{r^3} \, e^{-jkr} \, dV_{\mathrm{T}} \,. \tag{2.12}$$

Mathematically this formula is rather intricate due to the cross product. As an alternative we use the vector potential $\mathbf{a}$, which features the much simpler description[2]

$$\mathbf{a}(\mathbf{p}_{\mathrm{R}}) = \frac{\mu_0}{4\pi} \, \mathbf{j} \frac{e^{-jkr}}{r} \, dV_{\mathrm{T}} \,. \tag{2.13}$$

For the lengthy derivations of (2.12) and (2.13) we refer to [147, Cpt. 18 and 21]. The formulas incorporate the concept of *retarded time*: the field at position $\mathbf{p}_{\mathrm{R}}$ and time t is due to a source at $\mathbf{p}_{\mathrm{R}}$ at the earlier time $t - r/c$ because the effect propagates over the distance $r = \|\mathbf{p}_{\mathrm{R}} - \mathbf{p}_{\mathrm{T}}\|$ at the speed of light c. This means that the complex source phasor must be retarded, which is accomplished by multiplying with e^{-jkr}. This term uses the wavenumber $k = \frac{\omega}{c} = \frac{2\pi}{\lambda}$ at the considered frequency. The significance of this subtle detail to wireless engineering was emphasized by Ramo, "When retardation is neglected in the analysis of a circuit, the result will inevitably contain no possibility for radiation of energy" [154, Sec. 5-16].

---

[1]The notion of a voltage is problematic in the magnetic induction context where line integrals of $\mathbf{e}$ are path dependent due to non-zero curl $\nabla \times \mathbf{e}$. In the lumped element context this problem is fixed by assuming that magnetic fields are zero (or at least negligible) outside of the black box that represents an inductance, leading to a curl-free $\mathbf{e}$ between the terminals [147, Sec. 22-1]. We circumvented the issue by using a straight line $\mathcal{G}$ as integration path. This convenient but arbitrary choice is meaningful by the following argument. Consider the decomposition $\mathbf{e} = \mathbf{e}_{\mathrm{ind}} + \mathbf{e}_{\mathrm{wire}}$ where $\mathbf{e}_{\mathrm{ind}}$ is the electric field that would be observed in the absence of the wire and $\mathbf{e}_{\mathrm{wire}}$ is due to the charge density in the wire. The alternative voltage definition $V'_{\mathrm{R}} = -\int_{\mathcal{G}} \mathbf{e}_{\mathrm{wire}} \cdot d\boldsymbol{\ell}_{\mathrm{R}}$ is path independent because $\mathbf{e}_{\mathrm{wire}}$ is a conservative field. We note that $|V'_{\mathrm{R}} - V_{\mathrm{R}}| = |\int_{\mathcal{G}} \mathbf{e}_{\mathrm{ind}} \cdot d\boldsymbol{\ell}_{\mathrm{R}}| \le \max \|\mathbf{e}_{\mathrm{ind}}\| \cdot \ell_{\mathcal{G}}$ with gap width $\ell_{\mathcal{G}}$. A vanishing gap size $\ell_{\mathcal{G}} \to 0$ implies $|V'_{\mathrm{R}} - V_{\mathrm{R}}| \to 0$, although $|V'_{\mathrm{R}}|$ does not vanish. Hence, the approximation $V'_{\mathrm{R}} \approx V_{\mathrm{R}}$ is very accurate for a small gap. The discussion also shows that the induced voltage can be affected quite significantly by the shape of the feed wires and other termination circuitry (i.e. by the technical realization of a gap-closing curve $\mathcal{G}$).

[2]This formula holds under Lorenz gauge $\nabla \cdot \mathbf{a} = -j\omega\phi/c^2$, a popular choice for fixing the arbitrary divergence of $\mathbf{a}$ (which is left unspecified by the requirement $\nabla \times \mathbf{a} = \mathbf{b}$). The divergence of $\mathbf{a}$ is not relevant to our derivations because we use $\mathbf{a}$ only in almost-closed line integrals (cf. Footnote 1) whose values depends only on the curl of $\mathbf{a}$ (cf. Helmholtz decomposition and Stokes' theorem).

We consider a small element of a driven transmit wire. The element has directed length $d\boldsymbol{\ell}_{\mathrm{T}}$, volume dV_{T}, and carries a current i_{T}. The current density $\mathbf{j}$ may not be constant across the wire cross section (cf. skin effect). We note that (2.13) is linear in $\mathbf{j}$ and denote $\mathbf{j}_{\mathrm{avg}}$ for the mean current density in dV_{T}. If the wire is thin then $\mathbf{j}_{\mathrm{avg}}$ determines the contribution of this element to the vector potential $\mathbf{a}$ at a remote point $\mathbf{p}_{\mathrm{R}}$. This is subsequently assumed. We proceed by using the property $i_{\mathrm{T}}\, d\boldsymbol{\ell}_{\mathrm{T}} = \mathbf{j}_{\mathrm{avg}}\, dV_{\mathrm{T}}$ in (2.13). By superposition of many such small elements we find that the vector potential generated by a wire of geometry $\mathcal{C}_{\mathrm{T}}$ carrying a current i_{T} is

$$\mathbf{a}(\mathbf{p}_{\mathrm{R}}) = \frac{\mu_0}{4\pi} \int_{\mathcal{C}_{\mathrm{T}}} i_{\mathrm{T}}\, e^{-jkr}\, \frac{d\boldsymbol{\ell}_{\mathrm{T}}}{r} \ . \tag{2.14}$$

As outlined, we are ultimately interested in the effect of the transmit current i_{T} and the resulting vector potential (2.14) on the receive wire $\mathcal{C}_{\mathrm{R}}$. By using $\mathbf{a}$ from (2.14) and the law of induction in the form $v_{\mathrm{R}} = j\omega \int_{\mathcal{C}_{\mathrm{R}}} \mathbf{a} \cdot d\boldsymbol{\ell}_{\mathrm{R}}$ from (2.11) we obtain the induced voltage between the receive-wire terminals due to i_{T} running through $\mathcal{C}_{\mathrm{T}}$,

$$v_{\mathrm{R}} = \frac{j\omega\mu_0}{4\pi} \int_{\mathcal{C}_{\mathrm{R}}} \int_{\mathcal{C}_{\mathrm{T}}} i_{\mathrm{T}}\, e^{-jkr}\, \frac{d\boldsymbol{\ell}_{\mathrm{T}} \cdot d\boldsymbol{\ell}_{\mathrm{R}}}{r} \ . \tag{2.15}$$

Within the laws of classical physics and special relativity this is a general law for harmonic signals, a thin transmit wire[3] along the curve $\mathcal{C}_{\mathrm{T}}$, and a receive wire with just a small gap along the closed curve $\mathcal{C}_{\mathrm{R}}$. Regarding the integrand, note that i_{T} is a function of $\mathbf{p}_{\mathrm{T}} \in \mathcal{C}_{\mathrm{T}}$ while the distance $r = \|\mathbf{p}_{\mathrm{R}} - \mathbf{p}_{\mathrm{T}}\|$ depends on both $\mathbf{p}_{\mathrm{T}} \in \mathcal{C}_{\mathrm{T}}$ and $\mathbf{p}_{\mathrm{R}} \in \mathcal{C}_{\mathrm{R}}$.

2.1.3   The Low-Frequency Case

We now consider a transmit coil that is electrically small, i.e. its wire length is small compared to the employed wavelength ($\ell_{\mathrm{T}} \ll \lambda$). As a consequence, i_{T} is constant over $\mathcal{C}_{\mathrm{T}}$ according to Kirchhoff's current law for low-frequency operation on circuits. This is backed up by the results of Storer [155, Fig. 3] which show that the current distribution over a thin-wire single-turn circular loop is approximately constant for $\ell_{\mathrm{T}} \leq \lambda/10$. The same rough criterion $\ell_{\mathrm{T}} \leq \lambda/10$ is stated by Balanis [62, Sec. 5.3.2].

---

[3]In stating (2.15) we silently neglected the charge density ρ over the transmit wire which, by the charge conservation law [147, Sec. 13-2], necessarily arises if the current $i_{\mathrm{T}}(\mathbf{p}_{\mathrm{T}})$ varies over this wire. This ρ of course causes an electric potential ϕ [147, Eq. 21.15] which may affect the voltage v_{R} between the receive coil terminals. We neglect the effect because the voltage change goes to zero when the terminals are close (cf. Footnote 1), hence v_{R} is dominated by magnetic induction for a small gap.

If i_T is indeed spatially constant then it can be pulled out of the integral (2.15) and the equation takes the form $v_R = Z_{RT} i_T$. The mutual impedance Z_{RT} is given by the following proposition, which summarizes the preceding exposition.

Proposition 2.1. *The mutual impedance between two wires, evaluated at radial frequency $\omega = 2\pi f$, is given by the double line integral*

$$Z_{RT} = \frac{j\omega\mu_0}{4\pi} \int_{\mathcal{C}_R} \int_{\mathcal{C}_T} e^{-jkr} \frac{d\boldsymbol{\ell}_T \cdot d\boldsymbol{\ell}_R}{r} \tag{2.16}$$

under the conditions:

1. *The current-carrying transmit wire along the curve $\mathcal{C}_T$ is thin.*

2. *The receive wire along curve $\mathcal{C}_R$ is closed except for a small gap for the terminals.*

3. *The transmit wire is electrically small, i.e. the wire length is small compared to the wavelength λ to ensure an approximately constant spatial current distribution.*

The formula (2.16) is reciprocal in structure since the same formula applies when the roles of transmitter and receiver are exchanged. This holds for all coupling formulas presented in the following. This reciprocity is a general property of linear antennas [156] and electromagnetics [52, Sec. 1.9] rather than a magneto-inductive peculiarity.

2.1.4 The Magnetoquasistatic Regime

We consider the interesting special case $kr \ll 1$, which occurs when the link distance r is much smaller than a wavelength or, likewise, when the operating frequency is very small. In this case one can neglect the retardation term by arguing $e^{-jkr} \approx e^{j0} = 1$. In consequence, the mutual impedance (2.16) takes the purely imaginary value

$$Z_{RT} \approx j\omega M\,, \qquad\qquad M = \frac{\mu_0}{4\pi} \int_{\mathcal{C}_R} \int_{\mathcal{C}_T} \frac{d\boldsymbol{\ell}_T \cdot d\boldsymbol{\ell}_R}{r} \tag{2.17}$$

whereby the real-valued *mutual inductance* M does not depend on frequency. This equation is known as Neumann formula [157] and describes inductive coupling between two wires in terms of their geometries and relative arrangement. An attempt of an intuitive explanation was made by Feynman, "It depends on a kind of average separation of the two circuits, with the average weighted most for parallel segments of the two coils" [147, Sec. 17-6].

23

Formula (2.17) is exact in the magnetoquasistatic (MQS) physical system where Maxwell's equation (2.5) is changed from $\nabla \times \mathbf{b} = \mu_0 \left(\mathbf{j} + j\omega\,\varepsilon_0 \mathbf{e} \right)$ to just $\nabla \times \mathbf{b} = \mu_0 \mathbf{j}$, i.e. one discards time-variant electric fields as cause of solenoidal magnetic fields. This eliminates the possibility of coupling through radiated waves because the crucial interplay of $\mathbf{e}$ and $\mathbf{b}$ was discarded.

2.1.5 Small Loops, Circular Loops, Magnetic Dipoles

For certain transmit coil shapes the generated magnetic field has special geometrical and mathematical structure. In particular, we consider a coil with $\mathring{N}_{\mathrm{T}}$ turns as illustrated in Fig. 2.2. We require that the area enclosed by each turn is describable by a surface element whose orientation is consistent across all turns. Hence the turns must be reasonably flat but the shape of their outline can be arbitrary. This class of coils comprises, for example, solenoids with a small pitch angle and spider web coils. We work with the following geometric quantities:

- $\mathbf{p}_{\mathrm{T}}$ denotes the loop center position.

- $\mathbf{o}_{\mathrm{T}}$ is a unit vector describing the loop axis orientation which is orthogonal to the flat turns (the right-hand rule determines the sign).

- $\mathbf{p}_{\mathrm{R}}$ is the reference point where we want to determine the magnetic field.

- $r = \|\mathbf{p}_{\mathrm{R}} - \mathbf{p}_{\mathrm{T}}\|$ is the distance to the reference point.

- $\mathbf{u} = \frac{1}{r}(\mathbf{p}_{\mathrm{R}} - \mathbf{p}_{\mathrm{T}})$ is a unit vector in direction of the reference point.

- A_{T} is the mean area enclosed by the wire turns.

Proposition 2.2. *Consider an electrically small coil that carries a current i_{T} and has $\mathring{N}_{\mathrm{T}}$ wire turns. Each turn roughly sits in a 2D plane and the perpendicular orientation $\mathbf{o}_{\mathrm{T}}$ is consistent across all turns. Then the generated magnetic field (unit tesla) in complex phasor representation is, with good approximation, given by*

$$\mathbf{b} = \frac{\mu_0 A_{\mathrm{T}} \mathring{N}_{\mathrm{T}} k^3}{2\pi}\, e^{-jkr} \left(\left(\frac{1}{(kr)^3} + \frac{j}{(kr)^2} \right) \boldsymbol{\beta}_{\mathrm{NF}} + \frac{1}{2kr}\,\boldsymbol{\beta}_{\mathrm{FF}} \right) i_{\mathrm{T}}\,. \qquad (2.18)$$

The formula is accurate when r is large compared to the coil dimensions. It is exact for a magnetic dipole or a circular single-turn loop with $\mathbf{p}_{\mathrm{R}}$ outside the loop. The formula

24

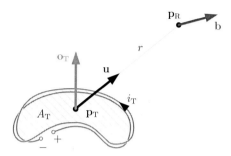

Figure 2.2: A coil with multiple flat turns (here $\mathring{N}_T = 2$) driven by a current i_T. The coil center is located at position $\mathbf{p}_T$. The sketch shows the relevant geometric quantities for calculating the magnetic field $\mathbf{b}$ at a reference point $\mathbf{p}_R$. Similar illustrations are found in [151, Fig. 2.10] and [147, Fig. 14-6].

uses the unitless field vector quantities $\boldsymbol{\beta}_{NF}$ and $\boldsymbol{\beta}_{FF}$, which we call the scaled near field and the scaled far field, respectively. They are given by[4]

$$\boldsymbol{\beta}_{NF} = \mathbf{F}_{NF}\,\mathbf{o}_T\,, \qquad \mathbf{F}_{NF} = \frac{1}{2}\Big(3\mathbf{u}\mathbf{u}^T - \mathbf{I}_3\Big)\,, \qquad (2.19)$$

$$\boldsymbol{\beta}_{FF} = \mathbf{F}_{FF}\,\mathbf{o}_T\,, \qquad \mathbf{F}_{FF} = \mathbf{I}_3 - \mathbf{u}\mathbf{u}^T\,. \qquad (2.20)$$

The statement is derived in Appendix A, based on an existing trigonometric field description for a circular loop. The formula is accurate at larger distance r because there the field is accurately described by a magnetic dipole at $\mathbf{p}_T$ with dipole moment $A_T\mathring{N}_T i_T \mathbf{o}_T$. The extension to multi-turn solenoids follows from superposition of multiple single-turn loops (an appropriate model when the pitch angle is small) and the fact that, for large r, the offset between the coil center $\mathbf{p}_T$ and the individual turn centers becomes negligible. Likewise, the proposition extends to arbitrary outline shapes because the current around A_T can be modeled equivalently as superposition of many small loops distributed across A_T, each carrying i_T and canceling each other on the interior (a common argument in the context of Stokes' theorem, cf. [147, Fig. 3-9]). The extensions also follow by requiring a reciprocal mutual impedance between a small

---

[4]It shall be noted that the linear transforms $\mathbf{F}_{NF}, \mathbf{F}_{FF} \in \mathbb{R}^{3\times3}$ from (2.19) and (2.20) depend on the transmitter-to-receiver direction $\mathbf{u}$ because our formalism allows for an arbitrary choice of coordinate system. They would be constant if the coordinate system was fixed with respect to $\mathbf{u}$ (e.g. by setting $\mathbf{u} = [1\ 0\ 0]^T$). Another noteworthy aspect is that the linear transform $\boldsymbol{\beta}_{FF} = \mathbf{F}_{FF}\,\mathbf{o}_T = \mathbf{o}_T - \mathbf{u}(\mathbf{u}^T\mathbf{o}_T)$ simply removes the $\mathbf{u}$-component from $\mathbf{o}_T$, like a step of a Gram-Schmidt process. This corresponds to the fact that transverse electromagnetic waves have no radial field component. Hence $\boldsymbol{\beta}_{FF}$ vanishes when $\mathbf{u} \approx \mathbf{o}_T$, corresponding to the radiation pattern zero on the axis of a circular loop antenna, which can be seen in [62, Fig. 5.8(a)].

circular coil and a coil of more complicated shape (cf. Proposition 2.4 later on).

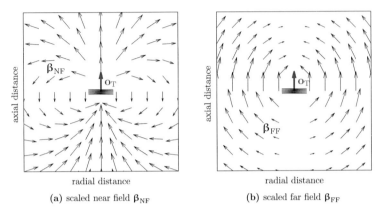

(a) scaled near field β_{NF} (b) scaled far field β_{FF}

Figure 2.3: Scaled near and far field illustrated as vector fields around a transmitting coil, whose axis orientation is described by unit vector $\mathbf{o}_{\mathrm{T}}$. They are scaled in the sense that no path loss occurs and their maximum magnitude is 1. In particular, $\beta_{\mathrm{NF}} = 1$ holds on the coil axis ($\mathbf{u} = \pm\mathbf{o}_{\mathrm{T}}$) and $\beta_{\mathrm{FF}} = 1$ holds in the plane where $\mathbf{u}$ is orthogonal to $\mathbf{o}_{\mathrm{T}}$. On this plane, β_{NF} attains its smallest magnitude $\beta_{\mathrm{NF}} = \frac{1}{2}$. The far field fades to $\beta_{\mathrm{FF}} = 0$ on the coil axis.

The introduced vector fields $\boldsymbol{\beta}_{\mathrm{NF}}$ and $\boldsymbol{\beta}_{\mathrm{FF}}$ have rotational symmetry around the transmit coil axis $\mathbf{o}_{\mathrm{T}}$. They are unitless and scaled such they have no path loss: their maximum magnitude is 1. They are illustrated and explained by Fig. 2.3.

Proposition 2.3. *The magnitudes of the scaled near field and scaled far field are*

$$\beta_{\mathrm{NF}} := \|\boldsymbol{\beta}_{\mathrm{NF}}\| = \frac{1}{2}\sqrt{1 + 3(\mathbf{u}^{\mathrm{T}}\mathbf{o}_{\mathrm{T}})^2}\,, \qquad \frac{1}{2} \le \beta_{\mathrm{NF}} \le 1\,, \qquad (2.21)$$

$$\beta_{\mathrm{FF}} := \|\boldsymbol{\beta}_{\mathrm{FF}}\| = \sqrt{1 - (\mathbf{u}^{\mathrm{T}}\mathbf{o}_{\mathrm{T}})^2}\,, \qquad 0 \le \beta_{\mathrm{FF}} \le 1\,. \qquad (2.22)$$

Proof. Many equivalent statements are found in standard literature on dipoles and dipole antennas, e.g. [147, Eq. 6.14], so this shall just provide a short derivation in our notation. We note that $\beta_{\mathrm{NF}}^2 = \boldsymbol{\beta}_{\mathrm{NF}}^{\mathrm{T}}\boldsymbol{\beta}_{\mathrm{NF}} = \mathbf{o}_{\mathrm{T}}^{\mathrm{T}}\mathbf{F}_{\mathrm{NF}}^2\,\mathbf{o}_{\mathrm{T}}$ and calculate $\mathbf{F}_{\mathrm{NF}}^2 = \frac{1}{4}(\mathbf{I}_3 + 3\mathbf{u}\mathbf{u}^{\mathrm{T}})$. Hence $\beta_{\mathrm{NF}}^2 = \frac{1}{4}(\mathbf{o}_{\mathrm{T}}^{\mathrm{T}}\mathbf{I}_3\mathbf{o}_{\mathrm{T}} + 3(\mathbf{o}_{\mathrm{T}}^{\mathrm{T}}\mathbf{u})(\mathbf{u}^{\mathrm{T}}\mathbf{o}_{\mathrm{T}})) = \frac{1}{4}(1 + 3(\mathbf{u}^{\mathrm{T}}\mathbf{o}_{\mathrm{T}})^2)$. Taking the square root yields the formula for β_{NF}. For β_{FF} the proof is analogous and facilitated by $\mathbf{F}_{\mathrm{FF}}^2 = \mathbf{F}_{\mathrm{FF}}$. The value ranges are a trivial consequence of the magnitude formulas. $\qquad\square$

In fact one can express $\beta_{FF} = \sin\theta_T$ where θ_T is the angle between $\mathbf{u}$ and $\mathbf{o}_T$, i.e. $\mathbf{u}^T\mathbf{o}_T = \cos\theta_T$. For β_{NF} no noteworthy insight is obtained from a trigonometric formulation. We use the vector formulation throughout this thesis because it provides a convenient interface to the linear-algebraic approach to communication theory as exercised in [158, Cpt. 2] and [90, 159] and this thesis (in particular later in Sec. 2.3.2 and Sec. 3.8).

To comprehend the meaning of a complex field vector such as $\mathbf{b}$ in (2.18), recall from (2.6) that the phasor $\mathbf{b}$ represents an oscillating field $\vec{B}(t) = \sqrt{2}\,\text{Re}\{\mathbf{b}\}\cos(\omega t) - \sqrt{2}\,\text{Im}\{\mathbf{b}\}\sin(\omega t)$. If $\text{Re}\{\mathbf{b}\}$ and $\text{Im}\{\mathbf{b}\}$ are linearly independent then $\vec{B}$ oscillates on an ellipsis. For linearly dependent parts, the ellipsis degenerates to a line. In Cpt. 4 we will see that this aspect has significant implications for the outage probability of a link between arbitrarily oriented coils.

Proposition 2.4. *Consider a receive coil with center position $\mathbf{p}_R$ and wire geometry such that the area enclosed by the turns can be meaningfully described by flat surfaces with equal orientations. The transmit coil geometry and distance are such that Proposition 2.2 holds. Then the mutual impedance between the coils is approximated by*

$$Z_{RT} = \frac{j\omega\mu_0 A_T \mathring{N}_T A_R \mathring{N}_R\, k^3}{2\pi} e^{-jkr}\left(\left(\frac{1}{(kr)^3} + \frac{j}{(kr)^2}\right)J_{NF} + \frac{1}{2kr}\,J_{FF}\right) \qquad (2.23)$$

which uses the (unitless) alignment factors

$$J_{NF} = \mathbf{o}_R^T\boldsymbol{\beta}_{NF} = \mathbf{o}_R^T\,\mathbf{F}_{NF}\,\mathbf{o}_T\ , \qquad J_{NF} \in [-\beta_{NF}, \beta_{NF}] \subseteq [-1, 1]\ , \qquad (2.24)$$

$$J_{FF} = \mathbf{o}_R^T\boldsymbol{\beta}_{FF} = \mathbf{o}_R^T\,\mathbf{F}_{FF}\,\mathbf{o}_T\ , \qquad J_{FF} \in [-\beta_{FF}, \beta_{FF}] \subseteq [-1, 1]\ . \qquad (2.25)$$

Proof. If the receive-wire geometry can be meaningfully represented by flat surfaces with area A_R enclosed by $\mathring{N}_R$ turns and with equal orientation $\mathbf{o}_R$, then the magnetic flux $\Phi = \iint_{A_R} \mathbf{b} \cdot d\mathbf{s}_R$ is approximated by the scalar product $\Phi \approx A_R\mathring{N}_R\mathbf{o}_R^T\,\mathbf{b}(\mathbf{p}_R)$. Combining this with the $\mathbf{b}$-field description (2.18) and $Z_{RT} = v_R/i_T = j\omega\Phi/i_T$ yields the proposition. The value range of J_{NF} (analogously, of J_{FF}) follows from the Cauchy–Schwarz inequality $|J_{NF}| = |\mathbf{o}_R^T\boldsymbol{\beta}_{NF}| \leq \|\mathbf{o}_R\| \cdot \|\boldsymbol{\beta}_{NF}\| = 1 \cdot \beta_{NF}$. $\qquad\square$

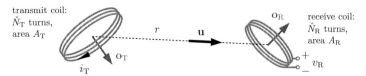

transmit coil: $\overset{\circ}{N}_T$ turns, area A_T

r $\mathbf{u}$

$\mathbf{o}_R$ receive coil: $\overset{\circ}{N}_R$ turns, area A_R

$\mathbf{o}_T$

i_T

$+$ v_R $-$

Figure 2.4: Magneto-inductive link between two coils. The unit vector $\mathbf{u}$ points from transmitter to receiver while the unit vectors $\mathbf{o}_T$ and $\mathbf{o}_R$ describe the axis orientations of the transmit and receive coils, respectively. The relation $v_R = Z_{RT} i_T$ is characterized by the mutual impedance Z_{RT}, which is described by Proposition 2.4.

Proposition 2.5. *Under the same conditions as in Proposition 2.4, the mutual inductance between two coils is accurately approximated by[5]*

$$M = \frac{\mu_0 A_T \overset{\circ}{N}_T A_R \overset{\circ}{N}_R}{2\pi} \frac{J_{NF}}{r^3} \,. \tag{2.26}$$

Proof. Require $M = Z_{RT}/(j\omega)$ according to (2.17). Substitute (2.23) for Z_{RT} and, with the resulting expression, form the low-frequency limit $k \to 0$ to obtain the formula. $\square$

A necessary criterion for (2.23) being accurate is that $\mathbf{b}(\mathbf{p}_R)$ is representative of the mean-$\mathbf{b}$ over the coil. This holds when the dimensions of the receive coil are small compared to r. To summarize the criteria with a tangible statement, (2.23) is accurate when (i) both coils have dimensions much smaller than $\min\{r, \lambda\}$ and (ii) either coil geometry can be modeled in terms of a composition of flat single-turn loops with equal surface orientation. The quantities $A_T \overset{\circ}{N}_T$ and $A_R \overset{\circ}{N}_R$ can be replaced by $\sum_{n=1}^{\overset{\circ}{N}_T} A_{T,n}$ and $\sum_{n=1}^{\overset{\circ}{N}_R} A_{R,n}$ when the enclosed areas are not equal, e.g. for a printed coil or a spiderweb coil. The formula (2.23) demonstrates reciprocity in the term $A_T \overset{\circ}{N}_T A_R \overset{\circ}{N}_R$ and in the bilinear forms $J_{NF} = \mathbf{o}_R^T \mathbf{F}_{NF} \mathbf{o}_T$ and $J_{FF} = \mathbf{o}_R^T \mathbf{F}_{FF} \mathbf{o}_T$ whereby the matrices $\mathbf{F}_{NF}, \mathbf{F}_{FF}$ are symmetric and invariant under a sign flip $\mathbf{u}' = -\mathbf{u}$. Example values and situations for the near- and far-field alignment factors are shown in Fig. 2.5.

---

[5]The quantity J was introduced by Kisseleff et al. [67, 160] for near-field links between arbitrarily oriented coils. They called J the polarization factor and describe it in terms of three trigonometric angles. These angles need to be measured with correct sign which is tricky in 3D. Our vector formulation is more straightforward to use (a sign error would, at worst, flip the sign of J and not lead to a completely different result) and is compatible with the linear algebra of communication theory. Our definition exhibits $J_{NF} \in [-1, 1]$ such that $10 \log_{10} J_{NF}^2$ characterizes the misalignment loss of a near-field link in dB (or rather $10 \log_{10} J_{NF}^4$ for RFID load modulation). The definition of [67, 160] uses a different value range $[-2, 2]$, their formula for M is however equivalent to (2.26).

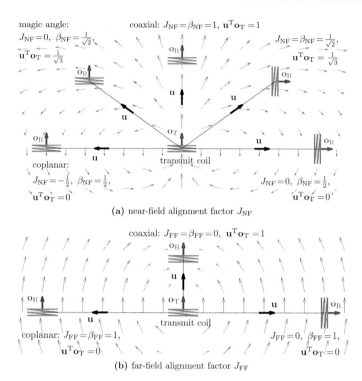

(a) near-field alignment factor J_{NF}

(b) far-field alignment factor J_{FF}

Figure 2.5: Example arrangements of coil pairs and the associated values of the near-field alignment factor $J_{NF} \in [-1,1]$ and the far-field alignment factor $J_{FF} \in [-1,1]$.

In the following we introduce an equivalent formulation of Z_{RT} in (2.23) which will prove very useful later in the thesis. It is a bilinear form of the two coil orientations

$$Z_{RT} = \mathbf{o}_R^T \, \mathbf{Z}_{3DoF} \, \mathbf{o}_T , \tag{2.27}$$

$$\mathbf{Z}_{3DoF} = Z_0 \left(\left(\frac{1}{(kr)^3} + \frac{j}{(kr)^2} \right) \mathbf{F}_{NF} + \frac{1}{2kr} \, \mathbf{F}_{FF} \right) , \tag{2.28}$$

$$Z_0 = \frac{j\omega\mu_0 A_T \mathring{N}_T A_R \mathring{N}_R \, k^3}{2\pi} e^{-jkr} . \tag{2.29}$$

The specific case of coaxial coil arrangement (i.e. $\mathbf{o}_{\mathrm{T}} = \mathbf{o}_{\mathrm{R}} = \mathbf{u}$) exhibits

$$Z_{\mathrm{RT}}^{\mathrm{coax}} = Z_0 \left(\frac{1}{(kr)^3} + \frac{j}{(kr)^2} \right) \qquad \text{through } J_{\mathrm{NF}} = 1, \, J_{\mathrm{FF}} = 0. \qquad (2.30)$$

Likewise, coplanar coil arrangement (i.e. $\mathbf{o}_{\mathrm{T}} = \mathbf{o}_{\mathrm{R}}$ are orthogonal to $\mathbf{u}$) exhibits

$$Z_{\mathrm{RT}}^{\mathrm{copl}} = \frac{Z_0}{2} \left(-\frac{1}{(kr)^3} - \frac{j}{(kr)^2} + \frac{1}{kr} \right) \qquad \text{through } J_{\mathrm{NF}} = -\tfrac{1}{2}, \, J_{\mathrm{FF}} = 1. \qquad (2.31)$$

These are in fact eigenvalues of $\mathbf{Z}_{3\mathrm{DoF}}$ because $\mathbf{Z}_{3\mathrm{DoF}}\mathbf{u} = Z_{\mathrm{RT}}^{\mathrm{coax}}\mathbf{u}$ and $\mathbf{Z}_{3\mathrm{DoF}}\mathbf{u}_{\perp} = Z_{\mathrm{RT}}^{\mathrm{copl}}\mathbf{u}_{\perp}$ for any $\mathbf{u}_{\perp}$ that is orthogonal to $\mathbf{u}$ (hence the eigenvalue $Z_{\mathrm{RT}}^{\mathrm{copl}}$ has a geometric multiplicity of two). This leads to the eigenvalue decomposition

$$\mathbf{Z}_{3\mathrm{DoF}} = \mathbf{Q}_{\mathbf{u}} \begin{bmatrix} Z_{\mathrm{RT}}^{\mathrm{coax}} & 0 & 0 \\ 0 & Z_{\mathrm{RT}}^{\mathrm{copl}} & 0 \\ 0 & 0 & Z_{\mathrm{RT}}^{\mathrm{copl}} \end{bmatrix} \mathbf{Q}_{\mathbf{u}}^{\mathrm{T}} \qquad (2.32)$$

where any orthogonal matrix $\mathbf{Q}_{\mathbf{u}} \in \mathbb{R}^{3\times 3}$ with $\mathbf{u}$ in the first column holds a suitable set of eigenvectors. As an illustrative example, if the coordinate system is fixed so that $\mathbf{u} = [1\ 0\ 0]^{\mathrm{T}}$ and the canonical choice $\mathbf{Q}_{\mathbf{u}} = \mathbf{I}_3$ is made, we obtain the diagonalized form $\mathbf{Z}_{3\mathrm{DoF}} = \mathrm{diag}(Z_{\mathrm{RT}}^{\mathrm{coax}}, Z_{\mathrm{RT}}^{\mathrm{copl}}, Z_{\mathrm{RT}}^{\mathrm{copl}})$. The same decomposition was stated for the pure near-field case in [95].

It shall be noted that formulas analogous to (2.23) and (2.27) hold for the mutual impedance between two electric dipoles (just with a different multiplier Z_0) due to electromagnetic duality [62, Sec. 3.7]. The coupling between a magnetic and an electric dipole could be calculated using the electric-field formula (A.4) from Appendix A.

Given a link with fixed kr and $\mathbf{u}$, an interesting question is the optimal choice of coil orientations $\mathbf{o}_{\mathrm{T}}, \mathbf{o}_{\mathrm{R}}$ such that $|Z_{\mathrm{RT}}|$ is maximized. The above eigenvalue decomposition reveals that coaxial arrangement is optimal for small $kr \leq kr_{\mathrm{th}}$ and that coplanar arrangement is optimal otherwise. The threshold is found to be (and was also stated in [161])

$$kr_{\mathrm{th}} = \sqrt{\frac{\sqrt{37}+5}{2}} = 2.3540 \qquad \Longleftrightarrow \qquad r_{\mathrm{th}} = 0.3747 \cdot \lambda. \qquad (2.33)$$

The case $kr = kr_{\mathrm{th}}$ exhibits $|Z_{\mathrm{RT}}^{\mathrm{coax}}| = |Z_{\mathrm{RT}}^{\mathrm{copl}}|$ and thus has the property $\mathbf{Z}_{3\mathrm{DoF}}\mathbf{Z}_{3\mathrm{DoF}}^{\mathrm{H}} = |Z_{\mathrm{RT}}^{\mathrm{coax}}|^2 \mathbf{I}_3 = |Z_{\mathrm{RT}}^{\mathrm{copl}}|^2 \mathbf{I}_3$. We will revisit this property in the context of spatial multiplexing over MIMO links.

For the different introduced formulas for Z_{RT} in this section, Appendix B discusses the mathematical correspondences between them. It furthermore shows the mathematical cause of the different propagation modes (including radiation) arising from electrically small coils.

2.1.6 Effect of Propagation Medium

So far we discussed formulas for the magnetic field and mutual impedance in the absence of a propagation medium, i.e. in free space. In the following we summarize the simple extension to a homogeneous medium that is furthermore linear and isotropic, with relative permeability μ_r and relative permittivity ϵ_r (hence $\mu = \mu_r\mu_0$ and $\epsilon = \epsilon_r\epsilon_0$). The phase velocity of wave propagation changes to $c_p = \frac{c}{\sqrt{\mu_r\epsilon_r}} = \frac{1}{\sqrt{\mu\epsilon}}$, whereby the speed of light c was the phase velocity in free space. [52, Sec. 1.4]

Let us first consider a lossless medium (a.k.a. non-conducting medium or dielectric). This case is incorporated into the formulas simply by replacing any occurrence of μ_0 by μ and by now calculating the wavenumber according to $k = \frac{\omega}{c_p} = \omega\sqrt{\mu\epsilon}$.

We now consider a lossy medium with a non-zero conductivity σ (unit $\frac{1}{\Omega\cdot m}$). Instead of the wavenumber, this case is characterized by the complex propagation constant [52, Eq. 1.52]

$$\gamma = j\omega\sqrt{\mu\epsilon\left(1 - \frac{j\sigma}{\omega\epsilon}\right)}. \tag{2.34}$$

Now the exponential function $e^{-\gamma r}$ replaces every occurrence of e^{-jkr} in the mutual impedance formulas. Hence, any field magnitude or link coefficient will be proportional to the term $e^{-\gamma r} = e^{-\mathrm{Re}(\gamma)r}e^{-j\mathrm{Im}(\gamma)r}$ whereby $\mathrm{Re}(\gamma)$ is the exponential decay rate. In other words, the medium conductivity causes an amplitude attenuation of e^{-1} over a distance of $\frac{1}{\mathrm{Re}(\gamma)}$ which is called the skin depth or depth of penetration. The term $e^{-j\mathrm{Im}(\gamma)r}$ on the other hand is associated with spatial frequency $\mathrm{Im}(\gamma)$ and phase velocity $c_p = \frac{\omega}{\mathrm{Im}(\gamma)}$. For good conductors ($\sigma \gg \omega\epsilon$) the more specific formulas $\gamma \approx (1+j)\sqrt{\frac{\omega\mu\sigma}{2}}$ and $\frac{1}{\mathrm{Re}(\gamma)} \approx \sqrt{\frac{2}{\omega\mu\sigma}}$ apply. We note that the depth of penetration decays with frequency according to $\frac{1}{\mathrm{Re}(\gamma)} \propto \sqrt{\frac{1}{\omega}}$, a manifestation of the material penetration advantage of low-frequency magnetic induction.

When the propagation medium is non-homogeneous, e.g., when the environment comprises various conducting and dielectric objects, then usually the propagation characteristics can not be described in closed form. This is the domain of wireless channel modeling. A notable exception is the field generated by a coil near a perfectly conducting half-space: the field is the sum of the free-space solutions for the actual coil location and for the coil location mirrored at the plane that separates the conducting half-space [61].

2.1.7 Iron Cores

The use of coils with a high-permeability ferromagnetic iron core can yield an immense increase of the magnetic flux density and thus a link improvement. At larger frequencies however core loses become drastic, e.g. at around $200\,\text{Hz}$ for laminated ferromagnetic cores [162] and around $300\,\text{kHz}$ for ferrite cores [163]. Furthermore the added mass, occupied volume and potential biocompatibility problems [164] of an iron core may be undesired from an application perspective (e.g. medical in-body applications). Iron cores are thus not considered any further by this dissertation; they are however certainly an interesting aspect for future work.

2.2 Coil Self-Impedance

A suitable model for the self-impedance of a coil is required to describe its electrical interaction with connected circuitry. The complex-valued self-impedance Z_{self} relates voltage v and current i at the coil terminals via $v = Z_{\text{self}}\, i$. We model its value as a function of frequency f in terms of the equivalent circuit shown in Fig. 2.6.

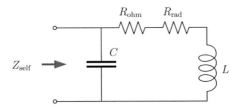

Figure 2.6: Equivalent circuit description of the self-impedance of an electrically small coil, taken from [62, Fig. 5.4]. It comprises four lumped elements: the self-inductance L, ohmic resistance R_{ohm}, radiation resistance R_{rad}, and self-capacitance C. The values of R_{ohm} and R_{rad} are frequency-dependent.

According to this equivalent circuit the self-impedance value is

$$Z_{\text{self}}(f) = \frac{\bar{Z}_{\text{self}}(f)}{1 + j\omega C \bar{Z}_{\text{self}}(f)} \tag{2.35}$$

where $\bar{Z}_{\text{self}}$ is the value when the self-capacitance C is neglected (e.g. at low f) or zero,

$$\bar{Z}_{\text{self}}(f) = R_{\text{ohm}}(f) + R_{\text{rad}}(f) + j\omega L \,. \tag{2.36}$$

The ohmic resistance R_{ohm}, radiation resistance R_{rad}, self-inductance L, and self-capacitance C all depend on the specific coil geometry; formulas will be given in the following subsection.

Tightly wound multi-turn coils usually exhibit a self-resonance frequency f_{res}. Thereby the interplay of inductance and inter-turn capacitance compensates the coil reactance, i.e. $\text{Im}(Z_{\text{self}}(f)) = 0$ at $f = f_{\text{res}}$. According to a circuit analysis,

$$\omega_{\text{res}} = 2\pi f_{\text{res}} = \sqrt{\frac{1}{LC} - \left(\frac{R}{L}\right)^2} \overset{\text{high Q}}{\approx} \frac{1}{\sqrt{LC}} \,. \tag{2.37}$$

The frequency-dependent resistance $R = R_{\text{ohm}} + R_{\text{rad}}$ prevents a direct evaluation of the precise formula. A simple workaround is to compute f_{res} with the approximation, evaluate R at this frequency and then compute a refined value of f_{res}. Most literature just uses the approximation because the relative error, given by $1 - \sqrt{1 - \frac{1}{Q^2}} \approx \frac{1}{2Q^2}$, is very small for reasonably large Q. At $f = f_{\text{res}}$ the self-impedance takes the large resistive value $Z_{\text{self}} = \frac{L}{RC} = RQ^2$. The coil quality factor relating to self-resonance is

$$Q = \frac{\omega_{\text{res}} L}{R} = \frac{1}{\omega_{\text{res}} RC} = \frac{f_{\text{res}}}{B_{\text{3dB}}} \,. \tag{2.38}$$

The same Q-factor formula applies when resonance is realized at a different f_{res} by connecting the coil with a serial or parallel capacitance (or a more involved matching network). In this context please also note that R is f-dependent.

Caution is advised when an equivalent circuit model is used to describe $Z_{\text{self}}(f)$ of a multi-turn coil near and above f_{res} because the effect of C is associated with a distributed charge density and a spatially varying current over the wire, i.e. the coil can not be considered electrically small. A proper description of these effects would require a transmission line model with distributed inductance and capacitance [69, 165], and one would have to consider the field generated by the charge density. Put differently, the equivalent circuit model of $Z_{\text{self}}(f)$ is certainly appropriate if f is appreciably smaller than f_{res} and also $\lambda = \frac{c}{f}$ is much longer than the wire.

The coil self-capacitance will not be depicted explicitly in circuit diagrams throughout this thesis in order to keep them simple and avoid confusion with matching circuits; it is however considered in all numerical results. The presented self-impedance model is compared to measurements later in Fig. 7.13 of Sec. 7.5.

2.2.1 Formulas for Circular Single-Layer Solenoid Geometry

The general geometry is shown in Fig. 2.7a. The electrical properties are characterized by the coil diameter D_c, axial coil length ℓ_c, wire diameter D_w, wire length $\ell_w = \sqrt{(\pi D_c \mathring{N})^2 + \ell_c^2}$, turn number $\mathring{N}$, enclosed area $A = \frac{\pi}{4} D_c^2$, the permeability μ of the surrounding medium and the conductivity σ of the wire material (e.g., $\sigma \approx 6 \cdot 10^7 \frac{1}{\Omega \cdot m}$ for copper).

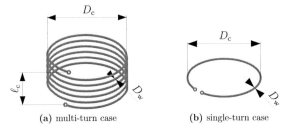

(a) multi-turn case　　　　　(b) single-turn case

Figure 2.7: Circular single-layer solenoid coil geometry (feed wires are not considered).

The ohmic resistance of a single-layer solenoid coil is given by [62, Sec. 5.2.3]

$$R_{\text{ohm}} = \frac{\ell_w}{\sigma A_\delta} \left(1 + \frac{R_p}{R_0} \right) \tag{2.39}$$

where $A_\delta = \pi (\frac{D_w}{2})^2 - \pi (\frac{D_w}{2} - \delta)^2 = \pi \delta (D_w - \delta)$ is the wire cross-sectional area where current effectively flows. It is determined by the current skin depth $\delta = \min\{\frac{D_w}{2}, \sqrt{\frac{2}{\omega \mu \sigma}}\}$ (cf. depth of penetration $\frac{1}{\text{Re}(\gamma)}$ in Sec. 2.1.6) [26, Eq. 3.26]. If the skin depth is $\delta = \sqrt{\frac{2}{\omega \mu \sigma}}$ and appreciably larger than $\frac{D_w}{2}$ (i.e. for $f > \frac{4}{\pi D_w^2 \mu \sigma}$) then $A_\delta \approx \pi D_w \delta$ and the ohmic resistance scales like $R_{\text{ohm}} \propto \frac{1}{A_\delta} \propto \frac{1}{\delta} \propto \sqrt{f}$ due to the skin effect. The ratio $\frac{R_p}{R_0} \geq 0$ describes the relative increase of the ohmic resistance due to the proximity effect, i.e. the added resistance due to the interaction of magnetic fields by nearby coil turns. Its value as a function of turn spacing and turn number is given by Fig. 2.8.

The radiation resistance is given by [62, Sec. 5.2.3]

$$R_{\text{rad}} = \sqrt{\frac{\mu}{\epsilon}} \frac{A^2 k^4 \mathring{N}^2}{6\pi} = \frac{1}{3} \mu k^3 f A^2 \mathring{N}^2 \tag{2.40}$$

for an electrically small coil, i.e. the formula is valid for about $\ell_w \leq \frac{\lambda}{10}$ or rather

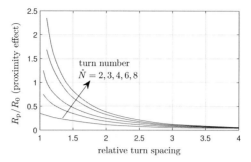

Figure 2.8: Relative increase of the ohmic coil resistance due to the proximity effect, plotted versus relative turn spacing $\ell_c/(D_w\mathring{N})$ for different turn numbers. Neighboring turns touch each other at a relative turn spacing of 1. This is a reproduction of the numerical results in [62, Fig. 5.3] which originate from [166]; please consult these sources for the most accurate results.

$f \le \frac{c}{10\ell_w}$. Note that $R_{\text{rad}} \propto f^4$ in this regime since $k = \frac{2\pi f}{c}$.

The self-inductance L in unit henry is given by [167]

$$L = \frac{\mu\mathring{N}^2 A}{3\pi\ell_c}\left(\frac{4\ell_c\tilde{\ell}}{D_c^2}(F - E) + \frac{4\tilde{\ell}}{\ell_c}E - \frac{4D_c}{\ell_c}\right) \qquad (2.41)$$

whereby $\tilde{\ell} = \sqrt{\ell_c^2 + D_c^2}$. Furthermore, F and E are the elliptic integrals of the first and second kind, evaluated at $D_c/\tilde{\ell}$. It shall be noted that the formula (2.41) specifies the so-called external inductance; the actual coil inductance is the sum of external inductance and the internal inductance of the wire given by $L_{\text{int}} = \frac{\ell_w}{\pi D_w}\sqrt{\frac{\mu}{2\omega\sigma}}$ [62, Eq. 5-38]. This contribution is often negligibly small.

For the coil self-capacitance with unit farad we use the empirical formula [165, Sec. 5.3] (an adaptation of [168]),

$$C = \frac{4\,\epsilon\,\ell_c}{\pi a^2}\left(1 + 1.78 \cdot S\right) \qquad (2.42)$$

where ϵ is the medium permittivity, $a = \pi D_c\mathring{N}/\ell_w$ is the cosine of the turn pitch angle, and $S = 0.71744\left(\frac{D_c}{\ell_c}\right) + 0.93305\left(\frac{D_c}{\ell_c}\right)^{1.5} + 0.106\left(\frac{D_c}{\ell_c}\right)^2$. Clearly C increases with the coil flatness $\frac{D_c}{\ell_c}$. The formula is valid for about $\frac{D_c}{\ell_c} \le 5$, i.e. not for very flat coils.

An interesting special case is a coil with just a single flat circular turn (i.e. $\mathring{N} = 1$, $\ell_c = 0$, $\ell_w = \pi D_c$) as shown in Fig. 2.7b. For this case the simpler inductance formula $L = \frac{1}{2}\mu D_c\left(\log(\frac{8D_c}{D_w}) - 2\right)$ is available [62, Eq. 5-37a]. Here, self-capacitance is not a relevant concept because of the absence of tightly spaced turns, hence $C = 0$ can be assumed. In consequence, the coil will not be self-resonant at any frequency where it is electrically small. Likewise, there is no proximity effect and so $\frac{R_p}{R_0} = 0$.

We conclude the exposition on single-layer solenoids with the evaluation in Fig. 2.9 of $\mathrm{Re}(Z_{\mathrm{self}})$ and the realizable Q-factor (which multiplies the maximum power gain over a weakly-coupled link, as we will see later) over frequency.

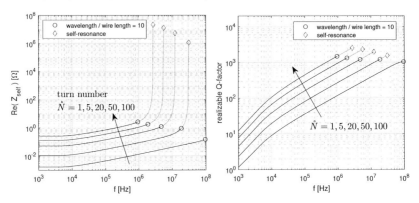

Figure 2.9: Evolution of selected electrical properties of single-layer solenoid coils over frequency f, plotted for varying turn number $\mathring{N}$. The coil diameter $D_{\mathrm{c}} = 100\,\mathrm{mm}$, the wire diameter $D_{\mathrm{w}} = 2\,\mathrm{mm}$, and the coil length $\ell_{\mathrm{c}} = \mathring{N} \cdot 4\,\mathrm{mm}$, hence the relative turn spacing $\frac{\ell_{\mathrm{c}}}{\mathring{N} D_{\mathrm{w}}} = 2$. The dotted line indicates that the coil may not be considered electrically small at such f. We observe that the maximum Q-factor that can be realized with an electrically small coil is similar for all $\mathring{N}$ (but the associated f depends heavily on $\mathring{N}$).

2.3 Coil Interaction

We consider an arbitrary arrangement of N coils. The coil ports are characterized by voltage v_n and current i_n at the coil terminals $n = 1, \ldots, N$, as illustrated in Fig. 2.10.

The m-th port voltage is the superposition of the self-induced voltage $Z_{m,m} i_m$ and the induced voltages $Z_{m,n} i_n$ due to the currents in all other coils $n \neq m$. In short,

$$
\begin{bmatrix} v_1 \\ v_2 \\ \vdots \\ v_N \end{bmatrix} = \begin{bmatrix} Z_{1,1} & Z_{1,2} & \cdots & Z_{1,N} \\ Z_{2,1} & Z_{2,2} & & \vdots \\ \vdots & & \ddots & \vdots \\ Z_{N,1} & \cdots & \cdots & Z_{N,N} \end{bmatrix} \cdot \begin{bmatrix} i_1 \\ i_2 \\ \vdots \\ i_N \end{bmatrix} \tag{2.43}
$$

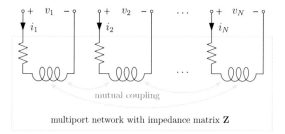

$$multiport\ network\ with\ impedance\ matrix\ \mathbf{Z}$$

Figure 2.10: Mutual coupling between a set of coils modeled in terms of an N-port network with impedance matrix $\mathbf{Z}$. This model was also employed by [26, 169].

or $\mathbf{v} = \mathbf{Zi}$ in matrix notation. The $N \times N$ impedance matrix $\mathbf{Z}$ characterizes the N-port network between all N coil terminals.[6] This multiport description is appropriate because electrically small coils fulfill the port condition (i_n equals the exiting current at the opposite terminal of the n-th port). This impedance-matrix approach for magnetic induction has also been used by [26, 70, 87].

This multiport network between the N coils is passive and reciprocal [26], i.e. it exhibits a symmetric (not hermitian) impedance matrix $\mathbf{Z} = \mathbf{Z}^{\mathrm{T}}$ (every passive network is reciprocal unless it contains certain special materials such as plasmas [170]). Also, for every passive network the real part $\mathrm{Re}(\mathbf{Z})$ is a positive semidefinite matrix, i.e. the active power into the network $\mathrm{Re}(\mathbf{i}^{\mathrm{H}}\mathbf{Zi}) = \mathbf{i}^{\mathrm{H}}\mathrm{Re}(\mathbf{Z})\mathbf{i} \geq 0$ for all current vectors $\mathbf{i}$.

2.3.1 Partially Terminated Multiport Network

Now consider that the N coils are partitioned according to $N = N_{\mathrm{p}} + N_{\mathrm{s}}$ with N_{p} coils on the primary side and N_{s} coils on the secondary side. According to (2.43) their electrical signals are related by

$$\begin{bmatrix} \mathbf{v}_{\mathrm{p}} \\ \mathbf{v}_{\mathrm{s}} \end{bmatrix} = \begin{bmatrix} \mathbf{Z}_{\mathrm{p}} & \mathbf{Z}_{\mathrm{p,s}} \\ \mathbf{Z}_{\mathrm{s,p}} & \mathbf{Z}_{\mathrm{s}} \end{bmatrix} \cdot \begin{bmatrix} \mathbf{i}_{\mathrm{p}} \\ \mathbf{i}_{\mathrm{s}} \end{bmatrix}. \tag{2.44}$$

---

[6]If the n-th coil is open circuited then, in this formulation, $i_n = 0$ (no current can flow through the coil) and the voltages v_m for $m \neq n$ are thus unaffected by the n-th coil. Thereby we however implicitly neglect the effect that an open-circuited coil can have on the rest of the network near its self-resonance frequency [69, 165]. We neglect this aspect because we mostly consider operation well below the self-resonance frequency and because of the mathematical difficulty of describing the current distribution across a self-resonant coil, as described in Sec. 2.2 (and emphasized by the heuristic choice of a sinusoidal current distribution in [69]).

If the secondary-side coils do not affect the magnetic field at the primary side (e.g. if they are open circuited or only weakly coupled to the primary side) then $\mathbf{v}_p = \mathbf{Z}_p\,\mathbf{i}_p$ holds on the primary side. We are however interested in the relation $\mathbf{v}_p = \mathbf{Z}'_p\,\mathbf{i}_p$ in the presence of terminated secondary-side coils whose induced currents do affect the magnetic field and thus also affect $\mathbf{Z}'_p$. For example, the termination loads could be resonance capacitors. We can capture this aspect with the following formula.

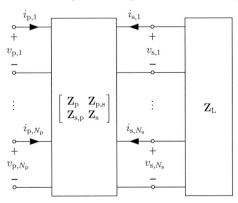

Figure 2.11: Partially loaded multiport network: terminating the N_s secondary-side ports with $\mathbf{Z}_L$ reduces the $(N_p + N_s)$-port network to a N_p-port network between the primary-side ports. Proposition 2.6 states that the primary-side electrical signals are related by $\mathbf{v}_p = \mathbf{Z}'_p\,\mathbf{i}_p$ through the impedance matrix $\mathbf{Z}'_p = \mathbf{Z}_p - \mathbf{Z}_{p,s}\left(\mathbf{Z}_s + \mathbf{Z}_L\right)^{-1}\mathbf{Z}_{s,p}$. The formula previously appeared in [100] and [26, Sec. 2.2].

Proposition 2.6. *Consider a network with $N_p + N_s$ ports that are partitioned into N_p primary-side ports and N_s secondary-side ports as in (2.44). When the secondary-side ports are terminated by an N_s-port network with impedance matrix $\mathbf{Z}_L \in \mathbb{C}^{N_s \times N_s}$, as shown in Fig. 2.11, then the port relation at the primary side is*

$$\mathbf{v}_p = \mathbf{Z}'_p\,\mathbf{i}_p\,, \qquad \mathbf{Z}'_p = \mathbf{Z}_p - \mathbf{Z}_{p,s}\left(\mathbf{Z}_s + \mathbf{Z}_L\right)^{-1}\mathbf{Z}_{s,p}\,. \qquad (2.45)$$

Proof. The second row of (2.44) is the equation $\mathbf{v}_s = \mathbf{Z}_{s,p}\mathbf{i}_p + \mathbf{Z}_s\mathbf{i}_s$ but $\mathbf{v}_s$ and $\mathbf{i}_s$ are also related by $\mathbf{v}_s = -\mathbf{Z}_L\mathbf{i}_s$ through the load (Ohm's law). Combined they give $\mathbf{i}_s = -(\mathbf{Z}_s + \mathbf{Z}_L)^{-1}\mathbf{Z}_{s,p}\mathbf{i}_p$. We use this expression in the first row of (2.44) to obtain $\mathbf{v}_p = \mathbf{Z}_p\mathbf{i}_p + \mathbf{Z}_{p,s}\mathbf{i}_s = (\mathbf{Z}_p - \mathbf{Z}_{p,s}(\mathbf{Z}_s + \mathbf{Z}_L)^{-1}\mathbf{Z}_{s,p})\,\mathbf{i}_p$. $\qquad\square$

Let us ponder the implications of Proposition 2.6 for inductive coupling in the magnetoquasistatic regime. There, $\mathbf{Z}_{p,s} = j\omega\mathbf{M}$ and $\mathbf{Z}_{s,p} = j\omega\mathbf{M}^{\mathrm{T}}$, whereby $\mathbf{M} \in \mathbb{R}^{N_p \times N_s}$

holds the mutual inductances between primary and secondary side. Subsequently, (2.45) gives $\mathbf{Z}'_\mathrm{p} = \mathbf{Z}_\mathrm{p} + \omega^2 \mathbf{M}(\mathbf{Z}_\mathrm{s} + \mathbf{Z}_\mathrm{L})^{-1}\mathbf{M}^\mathrm{T}$.

We want to study the case shown in Fig. 2.12 with a single secondary-side coil ($N_\mathrm{s} = 1$). The coil is loaded with a resonant reactive load $Z_\mathrm{L} = -j\mathrm{Im}(Z_\mathrm{s})$, giving $\mathbf{Z}'_\mathrm{p} = \mathbf{Z}_\mathrm{p} + \frac{\omega^2}{\mathrm{Re}(Z_\mathrm{s})}\mathbf{m}\mathbf{m}^\mathrm{T}$. Consider the active power P into the primary side for given port currents $\mathbf{i}_\mathrm{p}$. We find that $P = \mathbf{i}_\mathrm{p}^\mathrm{H}\mathrm{Re}(\mathbf{Z}'_\mathrm{p})\mathbf{i}_\mathrm{p} = \mathbf{i}_\mathrm{p}^\mathrm{H}\mathrm{Re}(\mathbf{Z}_\mathrm{p})\mathbf{i}_\mathrm{p} + \frac{\omega^2}{\mathrm{Re}(Z_\mathrm{s})}|\mathbf{m}^\mathrm{T}\mathbf{i}_\mathrm{p}|^2$ whereby both summands are positive since $\mathrm{Re}(\mathbf{Z}_\mathrm{p})$ is positive semidefinite. Thus, the presence of the terminated secondary-side coil increases the active power into the primary-side ports for fixed currents $\mathbf{i}_\mathrm{p}$. The increase $\frac{\omega^2}{\mathrm{Re}(Z_\mathrm{s})}|\mathbf{m}^\mathrm{T}\mathbf{i}_\mathrm{p}|^2$ is strictly positive unless $\mathbf{i}_\mathrm{p}$ and $\mathbf{m}$ are orthogonal (associated with zero net induced voltage at the secondary side).

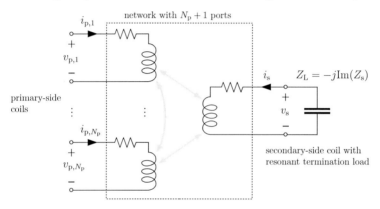

Figure 2.12: Example of a partially loaded ($N_\mathrm{p}+1$)-port network between N_p primary-side coils and one secondary-side coil with a resonant load termination. We are interested in the impedance matrix $\mathbf{Z}'_\mathrm{p}$ between the remaining N_p ports.

When either side is equipped with just a single coil, then $Z'_\mathrm{p} = Z_\mathrm{p} + \frac{\omega^2 M^2}{Z_\mathrm{s}+Z_\mathrm{L}}$, which becomes $Z'_\mathrm{p} = Z_\mathrm{p} + \frac{\omega^2 M^2}{\mathrm{Re}(Z_\mathrm{s})}$ when the load is an ideal resonance capacitor. This change in encountered primary-side coil impedance is the essential mechanism that enables load modulation at passive RFID tags [63,171]. The mechanism has also been proposed for improving passive tag localization in [26,172].

2.3.2 Z between Colocated Arrays: Dipole Formula

We consider a MIMO link described by the center-to-center distances $r_{m,n}$ and direction vectors $\mathbf{u}_{m,n}$ between transmit and receive coils. Thereby we use a receive-coil index $m = 1,\ldots,N_\mathrm{R}$ and a transmit-coil index $n = 1,\ldots,N_\mathrm{T}$. If the array sizes are small compared to the average $r_{m,n}$ then $r_{m,n}$ and $\mathbf{u}_{m,n}$ are near-constant over all m, n and the link geometry can be described by a single distance r and a single direction $\mathbf{u}$. The argument is exact for colocated arrays of coils. If furthermore the transmit coils have flat turns (cf. Proposition 2.2) and r is appreciably larger than the coil dimensions, then the transmitter-to-receiver mutual impedances are given by the $N_{\mathrm{R}} \times N_{\mathrm{T}}$ matrix

$$\mathbf{Z}_{\mathrm{RT}} = \mathbf{O}_{\mathrm{R}}^{\mathrm{T}} \mathbf{Z}_{\mathrm{3DoF}} \mathbf{O}_{\mathrm{T}} \,. \tag{2.46}$$

This follows from a simple extension of (2.27) to the case at hand. The matrices

$$\mathbf{O}_{\mathrm{R}} = [\mathbf{o}_{\mathrm{R},1} \ldots \mathbf{o}_{\mathrm{R},N_{\mathrm{R}}}] \in \mathbb{R}^{3 \times N_{\mathrm{R}}} \,, \tag{2.47}$$

$$\mathbf{O}_{\mathrm{T}} = [\mathbf{o}_{\mathrm{T},1} \ldots \mathbf{o}_{\mathrm{T},N_{\mathrm{T}}}] \in \mathbb{R}^{3 \times N_{\mathrm{T}}} \tag{2.48}$$

hold all the coil orientations. We note that $\mathrm{rank}(\mathbf{Z}_{\mathrm{RT}}) \leq \min\{3, N_{\mathrm{T}}, N_{\mathrm{R}}\}$ because the matrix $\mathbf{Z}_{\mathrm{3DoF}}$ has dimension 3×3.

Chapter 3

General Modeling and Analysis of Magneto-Inductive Links

This chapter introduces the system model which is used throughout this thesis, which is based on the properties from Cpt. 2. In particular, we state a general multiple-input multiple-output (MIMO) signal model for magnetic induction links with an arbitrary number of transmit and receive coils in Sec. 3.1 and the accompanying noise model in Sec. 3.2, and clarify some formalities like noise whitening in Sec. 3.3. Based thereon we state the limits on the achievable power transfer efficiency and data rate (narrow- and broadband cases) in Sec. 3.4 and Sec. 3.5, respectively. Sec. 3.6 covers necessary matching network aspects for system modeling. We then discuss important special cases and mathematical structure in Sec. 3.7, spatial degrees of freedom in Sec. 3.8, and formalize the limits of communication via load modulation in Sec. 3.9.

3.1 Signal Propagation and Transmit Power

Consider a setup with N_T transmit coils for forward transmission (not load modulation with passive tags) and N_R receiver coils. We employ the notion of (2.43) and describe the electrical interaction between those $N_T + N_R$ coils in terms of the impedance matrix

$$\mathbf{Z}_C = \begin{bmatrix} \mathbf{Z}_{C:T} & \mathbf{Z}_{C:TR} \\ \mathbf{Z}_{C:RT} & \mathbf{Z}_{C:R} \end{bmatrix} \in \mathbb{C}^{(N_T+N_R)\times(N_T+N_R)} \cdot \Omega \tag{3.1}$$

with unit Ω (ohm). Hereby, the $N_T \times N_T$ matrix $\mathbf{Z}_{C:T}$ describes the transmit coils' interaction if the receive coils are absent or open circuited or if $\mathbf{Z}_{C:RT} = \mathbf{0}$; the analogous statement holds for the $N_R \times N_R$ matrix $\mathbf{Z}_{C:R}$. The diagonal elements of $\mathbf{Z}_{C:T}$ and $\mathbf{Z}_{C:R}$ hold the coil self-impedances which are described by (and can be calculated with) Sec. 2.2. All other elements are mutual impedances and described by Sec. 2.1. Note that the transmitter-receiver role assignment does not affect the electrical description: any current induces voltages in all coils, irrespective of their roles in the wireless application.

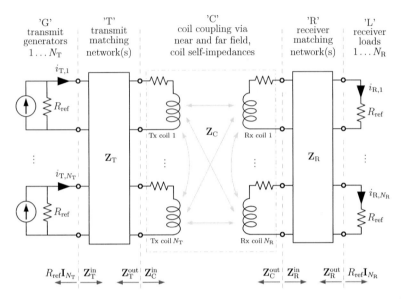

Figure 3.1: Circuit abstraction of N_T transmitting and N_R receiving loop antennas (coils) for wireless communication, power transfer, or localization. The circuit model specifies the relationship $\mathbf{i}_R = \mathbf{G}\mathbf{i}_T$ between the currents $\mathbf{i}_T$ delivered by the transmit generators and the currents $\mathbf{i}_R$ into the receiver loads (e.g. LNA inputs). The multiport matching networks are just an abstraction of any desired matching strategy, e.g., fully connected multiport designs or individual two-port networks per coil-load pair. This model is inspired by [53, Fig. 2].

In a wireless application the coils are terminated by transmit and receive circuits at the respective ports. Fig. 3.1 shows a general circuit description[1] of this circumstance. It is divided into five stages:

G: transmit generators

T: transmitter matching network(s)

C: coils, coupling, channel

R: receiver matching network(s)

L: receiver loads

---

[1]The modeling approach with the three electrical networks, characterized by their impedance matrices, and the subsequent formula(3.5) for signal propagation from multiple transmit generators to multiple receiver loads are according to the multiport circuit theory of communication by Ivrlac and Nossek [53, 170, 173]. The importance of circuit aspects for MIMO systems with coupled arrays was studied earlier by Wallace and Jensen [101] whose formalization uses scattering parameters. For rich introductory information and details on the employed concepts please consult these sources.

The transmit matching network connects transmit generators to the transmit coils while the receive matching network connects the receive coils to the receiver loads. We characterize the matching networks in terms of their impedance matrices

$$\mathbf{Z}_T = \begin{bmatrix} \mathbf{Z}_{T:G} & \mathbf{Z}_{T:CG}^T \\ \mathbf{Z}_{T:CG} & \mathbf{Z}_{T:C} \end{bmatrix} \in \mathbb{C}^{2N_T \times 2N_T} \cdot \Omega, \tag{3.2}$$

$$\mathbf{Z}_R = \begin{bmatrix} \mathbf{Z}_{R:C} & \mathbf{Z}_{R:LC}^T \\ \mathbf{Z}_{R:LC} & \mathbf{Z}_{R:L} \end{bmatrix} \in \mathbb{C}^{2N_R \times 2N_R} \cdot \Omega \tag{3.3}$$

whereby the notation $\mathbf{Z}_{T:CG}$ qualifies the block of $\mathbf{Z}_T$ which relates generators (G) to coils (C). The multiport networks $\mathbf{Z}_T$ and $\mathbf{Z}_R$ shall be considered as an abstraction that can account for any desired matching strategy (resonant power matching or noise matching) in any desired constellation (co-located coil arrays with multiport matching networks, distributed nodes with individual two-port networks, SISO, SIMO and MISO cases). This will be the subject of Sec. 3.6; to get an idea about possible matching network architectures see Fig. 3.4 to 3.6. The receiver loads are either the inputs to low-noise amplifiers (LNAs) of a communication receiver or tank circuits of a power receiver. We assume that each generator and each receiver load has the reference impedance $R_{ref} = 50\,\Omega$ at any frequency.

In the following we present a general model for signal propagation in noise,

$$\mathbf{i}_R = \mathbf{G}\,\mathbf{i}_T + \mathbf{i}_N \tag{3.4}$$

from the transmit generator currents $\mathbf{i}_T = [i_{T,1} \ \ldots \ i_{T,N_T}]^T$ to the receiver load currents $\mathbf{i}_R = [i_{R,1} \ \ldots \ i_{R,N_R}]^T$, distorted by noise $\mathbf{i}_N$. The model can serve for any magneto-inductive evaluation (communication, power transfer, or sensing) and entails any conceivable SISO, SIMO and MISO case. In the process we establish the relation between the currents and transmit power as well as received power. This shall allow for the formulation of physically meaningful power constraints and power transfer efficiencies.

In the power transfer context we are concerned with the noiseless model $\mathbf{i}_R = \mathbf{G}\mathbf{i}_T$ whereby $\mathbf{i}_T$ and $\mathbf{i}_R$ are the complex phasors of AC currents at the evaluation frequency. Their relation is linear because the electric network is linear. In the communication context, where $\mathbf{i}_T$ and $\mathbf{i}_R$ are information-bearing, the quantities $\mathbf{i}_T, \mathbf{i}_R, \mathbf{i}_N$ are considered as symbol-discrete samples of band-limited complex baseband signals; relation to the actual AC signals is described in [37]. The symbol time index is implied.

Proposition 3.1. *The transmitter-to-receiver current gain* $\mathbf{G} \in \mathbb{C}^{N_R \times N_T}$ *is given by*

$$\mathbf{G} = (R_{\mathrm{ref}}\mathbf{I}_{N_R} + \mathbf{Z}_{\mathrm{R:L}})^{-1}\mathbf{Z}_{\mathrm{R:LC}}(\mathbf{Z}_R^{\mathrm{in}} + \mathbf{Z}_{\mathrm{C:R}})^{-1}\mathbf{Z}_{\mathrm{C:RT}}(\mathbf{Z}_C^{\mathrm{in}} + \mathbf{Z}_{\mathrm{T:C}})^{-1}\mathbf{Z}_{\mathrm{T:CG}} . \quad (3.5)$$

We use the impedance matrix encountered between input ports of a multiport network when the respective output ports are terminated,

$$\mathbf{Z}_T^{\mathrm{in}} = \mathbf{Z}_{\mathrm{T:G}} - \mathbf{Z}_{\mathrm{T:CG}}^{\mathrm{T}}\left(\mathbf{Z}_{\mathrm{T:C}} + \mathbf{Z}_C^{\mathrm{in}}\right)^{-1}\mathbf{Z}_{\mathrm{T:CG}} , \quad (3.6)$$

$$\mathbf{Z}_C^{\mathrm{in}} = \mathbf{Z}_{\mathrm{C:T}} - \mathbf{Z}_{\mathrm{C:TR}}\left(\mathbf{Z}_{\mathrm{C:R}} + \mathbf{Z}_R^{\mathrm{in}}\right)^{-1}\mathbf{Z}_{\mathrm{C:RT}} , \quad (3.7)$$

$$\mathbf{Z}_R^{\mathrm{in}} = \mathbf{Z}_{\mathrm{R:C}} - \mathbf{Z}_{\mathrm{R:LC}}^{\mathrm{T}}\left(\mathbf{Z}_{\mathrm{R:L}} + R_{\mathrm{ref}}\mathbf{I}_{N_R}\right)^{-1}\mathbf{Z}_{\mathrm{R:LC}} . \quad (3.8)$$

Note the intuitive role of the mutual impedance matrix $\mathbf{Z}_{\mathrm{C:RT}}$, which relates transmit and receive coils, at the core of $\mathbf{G}$ in (3.5).

For later use we shall also state the impedance matrices encountered at output ports in the case of terminated input ports; they are given by

$$\mathbf{Z}_T^{\mathrm{out}} = \mathbf{Z}_{\mathrm{T:C}} - \mathbf{Z}_{\mathrm{T:CG}}\left(\mathbf{Z}_{\mathrm{T:G}} + R_{\mathrm{ref}}\mathbf{I}_{N_T}\right)^{-1}\mathbf{Z}_{\mathrm{T:CG}}^{\mathrm{T}} , \quad (3.9)$$

$$\mathbf{Z}_C^{\mathrm{out}} = \mathbf{Z}_{\mathrm{C:R}} - \mathbf{Z}_{\mathrm{C:RT}}\left(\mathbf{Z}_{\mathrm{C:T}} + \mathbf{Z}_T^{\mathrm{out}}\right)^{-1}\mathbf{Z}_{\mathrm{C:TR}} , \quad (3.10)$$

$$\mathbf{Z}_R^{\mathrm{out}} = \mathbf{Z}_{\mathrm{R:L}} - \mathbf{Z}_{\mathrm{R:LC}}\left(\mathbf{Z}_{\mathrm{R:C}} + \mathbf{Z}_C^{\mathrm{out}}\right)^{-1}\mathbf{Z}_{\mathrm{R:LC}}^{\mathrm{T}} . \quad (3.11)$$

Proof. All input and output impedance matrices follow directly from successive application of Proposition 2.6. The relation (3.5) is given almost equivalently[2] by [53, Eq. 16]. To prove it we consider the transmit coil currents $\mathbf{i}_C^{\mathrm{in}}$ into $\mathbf{Z}_C^{\mathrm{in}}$ (out of $\mathbf{Z}_T$). We note that $\mathbf{i}_C^{\mathrm{in}} = (\mathbf{Z}_C^{\mathrm{in}} + \mathbf{Z}_{\mathrm{T:C}})^{-1}\mathbf{Z}_{\mathrm{T:CG}}\mathbf{i}_T$ follows from $\mathbf{v}_C^{\mathrm{in}} = \mathbf{Z}_{\mathrm{T:CG}}\mathbf{i}_T - \mathbf{Z}_{\mathrm{T:C}}\mathbf{i}_C^{\mathrm{in}}$ (as part of the port relation of $\mathbf{Z}_T$) together with $\mathbf{v}_C^{\mathrm{in}} = \mathbf{Z}_C^{\mathrm{in}}\mathbf{i}_C^{\mathrm{in}}$. Repeating the same concept for the current gain past $\mathbf{Z}_C$ and then one more time past $\mathbf{Z}_R$ yields $\mathbf{i}_L$ as described by (3.5). □

The active power into the receiver loads is $P_{\mathrm{R},m} = |i_{\mathrm{R},m}|^2 R_{\mathrm{ref}}$ where $R_{\mathrm{ref}} = 50\,\Omega$ is the load impedance. The received sum-power is therefore $P_R = \sum_{m=1}^{N_R} P_{\mathrm{R},m} = \|\mathbf{i}_R\|^2 R_{\mathrm{ref}}$. We are also interested in the relation between $\mathbf{i}_T$ and transmit power to formulate physically meaningful power constraints [53, 173], a key requirement for a meaningful model of an energy-limited sensor network. This requires special care.

---

[2]The channel matrix formula (3.5) is equivalent to [53, Eq. 16] apart from the fact that we do not rely on the unilateral assumption $\mathbf{Z}_{\mathrm{C:TR}} \approx \mathbf{0}$. This assumption is well-justified for weak Tx-Rx coupling (the usual case in radio communications) but not for strong links as required for efficient power transfer. Note that verifying the equivalence of some formulas requires cumbersome applications of the Woodbury matrix identity (equivalent descriptions may seem different at first glance).

Proposition 3.2. *The active power delivered by the n-th transmit generator is*

$$P_{\mathrm{T},n} = \mathrm{Re}(\mathbf{Z}_{\mathrm{T}}^{\mathrm{in}}\,\mathbf{i}_{\mathrm{T}}\mathbf{i}_{\mathrm{T}}^{\mathrm{H}})_{n,n}\,. \tag{3.12}$$

For a random information-bearing transmit signal we can formulate a constraint on the average active power $\mathbb{E}[P_{\mathrm{T},n}]$ *delivered by the n-th transmit generator:*

$$\mathrm{Re}(\mathbf{Z}_{\mathrm{T}}^{\mathrm{in}}\,\mathbf{Q}_{\mathbf{i}})_{n,n} \leq P_{\mathrm{T},n}^{\mathrm{avail}}\,, \tag{3.13}$$

$$\mathbf{Q}_{\mathbf{i}} = \mathbb{E}[\mathbf{i}_{\mathrm{T}}\mathbf{i}_{\mathrm{T}}^{\mathrm{H}}]\,. \tag{3.14}$$

Proof. For the N_{T}-port network with $\mathbf{Z}_{\mathrm{T}}^{\mathrm{in}}$ faced by the transmit generators, consider the n-th port voltage $v_{\mathrm{T},n}$ and port current $i_{\mathrm{T},n}$. By [174, Sec. 3.1.1.6] the active power into this port is $P_{\mathrm{T},n} = \mathrm{Re}(v_{\mathrm{T},n}i_{\mathrm{T},n}^{*}) = \mathrm{Re}(\mathbf{v}_{\mathrm{T}}\mathbf{i}_{\mathrm{T}}^{\mathrm{H}})_{n,n} = \mathrm{Re}(\mathbf{Z}_{\mathrm{T}}^{\mathrm{in}}\,\mathbf{i}_{\mathrm{T}}\mathbf{i}_{\mathrm{T}}^{\mathrm{H}})_{n,n}$, which proves (3.12). Furthermore, $\mathbb{E}_{\mathbf{i}_{\mathrm{T}}}[P_{\mathrm{T},n}] = \mathbb{E}_{\mathbf{i}_{\mathrm{T}}}[\mathrm{Re}(\mathbf{Z}_{\mathrm{T}}^{\mathrm{in}}\,\mathbf{i}_{\mathrm{T}}\mathbf{i}_{\mathrm{T}}^{\mathrm{H}})_{n,n}] = \mathrm{Re}(\mathbf{Z}_{\mathrm{T}}^{\mathrm{in}}\,\mathbb{E}_{\mathbf{i}_{\mathrm{T}}}[\mathbf{i}_{\mathrm{T}}\mathbf{i}_{\mathrm{T}}^{\mathrm{H}}])_{n,n}$. □

The transmit power formulation is made complicated by the fact that $P_{\mathrm{T},n}$ depends on the entire $\mathbf{i}_{\mathrm{T}}$: the n-th port voltage depends on all port currents, not just $i_{\mathrm{T},n}$. The simpler statement $P_{\mathrm{T},n} = \mathrm{Re}(Z_{\mathrm{T}}^{\mathrm{in}})_{n,n}\,|i_{\mathrm{T},n}|^2$ holds iff $\mathbf{Z}_{\mathrm{T}}^{\mathrm{in}}$ is diagonal, e.g. for a decoupled array. A per-generator constraint is synonymous with a per-node or per-antenna constraint in the case of a distributed transmit array, e.g., for transmit cooperation between distributed sensor nodes.

Proposition 3.3. *The active sum-power delivered by the transmit generators is*

$$P_{\mathrm{T}} = \mathbf{i}_{\mathrm{T}}^{\mathrm{H}}\,\mathrm{Re}(\mathbf{Z}_{\mathrm{T}}^{\mathrm{in}})\,\mathbf{i}_{\mathrm{T}}\,. \tag{3.15}$$

For information-bearing random transmit signals, we can formulate a constraint on the average active sum-power $\mathbb{E}[P_{\mathrm{T}}]$ *delivered by the transmit amplifiers,*

$$\mathbb{E}[\,\mathbf{i}_{\mathrm{T}}^{\mathrm{H}}\mathrm{Re}(\mathbf{Z}_{\mathrm{T}}^{\mathrm{in}})\,\mathbf{i}_{\mathrm{T}}\,] \leq P_{\mathrm{T}}^{\mathrm{avail}}\,. \tag{3.16}$$

Proof. The same statement is found in [53]; we derive it for completeness and intuition: $P_{\mathrm{T}} = \sum_{n=1}^{N_{\mathrm{T}}} P_{\mathrm{T},n} = \sum_{n=1}^{N_{\mathrm{T}}} \mathrm{Re}(\mathbf{Z}_{\mathrm{T}}^{\mathrm{in}}\,\mathbf{i}_{\mathrm{T}}\mathbf{i}_{\mathrm{T}}^{\mathrm{H}})_{n,n} = \mathrm{tr}(\mathrm{Re}(\mathbf{Z}_{\mathrm{T}}^{\mathrm{in}}\,\mathbf{i}_{\mathrm{T}}\mathbf{i}_{\mathrm{T}}^{\mathrm{H}})) = \mathrm{Re}(\mathrm{tr}(\mathbf{Z}_{\mathrm{T}}^{\mathrm{in}}\,\mathbf{i}_{\mathrm{T}}\mathbf{i}_{\mathrm{T}}^{\mathrm{H}})) = \mathrm{Re}(\mathrm{tr}(\mathbf{i}_{\mathrm{T}}^{\mathrm{H}}\mathbf{Z}_{\mathrm{T}}^{\mathrm{in}}\mathbf{i}_{\mathrm{T}})) = \mathrm{Re}(\mathbf{i}_{\mathrm{T}}^{\mathrm{H}}\mathbf{Z}_{\mathrm{T}}^{\mathrm{in}}\mathbf{i}_{\mathrm{T}})$ where we used $P_{\mathrm{T},n}$ from (3.12), the cyclic permutation property of the trace, and that $\mathrm{tr}(z) = z$ for a scalar. We write $\mathbf{Z}_{\mathrm{T}}^{\mathrm{in}} = \mathbf{R} + j\mathbf{X}$ in terms of real and imaginary part. Because $\mathbf{Z}_{\mathrm{T}}^{\mathrm{in}}$ represents a passive reciprocal network, $\mathbf{R}$ and $\mathbf{X}$ are symmetric. Thus $\mathbf{i}_{\mathrm{T}}^{\mathrm{H}}\mathbf{R}\mathbf{i}_{\mathrm{T}}$ and $\mathbf{i}_{\mathrm{T}}^{\mathrm{H}}\mathbf{X}\mathbf{i}_{\mathrm{T}}$ are real-valued for any $\mathbf{i}_{\mathrm{T}} \in \mathbb{C}^{N_{\mathrm{T}}}$. We obtain $P_{\mathrm{T}} = \mathrm{Re}(\mathbf{i}_{\mathrm{T}}^{\mathrm{H}}\mathbf{R}\mathbf{i}_{\mathrm{T}} + j \cdot \mathbf{i}_{\mathrm{T}}^{\mathrm{H}}\mathbf{X}\mathbf{i}_{\mathrm{T}}) = \mathbf{i}_{\mathrm{T}}^{\mathrm{H}}\mathbf{R}\mathbf{i}_{\mathrm{T}}$, equivalent to (3.15). □

3.2 Noise Statistics

The statistics of the receiver noise vector $\mathbf{i_N}$ affect the communication performance of a link in a crucial way. Hence, an adequate noise model is required for subsequent studies. We employ the well-established assumption of a circularly-symmetric complex Gaussian distribution [90, Appendix A]

$$\mathbf{i_N} \sim \mathcal{CN}(\mathbf{0}, \mathbf{K_i})$$ (3.17)

considered over a narrow frequency band of width Δ_f. The remainder of the section shall describe the employed model for the covariance matrix $\mathbf{K_i} = \mathbb{E}[\mathbf{i_N} \mathbf{i_N^H}]$ of the noise currents $\mathbf{i_N}$ into the receiver loads. In particular we use the model

$$\mathbf{K_i} = \mathbb{E}[\mathbf{i_N} \mathbf{i_N^H}] = \frac{\sigma_{\mathrm{iid}}^2}{R_{\mathrm{ref}}^2} \mathbf{I}_{N_R} + \mathbf{Y_L} \Big(\mathbf{\Sigma_{\mathrm{therm}}} + \mathbf{\Sigma_{\mathrm{ext}}} + \mathbf{\Sigma_{\mathrm{LNA}}} \Big) \mathbf{Y_L^H}$$ (3.18)

which is analogous to [53,175]. Thereby $\sigma_{\mathrm{iid}}^2 \mathbf{I}_{N_R}$ accounts for uncorrelated noise between the different receivers which may arise at the low-noise amplifier (LNA) outputs, at later stages of the RF chain, or may model the limited resolution of the analog-to-digital converters. The standard deviation σ_{iid} characterizes an equivalent noise voltage at the LNA input. Furthermore $\mathbf{Y_L} = (R_{\mathrm{ref}} \mathbf{I}_{N_R} + \mathbf{Z_R^{out}})^{-1}$ is the serial admittance of the receiver load circuit and $R_{\mathrm{ref}} = 50\,\Omega$ is the receiver load impedance.

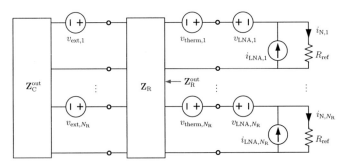

Figure 3.2: Circuit abstraction of considered receive-side noise sources: extrinsic noise, thermal noise, and noise added by the low-noise amplifiers (LNAs). The circuit determines their propagation to the receiver loads. The same model was given in [53].

Furthermore, the model accounts for thermal noise (the random motion of electrons in the receiver circuits) in $\mathbf{\Sigma_{\mathrm{therm}}}$, extrinsic noise due to picked up radiation in $\mathbf{\Sigma_{\mathrm{ext}}}$, and noise contributed by imperfect LNAs in $\boldsymbol{\Sigma}_{\mathrm{LNA}}$. These covariance matrices are associated with noise voltage sources that are in series with the receiver loads.

Passive reciprocal electrical networks have a convenient property: the covariance matrix of the *thermal noise* voltages (encountered at the ports when open circuited) is proportional to the real part of the network impedance matrix [176,177]. Hence, the thermal noise contribution by the network represented by $\mathbf{Z}_{\mathrm{R}}^{\mathrm{out}}$ is[3]

$$\boldsymbol{\Sigma}_{\mathrm{therm}} = 4k_{\mathrm{B}}T\Delta_f \,\mathrm{Re}(\mathbf{Z}_{\mathrm{R}}^{\mathrm{out}}) \tag{3.19}$$

with Boltzmann constant k_{B}, (actual) receiver temperature T, and the considered (narrow, single-sided) bandwidth Δ_f. Matrix $\boldsymbol{\Sigma}_{\mathrm{therm}}$ may be non-diagonal due to array coupling (non-diagonal $\mathbf{Z}_{\mathrm{C}}^{\mathrm{out}}$) and/or due to the matching network $\mathbf{Z}_{\mathrm{R}}$.

Extrinsic noise from background radiation and other unwanted field sources unrelated to the considered system is characterized by

$$\boldsymbol{\Sigma}_{\mathrm{ext}} = 4k_{\mathrm{B}}T_{\mathrm{A}}\Delta_f \,\mathbf{D}_{\mathrm{R}}\mathbf{R}_{\mathrm{rad}}^{\frac{1}{2}}\,\boldsymbol{\Phi}\,\mathbf{R}_{\mathrm{rad}}^{\frac{1}{2}}\,\mathbf{D}_{\mathrm{R}}^{\mathrm{H}} \tag{3.20}$$

where T_{A} is the antenna temperature (an established measure for the intensity of this phenomenon [150, Sec. 16.7]) and matrix $\mathbf{R}_{\mathrm{rad}} = \mathrm{diag}(R_1^{\mathrm{rad}}, \ldots, R_{N_R}^{\mathrm{rad}})$ holds the radiation resistances of the receive coils, which were described earlier in Sec. 2.2. The spatial correlation matrix $\boldsymbol{\Phi} \in \mathbb{C}^{N_R \times N_R}$ holds the correlation coefficients between the different induced receive voltages, fulfilling $\Phi_{m,m} = 1$ and $|\Phi_{m,n}| \leq 1$ for all m, n. The matrix $\mathbf{D}_{\mathrm{R}} = \mathbf{Z}_{\mathrm{R:LC}}(\mathbf{Z}_{\mathrm{R:C}} + \mathbf{Z}_{\mathrm{C}}^{\mathrm{out}})^{-1}$ is the voltage gain past the receiver matching network. For electrically large antennas, $\boldsymbol{\Sigma}_{\mathrm{ext}}$ usually dominates over $\boldsymbol{\Sigma}_{\mathrm{therm}}$ [53, Sec.II-E]. However, even for electrically small coils with $R_m^{\mathrm{rad}} \ll R_m^{\mathrm{ohm}}$, extrinsic noise can be dominant because the antenna temperature T_{A} can be exorbitantly large at low frequencies [178, Fig. 2].

For the noise added by each *low-noise amplifier* (LNA) we account with an established model [53, Sec. II-E] of two equivalent (and possibly correlated) noise sources: a voltage source v_{LNA} in series and a current source i_{LNA} in parallel to the load resistances and the ports of $\mathbf{Z}_{\mathrm{R}}^{\mathrm{out}}$. The resulting contribution to $\mathbf{i}_{\mathrm{N}}$ is given by [175, Eq. 7]

$$\boldsymbol{\Sigma}_{\mathrm{LNA}} = \sigma_i^2 \left(R_{\mathrm{N}}^2 \mathbf{I}_{N_R} + \mathbf{Z}_{\mathrm{R}}^{\mathrm{out}}(\mathbf{Z}_{\mathrm{R}}^{\mathrm{out}})^{\mathrm{H}} - 2R_{\mathrm{N}}\mathrm{Re}(\rho^* \mathbf{Z}_{\mathrm{R}}^{\mathrm{out}}) \right) \tag{3.21}$$

---

[3]If $\mathrm{Re}(\mathbf{Z}_{\mathrm{R}}^{\mathrm{out}})$ is significantly affected by the radiation resistances (i.e. at large f) then (3.19) is inadequate; for rigorous results one would have to single out the ohmic parts. However, since we do not expect noteworthy differences for electrically small coils and because the validity of the equivalent circuit model is in question in the discussed situation, we refrain from this detailed approach.

whereby the quantities σ_i^2, R_N, ρ are the LNA noise parameters which describe the statistics of the corresponding equivalent noise sources in Fig. 3.2. In detail, $\sigma_i^2 = \mathbb{E}[|i_{LNA}|^2]$ is the current variance, $R_N = \frac{1}{\sigma_i}\sqrt{\mathbb{E}[|v_{LNA}|^2]}$ the so-called noise resistance, and $\rho = \frac{1}{\sigma_i^2 R_N}\mathbb{E}[v_{LNA}i_{LNA}^*] \in \mathbb{C}$ the correlation coefficient which of course fulfills $|\rho| \leq 1$.

3.3   Useful Equivalent Models

The MIMO model $\mathbf{i_R} = \mathbf{G}\,\mathbf{i_T} + \mathbf{i_N}$ from (3.4) is in terms of electrical currents and thus offers intuition and a direct interface to physical power. Yet, certain equivalent models offer more convenient mathematical properties in various circumstances. To this effect, we consider the equivalent MIMO signal and noise model

$$\mathbf{y} = \mathbf{Hx} + \mathbf{w}\,, \qquad\qquad \mathbf{w} \sim \mathcal{CN}(\mathbf{0}, \mathbf{K}) \qquad (3.22)$$

which is related to the current model via the simple linear transformation and scaling

$$\mathbf{y} = R_{\text{ref}}^{\frac{1}{2}}\mathbf{i_R}\,, \quad \mathbf{H} = R_{\text{ref}}^{\frac{1}{2}}\mathbf{G}\,\text{Re}(\mathbf{Z}_T^{\text{in}})^{-\frac{1}{2}}\,, \quad \mathbf{x} = \text{Re}(\mathbf{Z}_T^{\text{in}})^{\frac{1}{2}}\mathbf{i_T}\,, \quad \mathbf{w} = R_{\text{ref}}^{\frac{1}{2}}\mathbf{i_N}\,. \quad (3.23)$$

This transformation was introduced by [53] with the intend of establishing consistency between information theory and active physical power. In particular, it fulfills the relations

$$P_R = \|\mathbf{y}\|^2\,, \qquad P_{R,m} = |y_m|^2\,, \qquad P_T = \|\mathbf{x}\|^2\,, \qquad \mathbf{K} = R_{\text{ref}}\mathbf{K_i} \qquad (3.24)$$

where $P_R = \|\mathbf{i_R}\|^2 R_{\text{ref}}$ is the received active sum-power, $P_{R,m} = |i_{R,m}|^2 R_{\text{ref}}$ the active power into the m-th receiver load, P_T the transmitted active sum-power (3.15), and $\mathbf{K_i}$ the covariance matrix of noise currents from (3.18). The vectors $\mathbf{x}, \mathbf{y}, \mathbf{w}$ have unit $\sqrt{\mathbf{W}}$. This model is useful for studying links with a colocated transmit array where the available sum-power can be shared among the generators. An important related fact is that x_n does *not* characterize the signals at the n-th transmit port and $|x_n|^2$ is *not equal*[4] to the per-generator transmit power (3.12) unless $\mathbf{Z}_T^{\text{in}}$ is diagonal. Per-generator power constraints must therefore be formulated strictly according to Proposition 3.2.

For communication-theoretic analyses the noise-whitened MIMO models

$$\bar{\mathbf{y}} = \bar{\mathbf{G}}\,\mathbf{i_T} + \bar{\mathbf{w}}\,, \qquad\qquad\qquad\qquad\qquad (3.25)$$

$$\bar{\mathbf{y}} = \bar{\mathbf{H}}\mathbf{x} + \bar{\mathbf{w}}\,, \qquad\qquad \mathbb{E}[\bar{\mathbf{w}}\bar{\mathbf{w}}^H] = \mathbf{I}_{N_R} \qquad (3.26)$$

are convenient because of the unit noise covariance matrix. They follow from a receive-side whitening operation $\bar{\mathbf{y}} = \mathbf{K}_\mathbf{i}^{-\frac{1}{2}} \mathbf{i}_\mathrm{R}$ or rather $\bar{\mathbf{y}} = \mathbf{K}^{-\frac{1}{2}} \mathbf{w}$. The associated channel matrices $\bar{\mathbf{G}} = \mathbf{K}_\mathbf{i}^{-\frac{1}{2}} \mathbf{G}$ and $\bar{\mathbf{H}} = \bar{\mathbf{G}} \operatorname{Re}(\mathbf{Z}_\mathrm{T}^\mathrm{in})^{-\frac{1}{2}}$ will be used for the calculation of achievable rates. In particular $\bar{\mathbf{H}}$ is the model of choice when a sum-power constraint applies and $\bar{\mathbf{G}}$ when per-generator power constraints apply.

3.4 Power Transfer Efficiency

For power transfer considerations we disregard noise and utilize the power-consistency properties of the model $\mathbf{y} = \mathbf{H}\mathbf{x}$ from (3.23), namely that $P_\mathrm{T} = \|\mathbf{x}\|^2$ and $P_\mathrm{R} = \|\mathbf{y}\|^2$. Hence we can define a power transfer efficiency (PTE)

$$\eta = \frac{P_\mathrm{R}}{P_\mathrm{T}} = \frac{\|\mathbf{H}\mathbf{x}\|^2}{\|\mathbf{x}\|^2} \tag{3.27}$$

from the transmit generators to the receiver loads. In the case of multiple transmit antennas, η depends on the direction of the transmit signal vector $\mathbf{x}$ and is upper-bounded by the largest eigenvalue of $\mathbf{H}^\mathrm{H}\mathbf{H}$,

$$\eta \le \eta_\mathrm{mrc} = \lambda_\mathrm{max}\!\left(\mathbf{H}^\mathrm{H}\mathbf{H}\right). \tag{3.28}$$

With the availability of channel state information at the transmitter (CSIT), the upper bound can be attained by setting $\mathbf{x} = \sqrt{P_\mathrm{T}}\,\frac{\mathbf{v}^*}{\|\mathbf{v}\|}$ where $\mathbf{v}$ is an eigenvector associated with the eigenvalue λ_max. This transmit beamforming scheme is called maximum-ratio combining [89]. If the transmit side uses just a single antenna then η is unaffected by

---

[4]In fact, given $\mathbf{Z}_\mathrm{T}^\mathrm{in}$ without special structure (e.g. diagonal), no linear transform $\mathbf{x} = \mathbf{A}\mathbf{i}_\mathrm{T}$ can establish the property $|x_n|^2 = P_{\mathrm{T},n}$ for all $\mathbf{i}_\mathrm{T} \in \mathbb{C}^{N_\mathrm{T}}$. This is proven in the following. First, we note that $|x_n|^2 = (\mathbf{x}\mathbf{x}^\mathrm{H})_{n,n} = (\mathbf{A}\mathbf{i}_\mathrm{T}\mathbf{i}_\mathrm{T}^\mathrm{H}\mathbf{A}^\mathrm{H})_{n,n}$. With $P_{\mathrm{T},n}$ from (3.12), we would require that for all $\mathbf{i}_\mathrm{T}$ and $n = 1\dots N_\mathrm{T}$ the equation $(\mathbf{A}\mathbf{i}_\mathrm{T}\mathbf{i}_\mathrm{T}^\mathrm{H}\mathbf{A}^\mathrm{H})_{n,n} = \operatorname{Re}(\mathbf{Z}_\mathrm{T}^\mathrm{in}\,\mathbf{i}_\mathrm{T}\mathbf{i}_\mathrm{T}^\mathrm{H})_{n,n}$ is fulfilled. We show that this requirement leads to a contradiction. We denote $\mathbf{A}^\mathrm{H} = [\mathbf{a}_1 \dots \mathbf{a}_{N_\mathrm{T}}]$ as well as $(\mathbf{Z}_\mathrm{T}^\mathrm{in})^\mathrm{H} = [\mathbf{z}_1 \dots \mathbf{z}_{N_\mathrm{T}}]$ and the requirement becomes $|\mathbf{a}_n^\mathrm{H}\mathbf{i}_\mathrm{T}|^2 = \operatorname{Re}(\mathbf{z}_n^\mathrm{H}\mathbf{i}_\mathrm{T}\,i_{\mathrm{T},n}^*)$. We consider transmit currents $\mathbf{i}_\mathrm{T} \ne \mathbf{0}$ with $i_{\mathrm{T},m} = 0$ for some transmitter m. The requirement for this transmitter $n = m$ is $|\mathbf{a}_m^\mathrm{H}\mathbf{i}_\mathrm{T}|^2 = 0$, i.e. $\mathbf{a}_m$ must be orthogonal to $\mathbf{i}_\mathrm{T}$. This must hold for any $\mathbf{i}_\mathrm{T}$ with $i_{\mathrm{T},m} = 0$ and thus the vector must have the structure $\mathbf{a}_m = [0 \dots 0\, A_{m,m}\, 0 \dots 0]^\mathrm{T}$. By repeating the argument for $m = 1 \dots N_\mathrm{T}$ we find that matrix $\mathbf{A}$ must be diagonal. On the other hand, $\|\mathbf{x}\|^2 = \|\mathbf{A}\mathbf{i}_\mathrm{T}\|^2$ must equal the sum power, i.e. $\mathbf{i}_\mathrm{T}^\mathrm{H}\mathbf{A}^\mathrm{H}\mathbf{A}\mathbf{i}_\mathrm{T} = \mathbf{i}_\mathrm{T}^\mathrm{H}\operatorname{Re}(\mathbf{Z}_\mathrm{T}^\mathrm{in})\mathbf{i}_\mathrm{T}$ must hold for all $\mathbf{Z}_\mathrm{T}^\mathrm{in}$ and $\mathbf{i}_\mathrm{T}$. Hence, $\mathbf{A}^\mathrm{H}\mathbf{A} = \operatorname{Re}(\mathbf{Z}_\mathrm{T}^\mathrm{in})$ must hold, but given a non-diagonal $\operatorname{Re}(\mathbf{Z}_\mathrm{T}^\mathrm{in})$ this can not be satisfied by a diagonal $\mathbf{A}$, which is a contradiction.

the transmit signal phase and takes the simpler form

$$\text{SIMO case:} \qquad \eta = \|\mathbf{h}\|^2, \qquad (3.29)$$

$$\text{SISO case:} \qquad \eta = |h|^2. \qquad (3.30)$$

The presented MIMO- and SIMO-case formulas are meaningful only when the receive-power quantity of interest actually conforms with the sum power $P_\mathrm{R} = \sum_{m=1}^{N_\mathrm{R}} P_{\mathrm{R},m}$. This is the case when either (i) we are interested in the sum power picked up by separate receivers (this will be the case in Cpt. 6) or (ii) when a colocated receiver is actually performing a summation of powers, e.g., if every receiver load charges an individual battery. Other approaches for power combining may lead to different results.[5]

3.5 Achievable Rates

We study the data rate over magneto-inductive wireless links in terms of achievable rates D and the channel capacity C (the largest achievable rate). A data rate is achievable if encoding and decoding functions for error-correcting codes exist such that the block error rate goes to zero with increasing block length at the given information rate [158, 180]. Near-capacity data rates can be realized with practical digital modulation schemes (for example 1024-QAM) in combination with a turbo code [181], low-density parity-check code [182], or polar code [183] for forward error correction.

3.5.1 Narrowband Rate Under Sum-Power Constraint

We are interested in the achievable data rate in bit/s over a narrow frequency band (bandwidth Δ_f) for reception in additive white Gaussian noise (AWGN) with the available transmit power $P_\mathrm{T}^\mathrm{avail}$ (which can be distributed arbitrarily among the generators). A famous result [159, 184] states: if the transmit signaling uses $\mathbf{x} \sim \mathcal{CN}(\mathbf{0}, \mathbf{Q_x})$ with statistically independent sampling for different symbol time indices and fulfills the

---

[5]A possible power combining strategy could employ coherent addition of the amplitudes y_m (e.g. after co-phasing them with analog phase shifters) which leads to the dependency $\eta \propto (\sum_{m=1}^{N_\mathrm{R}} |y_m|)^2$. This may suggest formidable PTE but, to the best of our knowledge, passive analog addition of signal amplitudes is possible only with severe losses. For example, a passive summing junction of resistors only allows for amplitude averaging, yielding $\eta = (\frac{1}{N_\mathrm{R}} \sum_{m=1}^{N_\mathrm{R}} |y_m|)^2 = \frac{1}{N_\mathrm{R}^2} (\sum_{m=1}^{N_\mathrm{R}} |y_m|)^2 \leq \frac{1}{N_\mathrm{R}} \eta_\mathrm{mrc}$. Componentwise lossless DC-conversion and a passive summing junction would give the same result. This poor performance is beaten even by receive-side selection combining which exhibits $\eta_\mathrm{sc} \geq \frac{1}{N_\mathrm{R}} \eta_\mathrm{mrc}$. Performing an actual addition of analog signals requires an amplifier [179, Sec. 4.2.4] whose power consumption however defeats the purpose of a wireless power receiver.

sum-power constraint $\mathrm{tr}(\mathbf{Q_x}) \leq P_{\mathrm{T}}^{\mathrm{avail}}$, then the information rate

$$D = \Delta_f \log_2 \det\left(\mathbf{I}_{N_{\mathrm{R}}} + \bar{\mathbf{H}}\mathbf{Q_x}\bar{\mathbf{H}}^{\mathrm{H}}\right) \tag{3.31}$$

in bit/s is achievable. Thereby $\bar{\mathbf{H}}$ is the noise-whitened channel matrix from (3.26). For example, $D = \Delta_f \log_2 \det(\mathbf{I}_{N_{\mathrm{R}}} + \frac{P_{\mathrm{T}}^{\mathrm{avail}}}{N_{\mathrm{T}}}\bar{\mathbf{H}}\bar{\mathbf{H}}^{\mathrm{H}})$ is achievable by allocating $\mathbf{Q_x} = \frac{P_{\mathrm{T}}^{\mathrm{avail}}}{N_{\mathrm{T}}}\mathbf{I}_{N_{\mathrm{T}}}$.

If $\bar{\mathbf{H}}$ is known to the transmitter then $\mathbf{Q_x}$ can be adapted to the channel. In particular, by allocating $\mathbf{Q_x} = \mathbf{V}\,\mathrm{diag}(P_1 \ldots P_{N_{\mathrm{T}}})\mathbf{V}^{\mathrm{H}}$ tailored to the singular value decomposition $\bar{\mathbf{H}} = \mathbf{U}\mathbf{\Sigma}\mathbf{V}^{\mathrm{H}}$, we find that the achievable rate takes the form

$$D = \sum_{m=1}^{N_{\mathrm{DoF}}} \Delta_f \log_2(1 + \lambda_m P_m)\,, \qquad \sum_{m=1}^{N_{\mathrm{DoF}}} P_m \leq P_{\mathrm{T}}^{\mathrm{avail}} \tag{3.32}$$

where $\lambda_m = (\mathbf{\Sigma}\mathbf{\Sigma}^{\mathrm{H}})_{m,m}$ are the eigenvalues of $\bar{\mathbf{H}}\bar{\mathbf{H}}^{\mathrm{H}}$. The spatial degrees of freedom $N_{\mathrm{DoF}} = \mathrm{rank}(\bar{\mathbf{H}}\bar{\mathbf{H}}^{\mathrm{H}}) \leq \min\{N_{\mathrm{T}}, N_{\mathrm{R}}\}$ quantify the number of usable parallel spatial channels. We assume that no power is assigned to useless channels, i.e. $P_m = 0$ for $m > N_{\mathrm{DoF}}$. The channel capacity C (the largest achievable rate D) is found by allocating power to the parallel channels according to the waterfilling rule [90, Eq. 5.43]

$$P_m = \max\{0, \mu - \tfrac{1}{\lambda_m}\}\,. \tag{3.33}$$

The quantity μ (colloquially called the "water level") can be computed by solving the equation $P_{\mathrm{T}}^{\mathrm{avail}} = \sum_{m=1}^{N_{\mathrm{DoF}}} \max\{0, \mu - \frac{1}{\lambda_m}\}$ for μ. This equation asserts that the power allocation P_m uses just the available sum power.

Allocating all power exclusively to the strongest spatial channel amounts to maximum-ratio combining on both ends and yields $D = \Delta_f \log_2(1 + \lambda_{\max}(\bar{\mathbf{H}}\bar{\mathbf{H}}^{\mathrm{H}})P_{\mathrm{T}}^{\mathrm{avail}})$. This rate is the channel capacity in the low-SNR case or if $N_{\mathrm{DoF}} = 1$. The situation $N_{\mathrm{DoF}} = 1$ of course comprises SIMO or MISO setups, associated with channel capacity $C = \Delta_f \log_2(1 + \|\bar{\mathbf{h}}\|^2 P_{\mathrm{T}}^{\mathrm{avail}})$, and SISO setups which exhibit $C = \Delta_f \log_2(1 + |\bar{h}|^2 P_{\mathrm{T}}^{\mathrm{avail}})$.

3.5.2 Broadband Rate Under Sum-Power Constraint

We consider a frequency band $f \in [f_{\mathrm{LO}}, f_{\mathrm{HI}}]$ whose bandwidth $B = f_{\mathrm{HI}} - f_{\mathrm{LO}}$ can be larger than the coherence bandwidth of the frequency-selective channel. We divide this wide band into N_f smaller sub-bands whereby their width $\Delta_f = \frac{B}{N_f}$ is chosen sufficiently small such that $\bar{\mathbf{H}}$ is approximately constant over each sub-band.[6] The

---

[6]A continuous-frequency formulation, as used in [67,85] for the SISO case, is avoided in order to maintain a simple notation (sums instead of integrals) and to highlight that the broadband MIMO

sub-band center frequencies are denoted f_k for $k = 1 \ldots N_f$. The associated noise-whitened channel matrices are denoted $\bar{\mathbf{H}}_k$. They describe N_f parallel MIMO channels

$$\bar{\mathbf{y}}_k = \bar{\mathbf{H}}_k \mathbf{x}_k + \bar{\mathbf{w}}_k \,, \qquad \bar{\mathbf{w}}_k \sim \mathcal{CN}(\mathbf{0}, \mathbf{I}_{N_{\mathrm{R}}}) \,, \qquad k = 1, \ldots, N_f \qquad (3.34)$$

whereby any pair $\bar{\mathbf{w}}_k$, $\bar{\mathbf{w}}_{k'}$ with $k \neq k'$ is statistically independent [90, Sec. 5.3.3]. Those parallel channels can be subsumed in a large abstract MIMO model with a block-diagonal channel matrix $\mathrm{diag}(\bar{\mathbf{H}}_1 \ldots \bar{\mathbf{H}}_{N_f})$ of dimensions $(N_{\mathrm{R}} N_f) \times (N_{\mathrm{T}} N_f)$ and noise covariance matrix $\mathbf{I}_{N_{\mathrm{R}} N_f}$. Hence (3.32) from Sec. 3.5.1 applies: any data rate

$$D = \sum_{k=1}^{N_f} \sum_{m=1}^{N_{\mathrm{DoF},k}} \Delta_f \log_2(1 + \lambda_{k,m} P_{k,m}) \,, \qquad \sum_{k=1}^{N_f} \sum_{m=1}^{N_{\mathrm{DoF},k}} P_{k,m} \leq P_{\mathrm{T}}^{\mathrm{avail}} \qquad (3.35)$$

is achievable. Thereby $\lambda_{k,m}$ is the m-th eigenvalue of $\bar{\mathbf{H}}_k \bar{\mathbf{H}}_k^{\mathrm{H}}$ and $N_{\mathrm{DoF},k} = \mathrm{rank}(\bar{\mathbf{H}}_k \bar{\mathbf{H}}_k^{\mathrm{H}})$. The channel capacity is found by allocating $P_{k,m}$ with the waterfilling rule (3.33).

As a special case we point out the SISO case $N_{\mathrm{T}} = N_{\mathrm{R}} = N_{\mathrm{DoF},k} = 1$ with achievable rate $D = \sum_{k=1}^{N_f} \Delta_f \log_2(1 + |\bar{h}_k|^2 P_k)$ subject to $\sum_{k=1}^{N_f} P_k \leq P_{\mathrm{T}}^{\mathrm{avail}}$. An illustrative example of the key quantities is given by Fig. 3.3. Clearly, the waterfilling solution and the resulting $\mathrm{SNR}(f)$ exhibit a systematic structure. In Appendix D we exploit this structure and solve the capacity problem in closed form for the SISO case with high-Q coils in AWGN. In particular, we find that transmit power should be allocated over a band $|f - f_{\mathrm{res}}| \leq B/2$ with $B = B_{3\mathrm{dB}} \left(\frac{6 |h_{\mathrm{res}}|^2 P_{\mathrm{T}}^{\mathrm{avail}}}{N_0 B_{3\mathrm{dB}}} \right)^{1/3}$. Thereby $B_{3\mathrm{dB}}$ is the 3-dB bandwidth of the resonant channel, N_0 the single-sided noise spectral density, and h_{res} the link coefficient at resonance (i.e. at the design frequency). The associated channel capacity is (in good approximation) given by $C = \frac{2}{\log(2)}(B - 2B_{3\mathrm{dB}} \arctan(\frac{B}{2B_{3\mathrm{dB}}}))$. Furthermore, the upper bound $C \leq \frac{|h_{\mathrm{res}}|^2 P_{\mathrm{T}}^{\mathrm{avail}}}{\log(2) N_0}$ is tight in the power-limited regime (Shannon limit) and the lower bound $C \geq \frac{2}{\log(2)}(B - \pi B_{3\mathrm{dB}})$ is tight for severe bandwidth-limitation. For details on the power allocation and modeling assumptions please consult Appendix D.

3.5.3 Narrowband Rate, Per-Generator Constraint

We consider the noise-whitened model $\bar{\mathbf{y}} = \bar{\mathbf{G}} \, \mathbf{i}_{\mathrm{T}} + \bar{\mathbf{w}}$ with $\bar{\mathbf{w}} \sim \mathcal{CN}(\mathbf{0}, \mathbf{I}_{N_{\mathrm{R}}})$ from (3.25). By Proposition 3.2, a constraint on the average power delivered by the n-th transmit generator reads $\mathrm{Re}(\mathbf{Z}_{\mathrm{T}}^{\mathrm{in}} \mathbf{Q}_{\mathbf{i}})_{n,n} \leq P_{\mathrm{T},n}^{\mathrm{avail}}$ with $\mathbf{Q}_{\mathbf{i}} = \mathbb{E}[\mathbf{i}_{\mathrm{T}} \mathbf{i}_{\mathrm{T}}^{\mathrm{H}}]$. By the argument of [185]

---

capacity problem is conceptionally equivalent to the narrowband problem: in both cases power is allocated optimally across parallel AWGN channels. Furthermore, a practical realization in terms of orthogonal frequency-division multiplexing (OFDM) would operate in a frequency-discrete fashion.

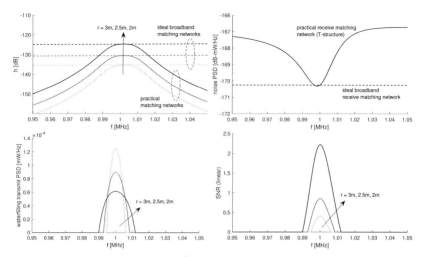

Figure 3.3: SISO link at different distances r between two coplanar coils with $1\,\mathrm{cm}$ diameter and 6 turns. The resonant design frequency is $1\,\mathrm{MHz}$ and the transmit power $1\,\mathrm{mW}$. The relation to the frequency-discrete quantities of Sec. 3.5.2 is as follows: over a small band Δ_f at f_k, the noise power spectral density (PSD) is $\frac{1}{\Delta_f}\mathbb{E}[|w_k|^2]$, the transmit PSD $\frac{1}{\Delta_f}P_k$, and the SNR $|\bar{h}_k|^2 P_k$. This numerical experiment assume a lumped element L-structure for transmit-side power matching and a T-structure for receive-side noise matching. The latter causes the noise PSD notch at the design frequency; the LNA noise correlation parameter was set to $\rho = 0.5 + 0.3j$ (details follow in Sec. 3.6).

in reference to [159], because the noise is Gaussian and the channel is known at the receiver, capacity is achieved with a transmit distribution $\mathbf{i}_\mathrm{T} \sim \mathcal{CN}(\mathbf{0}, \mathbf{Q_i})$ also in this case. The capacity-achieving transmit covariance matrix $\mathbf{Q_i}$ is found by solving the optimization problem

$$C = \max_{\mathbf{Q_i}} \Delta_f \log_2 \det \left(\mathbf{I}_{N_\mathrm{R}} + \bar{\mathbf{G}} \mathbf{Q_i} \bar{\mathbf{G}}^\mathrm{H} \right)$$

$$\text{subject to } \mathbf{Q_i} \succeq \mathbf{0},$$

$$\mathrm{Re}(\mathbf{Z}_\mathrm{T}^\mathrm{in} \mathbf{Q_i})_{n,n} \leq P_{\mathrm{T},n}^{\mathrm{avail}} \ \forall n \in \{1 \ldots N_\mathrm{T}\}. \tag{3.36}$$

This is a convex problem because the objective function and all constraints are convex (which can be assessed with the theory in [186]). Hence, it can be solved efficiently by established numerical methods [186]. A closed-form solution is unavailable.

A natural suboptimal approach is a diagonal allocation $\mathbf{Q_i} = \mathrm{diag}(\sigma_1^2 \ldots \sigma_{N_T}^2)$. It has the very convenient consequence that the average power constraint of the n-th generator becomes $\mathrm{Re}(Z_T^{\mathrm{in}})_{n,n} \, \sigma_n^2 \leq P_{T,n}^{\mathrm{avail}}$ which now concerns only the n-th port of $\mathbf{Z}_T^{\mathrm{in}}$. Thus, we can just set the current variances to $\sigma_n^2 = P_{T,n}^{\mathrm{avail}} / \mathrm{Re}(Z_T^{\mathrm{in}})_{n,n}$ in order to obtain the achievable rate

$$D = \Delta_f \log_2 \det \left(\mathbf{I}_{N_R} + \bar{\mathbf{G}} \mathbf{Q_i} \bar{\mathbf{G}}^{\mathrm{H}} \right), \qquad \mathbf{Q_i} = \underset{n=1\ldots N_T}{\mathrm{diag}} \left(\frac{P_{T,n}^{\mathrm{avail}}}{\mathrm{Re}(Z_T^{\mathrm{in}})_{n,n}} \right). \qquad (3.37)$$

3.5.4 Broadband Rate, Per-Generator Constraint

Analogous to Sec. 3.5.2, we consider parallel MIMO channels $\bar{\mathbf{y}}_k = \bar{\mathbf{G}}_k \mathbf{i}_{T,k} + \bar{\mathbf{w}}_k$ with $\bar{\mathbf{w}}_k \sim \mathcal{CN}(\mathbf{0}, \mathbf{I}_{N_R})$ for the narrow sub-bands $k = 1 \ldots N_f$. The data rate $D = \sum_{k=1}^{N_f} \Delta_f \log_2 \det \left(\mathbf{I}_{N_R} + \bar{\mathbf{G}}_k \mathbf{Q}_{i,k} \bar{\mathbf{G}}_k^{\mathrm{H}} \right)$ is achievable, whereby the covariance matrices $\mathbf{Q}_{i,k}$ of the transmit currents must fulfill $\sum_{k=1}^{N_f} \mathrm{Re}(\mathbf{Z}_T^{\mathrm{in}}(f_k) \mathbf{Q}_{i,k})_{n,n} \leq P_{T,n}^{\mathrm{avail}}$ for all generators $n = 1 \ldots N_T$. The capacity-achieving transmit covariance matrices $\mathbf{Q}_{i,k}$ have no known analytical solution (cf. the efforts in [187] regarding the broadband MISO capacity problem) and we do not attempt a numerical solution. Instead we will allocate the matrices $\mathbf{Q}_{i,k}$ with situation-specific heuristics when required.

3.6 Matching Strategies

This section discusses target values for the impedance matrices $\mathbf{Z}_T$ and $\mathbf{Z}_R$ of the transmit- and receive side matching network(s), respectively, and circuit designs which aim to attain these target values. There are different possible objectives for this design process; we consider maximizing the delivered power and maximizing the receive signal-to-noise ratio (SNR). Ideally, a matching network is composed of purely reactive elements and thus lossless; corresponding to symmetric and purely imaginary impedance matrices $\mathbf{Z}_T = j\mathbf{X}_T$ and $\mathbf{Z}_R = j\mathbf{X}_R$. [170]

The active power delivered by the transmit generators to the transmit antennas is maximized by *transmit-side power matching*. Likewise, the power delivered by the receive antennas to the receiver loads is maximized by *transmit-side power matching*. Ideally these strategies enforce (at the considered frequency)

$$\text{ideal transmit-side power matching} \quad \Longleftrightarrow \quad R_{\mathrm{ref}} \mathbf{I}_{N_T} = \mathbf{Z}_T^{\mathrm{in}}, \ \mathbf{Z}_T^{\mathrm{out}} = (\mathbf{Z}_C^{\mathrm{in}})^*, \qquad (3.38)$$

$$\text{ideal receive-side power matching} \quad \Longleftrightarrow \quad (\mathbf{Z}_C^{\mathrm{out}})^* = \mathbf{Z}_R^{\mathrm{in}}, \ \mathbf{Z}_R^{\mathrm{out}} = R_{\mathrm{ref}} \mathbf{I}_{N_R}. \qquad (3.39)$$

The transmit-side power matching condition (3.38) is achieved when the matching-network impedance matrix $\mathbf{Z}_T$ takes the specific value [53, Eq. 103]

$$\mathbf{Z}_T = \begin{bmatrix} \mathbf{Z}_{T:G} & \mathbf{Z}_{T:CG}^T \\ \mathbf{Z}_{T:CG} & \mathbf{Z}_{T:C} \end{bmatrix} = \begin{bmatrix} \mathbf{0}_{N_T} & \pm j R_{\mathrm{ref}}^{\frac{1}{2}} \operatorname{Re}(\mathbf{Z}_C^{\mathrm{in}})^{\frac{1}{2}} \\ \pm j R_{\mathrm{ref}}^{\frac{1}{2}} \operatorname{Re}(\mathbf{Z}_C^{\mathrm{in}})^{\frac{1}{2}} & -j\operatorname{Im}(\mathbf{Z}_C^{\mathrm{in}}) \end{bmatrix}. \tag{3.40}$$

Analogously, the receive-side power matching condition (3.39) is achieved for[7]

$$\mathbf{Z}_R = \begin{bmatrix} \mathbf{Z}_{R:C} & \mathbf{Z}_{R:LC}^T \\ \mathbf{Z}_{R:LC} & \mathbf{Z}_{R:L} \end{bmatrix} = \begin{bmatrix} -j\operatorname{Im}(\mathbf{Z}_C^{\mathrm{out}}) & \pm j R_{\mathrm{ref}}^{\frac{1}{2}} \operatorname{Re}(\mathbf{Z}_C^{\mathrm{out}})^{\frac{1}{2}} \\ \pm j R_{\mathrm{ref}}^{\frac{1}{2}} \operatorname{Re}(\mathbf{Z}_C^{\mathrm{out}})^{\frac{1}{2}} & \mathbf{0}_{N_R} \end{bmatrix}. \tag{3.41}$$

These rules, of course, also apply to the SISO case where all matrix blocks are scalars.

If both conditions (3.38) and (3.39) are fulfilled simultaneously, then the power transfer from the transmit generators to the receiver loads is maximized. If however the transmitter-receiver coupling is strong, $\mathbf{Z}_T^{\mathrm{out}}$ will affect $\mathbf{Z}_C^{\mathrm{out}}$ and $\mathbf{Z}_R^{\mathrm{in}}$ will affect $\mathbf{Z}_C^{\mathrm{in}}$ appreciably. In this case the value of $\mathbf{Z}_T$ affects the design of $\mathbf{Z}_R$ and vice versa. To our knowledge, such simultaneous matching problems have closed-form solutions only for the SISO case. This solution is presented in Appendix C in compact Z-parameter form, where we also state formulas for the necessary Z_T^{out} and Z_R^{in}. A possible heuristic approach to simultaneous matching of SIMO, MISO and MIMO links is an iterative round-robin alternation of transmitter matching (3.40) and receiver matching (3.41).

Maximizing the receive SNR under the employed noise model from Sec. 3.2 requires a different receiver matching strategy, called noise matching [53, Sec. IV.B]. Thereby, part of the strategy is to establish uncoupled outputs with

$$\mathbf{Z}_R^{\mathrm{out}} = Z_{\mathrm{opt}} \mathbf{I}_{N_R}, \qquad Z_{\mathrm{opt}} = R_N \left(\sqrt{1 - (\operatorname{Im}\rho)^2} + j\operatorname{Im}\rho \right) \tag{3.42}$$

whereby the SNR-optimal output impedance value Z_{opt} is determined by the noise parameters of the low-noise amplifier: the noise resistance R_N and correlation coefficient ρ given in Sec. 3.2. In particular, noise matching can be realized with a receive matching network whose impedance matrix $\mathbf{Z}_R$ has the value

$$\begin{bmatrix} \mathbf{Z}_{R:C} & \mathbf{Z}_{R:LC}^T \\ \mathbf{Z}_{R:LC} & \mathbf{Z}_{R:L} \end{bmatrix} = \begin{bmatrix} -j\operatorname{Im}(\mathbf{Z}_C^{\mathrm{out}}) & \pm j \sqrt{\operatorname{Re}(Z_{\mathrm{opt}})} \operatorname{Re}(\mathbf{Z}_C^{\mathrm{out}})^{\frac{1}{2}} \\ \pm j \sqrt{\operatorname{Re}(Z_{\mathrm{opt}})} \operatorname{Re}(\mathbf{Z}_C^{\mathrm{out}})^{\frac{1}{2}} & j\operatorname{Im}(Z_{\mathrm{opt}}) \mathbf{I}_{N_R} \end{bmatrix}. \tag{3.43}$$

---

[7]Formally, the off-diagonal $\pm$ can be chosen by preference but must be equal for both elements. A good strategy would be to choose the sign such that the matching network can be designed with mostly capacitors while inductors are avoided. In [53, Eq. 103] they employ a minus sign.

Satisfying the above conditions requires a fully connected multiport network; a possible design approach is illustrated in Fig. 3.4. Implementing such a network for arrays with $N_{\text{T}} > 1$ or $N_{\text{R}} > 1$ antennas however gives rise to a plethora of practical issues: the usable bandwidth may become very small or there may be significant ohmic losses or deviations from the target $\mathbf{Z}$-value due to component drift. Furthermore, capacitors or inductors with unrealistically large or small component values may be required. A possible resort is to use an individual two-port matching network per generator-coil pair or coil-load pair (this is also the only meaningful design for distributed arrays). This way, only diagonal blocks can be realized for $\mathbf{Z}_{\text{T}}$ and $\mathbf{Z}_{\text{R}}$. Finding their power- or SNR-optimal values is a complicated problem for coupled arrays (i.e. for non-diagonal $\mathbf{Z}_{\text{C}}^{\text{in}}$ or $\mathbf{Z}_{\text{C}}^{\text{out}}$). Possible approaches include numerical optimization or heuristics like matching just for the diagonal of $\mathbf{Z}_{\text{C}}^{\text{in}}$ or $\mathbf{Z}_{\text{C}}^{\text{out}}$. [103, 175]

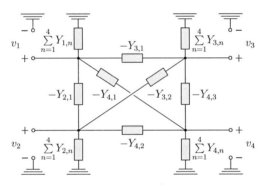

Figure 3.4: Approach for designing a $2N$-port matching network with desired electrical properties from a total of $N(2N+1)$ one-port elements (e.g., lumped elements) in Π-structure [188, 189]. The sketch shows a four-port network which can be used to match an array of two coils and two loads; the approach however transfers to larger N. The admittance values $Y_{m,n}$ are the entries of a symmetric admittance matrix $\mathbf{Y} = \mathbf{Z}^{-1}$ whereby $\mathbf{Z}$ is the desired symmetric impedance matrix at $f = f_{\text{design}}$. For a purely imaginary $\mathbf{Z} = j\mathbf{X}$ all elements of $\mathbf{Y}$ are purely imaginary as well, hence the network can be implemented with reactive lumped elements (capacitors and inductors).

A finished design then determines $\mathbf{Z}(f)$ for any f which, for $f \neq f_{\text{design}}$, will usually deviate from the desired electrical behavior.

The example in Fig. 3.7 shall give an idea of the effects and problems involved in matching a strongly coupled array. An important take-away message is that disregarding strong coupling to neighboring coils leads to very poor performance (we observe considerable resonant mode splitting). In contrary, good performance can be achieved by adapting the matching to the coupling conditions, even with the simple approach of one individual two-port network per coil-load pair. A full multiport design offers a systematic way to achieve optimal performance at the design frequency but may feature a small matching bandwidth. For a detailed discussion on the topic we refer to [103, 188, 189].

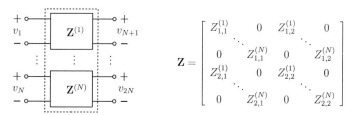

Figure 3.5: $2N$-port matching network design in terms of N two-port networks (one per coil-load pair). Compared to the fully connected design in Fig. 3.4 this design can be build with far less components (e.g. $2N$ lumped elements with L-structure two-port networks) and thus offers more robustness to component drift and a larger usable bandwidth. The drawback is that only impedance matrices $\mathbf{Z}$ with the shown structure can be realized, i.e. the strategies (3.40),(3.41),(3.43) can not be implemented for $N > 1$. This approach is mandatory for distributed arrays where no wired connection is possible across the array. [103]

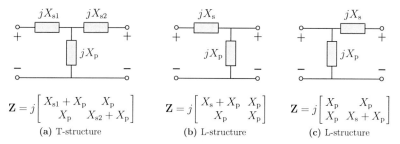

(a) T-structure   (b) L-structure   (c) L-structure

Figure 3.6: Useful structures for two-port matching network design out of reactive lumped elements (capacitances and inductances), accompanied by the realizable structure of the 2×2 impedance matrix. Finding the T-structure for a given symmetric $\mathbf{Z} = j\mathbf{X}$ such as the 2×2 versions of the rules (3.40),(3.41),(3.43) is straightforward: use lumped elements with reactance values $X_{\mathrm{p}} = X_{21}$, $X_{\mathrm{s1}} = X_{11} - X_{\mathrm{p}}$, and $X_{\mathrm{s2}} = X_{22} - X_{\mathrm{p}}$. The L-structures on the other hand allow for power-matching of an antenna to a load resistance (e.g. to $50\,\Omega$) with just two lumped elements; for the necessary formulas see [52, Sec. 5.1].

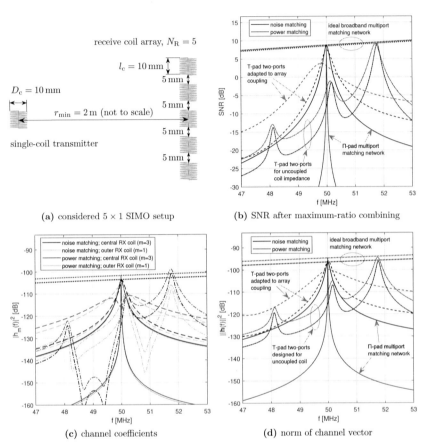

(a) considered 5×1 SIMO setup

(b) SNR after maximum-ratio combining

(c) channel coefficients

(d) norm of channel vector

Figure 3.7: Example of a SIMO link with a receive array of five strongly coupled coils and the spectral performance metrics resulting from different matching strategies. The experiment uses the same coil parameters and 50 MHz design frequency as Cpt. 5. The two-port networks adapted to the coupling conditions were obtained by re-matching each two-port network in five round-robin iterations. Any re-matching step considers the impedance $\tilde{Z}_m$ of the receive coil when all other coupled receive coils are terminated by their matched loads, given by $\tilde{Z}_m = 1/Y_{m,m} - (Z_{\mathrm{R}}^{\mathrm{in}})_{m,m}$ with $\mathbf{Y} = (\mathbf{Z}_{\mathrm{C}}^{\mathrm{out}} + \mathbf{Z}_{\mathrm{R}}^{\mathrm{in}})^{-1}$. The SNR results assume 1 mW of transmit power allocated uniformly over the shown 6 MHz band.

3.7 Channel Structure and Special Cases

The presented model may not be very accessible in its general form. It however attains simpler and more intuitive structure in various important special cases.

3.7.1 SISO Link with Simultaneous Power Matching

For a link between two coils, where a 2×2 impedance matrix $\mathbf{Z}_C$ describes the coil coupling, the simultaneous matching problem can be solved analytically. In Appendix C we derive the formula for the power transfer efficiency

$$\eta = |h|^2 = \frac{|Z_{C:RT}|^2}{R_T R_R \left(\sqrt{1 - \frac{\mathrm{Re}(Z_{C:RT})^2}{R_T R_R}} + \sqrt{1 + \frac{\mathrm{Im}(Z_{C:RT})^2}{R_T R_R}} \right)^2} \tag{3.44}$$

whereby we use the shorthand notation $R_T R_R := \mathrm{Re}(Z_{C:T}) \, \mathrm{Re}(Z_{C:R})$. In the magnetoquasistatic case $Z_{C:RT} = j\omega M$ where $\mathrm{Re}(Z_{C:RT}) = 0$, the expression (3.44) conforms with a well-known formula for inductive power transfer [81, Eq. 6], [190]. An important observation is that η in (3.44) increases when $\frac{|Z_{C:RT}|^2}{R_T R_R}$ increases, which can be seen by verifying $\partial \eta / \partial (\frac{\mathrm{Re}(Z_{C:RT})^2}{R_T R_R}) \geq 0$ and $\partial \eta / \partial (\frac{\mathrm{Im}(Z_{C:RT})^2}{R_T R_R}) \geq 0$. An important limiting case of (3.44) is obtained for weak transmitter-receiver coupling,

$$\frac{|Z_{C:RT}|^2}{R_T R_R} \ll 1 \qquad \Longrightarrow \qquad \eta = |h|^2 \approx \frac{|Z_{C:RT}|^2}{4 R_T R_R}. \tag{3.45}$$

Furthermore $\eta = 1$ for $\frac{\mathrm{Re}(Z_{C:RT})^2}{R_T R_R} = 1$ or $\frac{\mathrm{Im}(Z_{C:RT})^2}{R_T R_R} \to \infty$.

3.7.2 Unilateral Assumption for Weakly-Coupled Links

A very useful simplification is the so-called unilateral assumption [53, 170]

$$\text{set } \mathbf{Z}_{C:TR} = \mathbf{0} \qquad \Longrightarrow \qquad \mathbf{Z}_C^{\mathrm{in}} = \mathbf{Z}_{C:T}, \ \mathbf{Z}_C^{\mathrm{out}} = \mathbf{Z}_{C:R}. \tag{3.46}$$

It establishes that the transmit-side impedances are unaffected by the presence of the receiver and vice versa, which reflects the reality of weakly coupled links.[8] A very convenient consequence is that the transmit- and receive-side coils can be matched individually. Furthermore, the channel matrix becomes linear in $\mathbf{Z}_{C:RT}$ because the impedances on both ends (specifically $\mathbf{Z}_T^{in}, \mathbf{Z}_C^{in}, \mathbf{Z}_C^{out}, \mathbf{Z}_R^{out}$) are now unaffected by $\mathbf{Z}_{C:RT}$.

3.7.3 Weak Link with Ideal Matching

We employ the unilateral assumption, as is adequate for a weakly coupled link, and assume ideal power matching at both ends (i.e. the conditions (3.38) and (3.39) are fulfilled, either by design or by assumption). The channel matrix takes the form

$$\mathbf{G} = \mathbf{H} = \tfrac{1}{2} \operatorname{Re}(\mathbf{Z}_{C:R})^{-\frac{1}{2}} \mathbf{Z}_{C:RT} \operatorname{Re}(\mathbf{Z}_{C:T})^{-\frac{1}{2}} . \tag{3.47}$$

Similar expressions are common in the MIMO literature, e.g., in [191, Eq. 10], [192, Eq. 11] and [193, Eq. 6]. When noise matching is used at the receive-side instead of power matching then the same formula applies with a different scaling factor: $\frac{1}{2}$ is replaced by $\frac{\sqrt{R_{ref} \cdot \operatorname{Re}(Z_{opt})}}{R_{ref} + Z_{opt}}$.

In the SISO case, where $\mathbf{G}$ and $\mathbf{H}$ become the scalar coefficients g and h, the equation (3.47) takes the particularly simple form

$$g = h = \frac{Z_{C:RT}}{\sqrt{4R_T R_R}} \tag{3.48}$$

with coil resistances R_T and R_R and mutual impedance $Z_{C:RT}$. The PTE follows as $\eta = |h|^2 = \frac{|Z_{C:RT}|^2}{4R_T R_R}$, equivalent to the small-$Z_{C:RT}$ approximation of η in (3.44). A comparison suggests that the unilateral assumption is adequate if $|Z_{C:RT}|^2 \ll R_T R_R$.

An important special case is the SISO link where $Z_{C:RT}$ is according to the dipole formula (2.23), resulting in

$$h = \mathbf{o}_R^T \mathbf{H}_{3DoF} \mathbf{o}_T \tag{3.49}$$

whereby the central 3×3 matrix is just a scaled version of $\mathbf{Z}_{3DoF}$ from (2.28),

$$\mathbf{H}_{3DoF} = \frac{\mathbf{Z}_{3DoF}}{\sqrt{4R_T R_R}} = \alpha \left(\left(\frac{1}{(kr)^3} + \frac{j}{(kr)^2} \right) \mathbf{F}_{NF} + \frac{1}{2kr} \mathbf{F}_{FF} \right) , \tag{3.50}$$

$$\alpha = \frac{Z_0}{\sqrt{4R_T R_R}} = \frac{\mu_0 A_T \mathring{N}_T A_R \mathring{N}_R f k^3}{\sqrt{4R_T R_R}} j e^{-jkr} . \tag{3.51}$$

To understand the interface between this formalism and related introduced quantities

---

[8]Setting $\mathbf{Z}_{C:TR} = \mathbf{0}$ is a mathematically convenient and meaningful step, but violates the property $\mathbf{Z}_{C:TR} = \mathbf{Z}_{C:RT}^T$ inflicted by network reciprocity. For weakly coupled links this violation is negligible from a technical perspective while the obtained mathematical simplifications are significant. [170]

recall the definitions $\mathbf{F}_{\mathrm{NF}} = \frac{1}{2}(3\mathbf{u}\mathbf{u}^{\mathrm{T}} - \mathbf{I}_3)$ and $\mathbf{F}_{\mathrm{FF}} = \mathbf{I}_3 - \mathbf{u}\mathbf{u}^{\mathrm{T}}$, the scaled field vectors $\boldsymbol{\beta}_{\mathrm{NF}} = \mathbf{F}_{\mathrm{NF}}\mathbf{o}_{\mathrm{T}}$ and $\boldsymbol{\beta}_{\mathrm{FF}} = \mathbf{F}_{\mathrm{FF}}\mathbf{o}_{\mathrm{T}}$ from (2.19) and (2.20), and the alignment factors $J_{\mathrm{NF}} = \mathbf{o}_{\mathrm{R}}^{\mathrm{T}}\boldsymbol{\beta}_{\mathrm{NF}} \in [-1, 1]$ and $J_{\mathrm{FF}} = \mathbf{o}_{\mathrm{R}}^{\mathrm{T}}\boldsymbol{\beta}_{\mathrm{FF}} \in [-1, 1]$ from (2.24) and (2.25).

3.7.4 Channel Between Colocated Arrays

We consider the setup of Sec. 2.3.2 with a MIMO link between a transmit coil array and a receive coil array. Both arrays are either strictly colocated or their sizes are much smaller than the link distance. In (2.46) we showed that the formula $\mathbf{Z}_{\mathrm{C:RT}} = \mathbf{O}_{\mathrm{R}}^{\mathrm{T}}\mathbf{Z}_{\mathrm{3DoF}}\mathbf{O}_{\mathrm{T}}$ applies for the $N_{\mathrm{R}} \times N_{\mathrm{T}}$ transmitter-to-receiver mutual impedance matrix whereby $\mathbf{Z}_{\mathrm{3DoF}}$ is a 3×3 matrix given in (2.28), $\mathbf{O}_{\mathrm{R}} = [\mathbf{o}_{\mathrm{R},1} \ldots \mathbf{o}_{\mathrm{R},N_{\mathrm{R}}}]$ holds the receiver orientation vectors and $\mathbf{O}_{\mathrm{T}} = [\mathbf{o}_{\mathrm{T},1} \ldots \mathbf{o}_{\mathrm{T},N_{\mathrm{T}}}]$ holds the transmitter orientation vectors. If the link is furthermore weakly coupled (which is to be expected because the above formula requires a rather large link distance r; in principle however high-Q coils can have strong coupling at large r) and power-matched on both ends, then by (3.47) the channel matrix is

$$\mathbf{H} = \tfrac{1}{2} \operatorname{Re}(\mathbf{Z}_{\mathrm{C:R}})^{-\frac{1}{2}} \mathbf{O}_{\mathrm{R}}^{\mathrm{T}} \mathbf{Z}_{\mathrm{3DoF}} \, \mathbf{O}_{\mathrm{T}} \operatorname{Re}(\mathbf{Z}_{\mathrm{C:T}})^{-\frac{1}{2}} \qquad (3.52)$$

(noise matching gives the same result apart from a different scaling factor). For the channel rank we note that $\operatorname{rank}(\mathbf{H}) \leq 3$ because matrix $\mathbf{Z}_{\mathrm{3DoF}}$ has dimension 3×3.

As an interesting example consider that all transmit coils have the same orientation $\mathbf{o}_{\mathrm{T}}$ and all receive coils the same $\mathbf{o}_{\mathrm{R}}$, i.e. $\mathbf{O}_{\mathrm{T}} = \mathbf{o}_{\mathrm{T}} \mathbf{1}_{1 \times N_{\mathrm{T}}}$ and $\mathbf{O}_{\mathrm{R}}^{\mathrm{T}} = \mathbf{1}_{N_{\mathrm{R}} \times 1} \mathbf{o}_{\mathrm{R}}^{\mathrm{T}}$. The resulting channel matrix $\mathbf{H} = \frac{\mathbf{o}_{\mathrm{R}}^{\mathrm{T}}\mathbf{Z}_{\mathrm{3DoF}}\mathbf{o}_{\mathrm{T}}}{2} \cdot \operatorname{Re}(\mathbf{Z}_{\mathrm{C:R}})^{-\frac{1}{2}} \mathbf{1}_{N_{\mathrm{R}} \times N_{\mathrm{T}}} \operatorname{Re}(\mathbf{Z}_{\mathrm{C:T}})^{-\frac{1}{2}}$. To obtain some intuition let us assume uncoupled arrays with $\operatorname{Re}(\mathbf{Z}_{\mathrm{C:R}}) = R_{\mathrm{R}}\mathbf{I}_{N_{\mathrm{R}}}$ and $\operatorname{Re}(\mathbf{Z}_{\mathrm{C:T}}) = R_{\mathrm{T}}\mathbf{I}_{N_{\mathrm{T}}}$ (thereby ignoring that a colocated array will usually be coupled), resulting in $\mathbf{H} = h_{\mathrm{SISO}} \mathbf{1}_{N_{\mathrm{R}} \times N_{\mathrm{T}}}$ with $h_{\mathrm{SISO}} = \mathbf{o}_{\mathrm{R}}^{\mathrm{T}}\mathbf{H}_{\mathrm{3DoF}}\mathbf{o}_{\mathrm{T}}$. Clearly $\mathbf{H}$ has rank one which prohibits spatial multiplexing, yet an array gain can be achieved. With (3.28) we find that maximum-ratio beamforming on both ends yields a power transfer efficiency of $\eta = |h_{\mathrm{SISO}}|^2 \lambda_{\max}(\mathbf{1}_{N_{\mathrm{R}} \times N_{\mathrm{T}}} \mathbf{1}_{N_{\mathrm{T}} \times N_{\mathrm{R}}}) = |h_{\mathrm{SISO}}|^2 N_{\mathrm{T}} \cdot \lambda_{\max}(\mathbf{1}_{N_{\mathrm{R}} \times N_{\mathrm{R}}}) = |h_{\mathrm{SISO}}|^2 N_{\mathrm{T}} N_{\mathrm{R}}$ whereby the factor $N_{\mathrm{T}}N_{\mathrm{R}}$ is the array gain (cf. [159, Example 1]). Different results should be expected for strongly coupled arrays because some eigenvalues of $\operatorname{Re}(\mathbf{Z}_{\mathrm{C:T}})$ and $\operatorname{Re}(\mathbf{Z}_{\mathrm{C:R}})$ may be much larger than the individual coil resistances.

3.7.5   Weak 3×3 Link Between Orthogonal Colocated Arrays

We consider a 3×3 link under the assumptions of Sec. 3.7.4, but now the colocated $N_T = 3$ transmit coils have orthogonal axes and also the colocated $N_R = 3$ receive coils have orthogonal axes (tri-axial coils) as shown in Fig. 3.8. Formally, that means that $\mathbf{O}_T$ and $\mathbf{O}_R$ are orthogonal matrices. Such arrays are in fact uncoupled if the coils are electrically small [194], hence $\mathbf{Z}_{C:T}$ and $\mathbf{Z}_{C:R}$ are diagonal matrices. We consider the case where all coils in an array have the same parameters and thus $\mathbf{Z}_{C:T} = R_T \mathbf{I}_{N_T}$ and $\mathbf{Z}_{C:R} = R_R \mathbf{I}_{N_R}$. Hence (3.52) simplifies to

$$\mathbf{H} = \mathbf{O}_R^T \mathbf{H}_{3\text{DoF}} \mathbf{O}_T \in \mathbb{R}^{3\times 3} \qquad (3.53)$$

in complete analogy to (3.49). In more generality, $\mathbf{H} = a \cdot \mathbf{O}_R^T \mathbf{H}_{3\text{DoF}} \mathbf{O}_T$ with $a = 1$ for a power-matched receive array and $a = \frac{\sqrt{4R \cdot \text{Re}(Z_{\text{opt}})}}{R + Z_{\text{opt}}}$ for a noise-matched receive array.

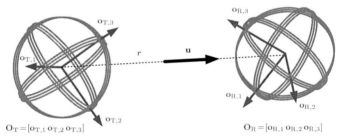

Figure 3.8: Magneto-inductive 3×3 MIMO link whereby each array consists of three colocated coils with orthogonal axes.

For this setup we can furthermore expect a scaled-identity noise covariance matrix $\mathbf{K} = \sigma^2 \mathbf{I}_3$: by inspecting (3.18) we find that Σ_{therm} and Σ_{LNA} are both scaled identity because $\mathbf{Z}_R^{\text{out}}$ is scaled identity. Likewise, the covariance matrix Σ_{ext} of extrinsically induced noise voltages is scaled identity for an orthogonal and uncoupled array if we assume i.i.d. impinging field components in x-, y- and z-direction.

Apart from communication, the 3×3 link described by (3.53) has powerful properties for localization. We note that the maximum eigenvalue fulfills $\lambda_{\max}(\mathbf{H}\mathbf{H}^H) = \lambda_{\max}(\mathbf{H}_{3\text{DoF}}\mathbf{H}_{3\text{DoF}}^H)$ and is given by either $|h^{\text{coax}}|^2 = |\alpha|^2 \left(\frac{1}{(kr)^6} + \frac{1}{(kr)^4} \right)$ in the near field or $|h^{\text{copl}}|^2$ in the far field. So after observing $\mathbf{H}$ through channel estimation, the equation $\frac{1}{(kr)^6} + \frac{1}{(kr)^4} = \frac{1}{|\alpha|^2} \lambda_{\max}(\mathbf{H}\mathbf{H}^H)$ can be solved for r to obtain a near-field case distance estimate (the approach is easily generalized with a case distinction). The approximation

$r \approx \frac{1}{k}|\alpha|^{\frac{1}{3}}(\lambda_{\max}(\mathbf{H}\mathbf{H}^{\mathrm{H}}))^{-\frac{1}{6}}$ applies in the near-field region and $r \approx \frac{|\alpha|}{2k}(\lambda_{\max}(\mathbf{H}\mathbf{H}^{\mathrm{H}}))^{-\frac{1}{2}}$ applies in the far-field region. This idea has been used by [59] for the magnetoquasistatic near-field case. However, this theory can also be used for direction finding: we can compute the transmitter-to-receiver direction $\mathbf{u}$ up a $\pm\mathbf{u}$ uncertainty from the eigenvalue decomposition of $\mathbf{H}\mathbf{H}^{\mathrm{H}}$ because a unit-length eigenvector associated with $\lambda_{\max}(\mathbf{H}\mathbf{H}^{\mathrm{H}})$ equals either $\mathbf{u}$ or $-\mathbf{u}$.

3.8 Spatial Degrees of Freedom

3.8.1 Colocated Coil Arrays

In Sec. 3.7.4 we showed that magneto-inductive links between colocated arrays exhibit a channel matrix with bounded rank($\mathbf{H}$) ≤ 3, hence a maximum of three parallel streams are available for spatial multiplexing. This is due to the three components b_x, b_y, b_z of the magnetic field $\mathbf{b}$ at the receive position, which can be used to convey three independent streams of information to the receiver. Spatial multiplexing over a direct link (i.e. without any multipath propagation) between colocated arrays has been considered for satellite communication which utilizes two orthogonal electric-field components (polarization) [195]. The result compares to a maximum of six spatial channels that are available between colocated arrays if both the magnetic field and the electric field are utilized [196].

The three spatial degrees of freedom are particularly accessible for the 3×3 link between orthogonal coil arrays discussed in Sec. 3.7.5. There we derived a noise-whitened channel matrix $\bar{\mathbf{H}} = \frac{a}{\sigma}\mathbf{O}_{\mathrm{R}}^{\mathrm{T}}\mathbf{H}_{\mathrm{3DoF}}\mathbf{O}_{\mathrm{T}}$. We rewrite $\bar{\mathbf{H}} = \frac{a}{\sigma}\mathbf{O}_{\mathrm{R}}^{\mathrm{T}}\mathbf{Q}_{\mathbf{u}}\,\mathrm{diag}(h^{\mathrm{coax}}, h^{\mathrm{copl}}, h^{\mathrm{copl}})\mathbf{Q}_{\mathbf{u}}^{\mathrm{T}}\mathbf{O}_{\mathrm{T}}$ using the eigenvalue decomposition (2.32) and the particular SISO link coefficients between coaxial or coplanar coil pairs,

$$h^{\mathrm{coax}} = \alpha\left(\frac{1}{(kr)^3} + \frac{j}{(kr)^2}\right), \qquad h^{\mathrm{copl}} = \frac{\alpha}{2}\left(-\frac{1}{(kr)^3} - \frac{j}{(kr)^2} + \frac{1}{kr}\right). \qquad (3.54)$$

Since $\mathbf{Q}_{\mathbf{u}}$ is an orthogonal matrix, $(\mathbf{O}_{\mathrm{R}}^{\mathrm{T}}\mathbf{Q}_{\mathbf{u}})$ and $(\mathbf{Q}_{\mathbf{u}}^{\mathrm{T}}\mathbf{O}_{\mathrm{T}})$ are orthogonal too and can be compensated by transmit beamforming and receive beamforming without affecting the transmit power or noise statistics (i.e. the array orientations are irrelevant). In consequence, the narrowband channel capacity is characterized by the equivalent noise-whitened channel matrix

$$\bar{\mathbf{H}}' = \frac{a}{\sigma} \begin{bmatrix} h^{\text{coax}} & & \\ & h^{\text{copl}} & \\ & & h^{\text{copl}} \end{bmatrix} \tag{3.55}$$

which reveals the available spatial channels: one with amplitude gain h^{coax} corresponding to a coaxial SISO link and two with channel gain h^{copl} corresponding to coplanar SISO links. One spatial channel is lost in the far field because h^{coax} decays faster than r^{-1}. An intuitive explanation is that the magnetic far field generated at the receiver position can only be realized on the plane orthogonal to the radial direction (two degrees of freedom), but not in radial direction. No such restriction applies in the near field, where all three degrees of freedom are available.

3.8.2 Distributed Coil Arrays

Spreading out the coils of the arrays involved in a magneto-inductive MIMO link can give rise to additional spatial degrees of freedom by increasing the channel rank. The circumstance is analogous to line-of-sight MIMO [197] which also aims for spatial multiplexing gains via spread arrays, possibly without any multipath propagation. We investigate the basic mechanism in terms of the eigenvalues $\lambda_m(\mathbf{HH}^{\text{H}})$ for magneto-inductive 2×2 links in Fig. 3.9 between two pairs of coplanar coils and in Fig. 3.10 between two pairs of coaxial coils, as a function of array spread (angle γ) and in near-, mid- and far-field regimes (the mid-field value $kr_{\text{th}} = 2.354$ is the threshold from (2.33)). For simplicity these numerical results assume uncoupled arrays in the sense that $\text{Re}(\mathbf{Z}_{\text{C:R}}) = R_{\text{R}}\mathbf{I}_{N_{\text{R}}}$ and $\text{Re}(\mathbf{Z}_{\text{C:T}}) = R_{\text{T}}\mathbf{I}_{N_{\text{T}}}$.

We observe that at $\alpha = 90°$ both eigenvalues attain $|h^{\text{coax}}|^2$ in the near-field cases, $|h^{\text{coax}}|^2 = |h^{\text{copl}}|^2$ in the specific mid-field cases and $|h^{\text{copl}}|^2$ in the coplanar far-field case. This is because $\alpha = 90°$ effectively gives rise to two parallel (non-interfering) SISO links with just those power gains. For the same cases we observe that λ_{max} is 4 times the 90°-value due to a array gain of $N_{\text{T}}N_{\text{R}} = 4$ for near-colocated arrays (see the example in Sec. 3.7.4). An exception is the coaxial far-field case where the channel fades for $\alpha = 0°$ and $\alpha = 90°$ because the far-field link vanishes for a coaxial pair of coils; appreciable eigenvalues however emerge near $\alpha = 55°$. Furthermore, note that the coaxial near-field case has equal eigenvalues at the magic angle $\alpha = \arccos(\frac{1}{\sqrt{3}}) = 54.74°$ and the coplanar near-field case at $\alpha = 90° - 54.74° = 35.26°$, because in these situations the diagonal links fade to zero (as seen earlier in Fig. 2.5).

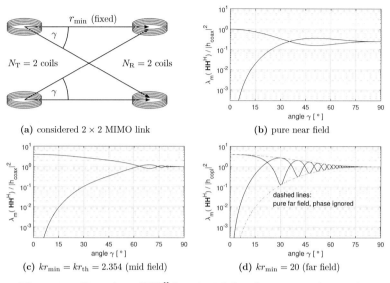

(a) considered 2 × 2 MIMO link

(b) pure near field

(c) $kr_{\min} = kr_{\text{th}} = 2.354$ (mid field)

(d) $kr_{\min} = 20$ (far field)

Figure 3.9: Eigenvalues of $\mathbf{HH}^{\text{H}}$ for a 2 × 2 link with pairwise coplanar coils.

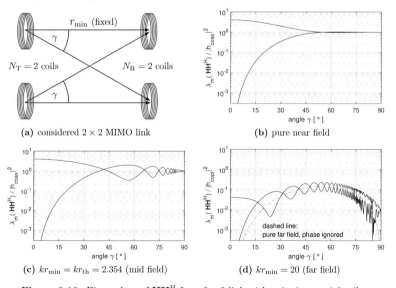

(a) considered 2 × 2 MIMO link

(b) pure near field

(c) $kr_{\min} = kr_{\text{th}} = 2.354$ (mid field)

(d) $kr_{\min} = 20$ (far field)

Figure 3.10: Eigenvalues of $\mathbf{HH}^{\text{H}}$ for a 2 × 2 link with pairwise coaxial coils.

65

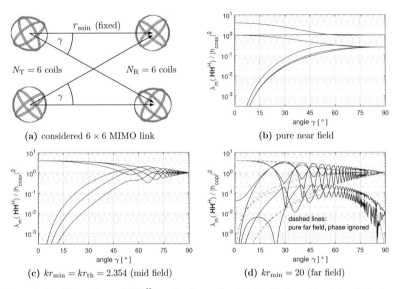

(a) considered 6×6 MIMO link

(b) pure near field

(c) $kr_{\min} = kr_{\mathrm{th}} = 2.354$ (mid field)

(d) $kr_{\min} = 20$ (far field)

Figure 3.11: Eigenvalues of $\mathbf{HH}^{\mathrm{H}}$ for the shown 6×6 link with four tri-axial coil clusters.

We proceed with a study of magneto-inductive 6×6 MIMO links of the kind shown in Fig. 3.11a. A first important observation concerns the structure of the channel matrix following from (3.53),

$$\mathbf{H} = \begin{bmatrix} \mathbf{O}_{\mathrm{R},1}^{\mathrm{T}} \mathbf{H}_{\mathrm{3DoF}}^{(1,1)} \mathbf{O}_{\mathrm{T},1} & \mathbf{O}_{\mathrm{R},1}^{\mathrm{T}} \mathbf{H}_{\mathrm{3DoF}}^{(1,2)} \mathbf{O}_{\mathrm{T},2} \\ \mathbf{O}_{\mathrm{R},2}^{\mathrm{T}} \mathbf{H}_{\mathrm{3DoF}}^{(2,1)} \mathbf{O}_{\mathrm{T},1} & \mathbf{O}_{\mathrm{R},2}^{\mathrm{T}} \mathbf{H}_{\mathrm{3DoF}}^{(2,2)} \mathbf{O}_{\mathrm{T},2} \end{bmatrix} = \underbrace{\begin{bmatrix} \mathbf{O}_{\mathrm{R},1}^{\mathrm{T}} & \\ & \mathbf{O}_{\mathrm{R},2}^{\mathrm{T}} \end{bmatrix}}_{\text{unitary}} \begin{bmatrix} \mathbf{H}_{\mathrm{3DoF}}^{(1,1)} & \mathbf{H}_{\mathrm{3DoF}}^{(1,2)} \\ \mathbf{H}_{\mathrm{3DoF}}^{(2,1)} & \mathbf{H}_{\mathrm{3DoF}}^{(2,2)} \end{bmatrix} \underbrace{\begin{bmatrix} \mathbf{O}_{\mathrm{T},1} & \\ & \mathbf{O}_{\mathrm{T},2} \end{bmatrix}}_{\text{unitary}}$$

(3.56)

which means that the orientations (or rather rotations) of the orthogonal coil clusters do not affect the channel eigenvalues, hence we do not have to specify them. The arguments extends to all MIMO links between arrays that consist of such orthogonal coil clusters. In the numerical results we again observe that two non-interfering 3×3 links emerge for large γ, each with one spatial channel that corresponds to a coaxial coil pair (which fades in the far-field, see Fig. 3.11d) and two spatial channels that correspond to coplanar coil pairs. For $\alpha = 0$ we observe the same three spatial channels again with an array gain of 4 in terms of power, except for the pure far-field case which misses the spatial channel corresponding to a coaxial SISO link (also at $\alpha = 90°$).

3.9 Limits of Cooperative Load Modulation

We consider the load modulation system in Fig. 3.12 where a number of tags transmit bits by switching between two different load terminations according to the currently transmitted bit. Meanwhile a receiving reader device attempts to decode those bits. The tags are cooperating in the sense that all tags transmit the same bit in each time slot (perfect synchronization is assumed). The reader drives the constant AC currents $\mathbf{i} = [i_1 \ \dots \ i_{N_R}]^T$ through its coil array and measures the resulting voltages $\mathbf{v} = [v_1 \ \dots \ v_{N_R}]^T$ across the same coils. The bits are decoded based on the observations $\mathbf{v}$. The electrical aspects of this approach were described in Proposition 2.6 and the accompanying examples. In this conceptual investigation we do not consider technical details of the reader circuits.

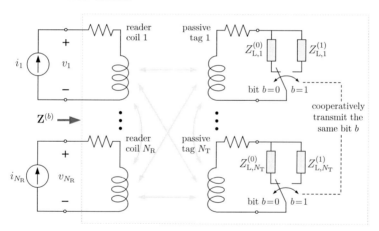

Figure 3.12: Circuit description of the considered cooperative load modulation scheme. We consider a total of N_T passive tags which cooperatively transmit the same bit stream (synchronization is assumed), which is decoded by a reader device with an array of N_R coils. The approach can be regarded as an idealized MIMO version of the SISO load modulation system in [63, Fig. 3.16]. Reasonable choices for the load impedances would be a resonance capacitance for all $Z_{L,n}^{(0)}$ and a vastly different load (e.g., some resistance) for all $Z_{L,n}^{(1)}$.

Let $\mathbf{Z}$ denote the impedance matrix of the reader device coil array. The core idea of load modulation is that the tag switching state affects $\mathbf{Z}$ and thus also the voltages

$\mathbf{v} = \mathbf{Z}\mathbf{i}$ [26, 63]. We denote $\mathbf{Z}^{(b)}$ for the value of $\mathbf{Z}$ when bit $b \in \{0, 1\}$ is transmitted. In doing so we assume that the bit rate f_b is sufficiently small for transients to die out between switching instants.[9] As a consequence of Proposition 2.6, the difference $\mathbf{Z}^{(1)} - \mathbf{Z}^{(0)}$ is usually very small because the tag-to-reader coupling applies twice. Hence we can describe the coarse electrical properties of the array with $\bar{\mathbf{Z}} = \frac{1}{2}(\mathbf{Z}^{(1)} + \mathbf{Z}^{(0)})$ instead of the momentary $\mathbf{Z}^{(b)}$.

We consider erroneous measurements $\mathbf{v}$ which might lead to bit errors, especially when $\mathbf{Z}^{(1)} - \mathbf{Z}^{(0)}$ is small. In particular we use the model

$$\mathbf{v} = \mathbf{Z}^{(b)}\mathbf{i} + \mathbf{v}_\mathrm{N}\,, \qquad \mathbf{v}_\mathrm{N} \sim \mathcal{CN}(\mathbf{0}, \mathbf{K}_\mathbf{v}) \tag{3.57}$$

whereby the Gaussian model for $\mathbf{v}_\mathrm{N}$ is motivated by the same arguments as in Sec. 3.2. The most optimistic model for the covariance matrix $\mathbf{K}_\mathbf{v}$ would be thermal noise only, i.e. $\mathbf{K}_\mathbf{v} = 4k_\mathrm{B}Tf_\mathrm{b}\,\mathrm{Re}(\bar{\mathbf{Z}})$ with Boltzmann constant k_B and temperature T. A meaningful more sophisticated model is $\mathbf{K}_\mathbf{v} = 4k_\mathrm{B}Tf_\mathrm{b}\,\mathrm{Re}(\bar{\mathbf{Z}}) + \sigma_v^2 \mathbf{I}_{N_\mathrm{R}} + \sigma_i^2 \bar{\mathbf{Z}}(\bar{\mathbf{Z}})^\mathrm{H}$, which also considers fluctuations in the measurement of $\mathbf{v}$ and imperfections in establishing the currents $\mathbf{i}$, which can certainly exceed thermal noise by orders of magnitude. For example, the values $\sigma_v = 1\,\mu\mathrm{V} + 10^{-5} \cdot \frac{\|\bar{\mathbf{Z}}\mathbf{i}\|}{\sqrt{N_\mathrm{R}}}$ and $\sigma_i = 10^{-5} \cdot \frac{\|\mathbf{i}\|}{\sqrt{N_\mathrm{R}}}$ could be reasonable for a capable reader.

Proposition 3.4. *The bit error probability of a maximum a-posteriori bit detector for equiprobable load-modulated bits is given in terms of the Q-function,*

$$\epsilon = Q(\sqrt{\mathrm{SNR}})\,, \qquad \mathrm{SNR} = \frac{1}{2}\,\|\mathbf{K}_\mathbf{v}^{-\frac{1}{2}}(\mathbf{Z}^{(1)} - \mathbf{Z}^{(0)})\mathbf{i}\|^2\,. \tag{3.58}$$

Proof. We apply noise whitening $\mathbf{y} = \mathbf{K}_\mathbf{v}^{-1/2}\mathbf{v}$ and obtain the model $\mathbf{y} = \mathbf{y}^{(b)} + \mathbf{w}$ with $\mathbf{w} \sim \mathcal{CN}(\mathbf{0}, \mathbf{I}_{N_\mathrm{R}})$. The two signal points $\mathbf{y}^{(b)} = \mathbf{K}_\mathbf{v}^{-1/2}\mathbf{Z}^{(b)}\mathbf{i} \in \mathbb{C}^{N_\mathrm{R}}$ have Euclidean distance $\sqrt{2\,\mathrm{SNR}} = \|\mathbf{y}^{(1)} - \mathbf{y}^{(0)}\|$. The quantity $\mathrm{Re}(\mathbf{a}^\mathrm{H}\mathbf{y})$ with $\mathbf{a} = \frac{\mathbf{y}^{(1)} - \mathbf{y}^{(0)}}{\|\mathbf{y}^{(1)} - \mathbf{y}^{(0)}\|}$ is a sufficient statistic for the detection of bit b [90, Eq. A-49]. Equivalently we consider the statistic $z = \sqrt{2}\cdot\mathrm{Re}(\mathbf{a}^\mathrm{H}(\mathbf{y} - \frac{1}{2}(\mathbf{y}^{(1)} - \mathbf{y}^{(0)})))$, yielding the signal model $z = \pm\sqrt{\mathrm{SNR}} + w_z$ with $w_z \sim \mathcal{N}(0, 1)$. The decision threshold $z = 0$ minimizes the error probability of bit detection if $b = 0$ and $b = 1$ are equiprobable [37], giving $\epsilon = Q(\sqrt{\mathrm{SNR}})$. $\qquad\square$

---

[9]In Appendix E we show how transients can be incorporated in the receive processing of load modulation systems, with the conclusion that neither the consideration nor the ignoring of transients is associated with significant gain or deterioration. Based on this insight we do not consider transients in the investigation at hand.

Proposition 3.5. *For a given source sum-power P, the* SNR*-maximizing current vector is given by the rule* $\mathbf{i} = \mathrm{Re}(\bar{\mathbf{Z}})^{-\frac{1}{2}}\mathbf{x}$ *whereby the vector* $\mathbf{x}$ *is an eigenvector of* $\mathrm{Re}(\bar{\mathbf{Z}})^{-\frac{\mathrm{H}}{2}}(\mathbf{Z}^{(1)} - \mathbf{Z}^{(0)})^{\mathrm{H}}\mathbf{K}_{\mathbf{v}}^{-1}(\mathbf{Z}^{(1)} - \mathbf{Z}^{(0)})\mathrm{Re}(\bar{\mathbf{Z}})^{-\frac{1}{2}}$ *that corresponds to the largest eigenvalue and has magnitude* $\|\mathbf{x}\| = \sqrt{P}$.

Proof. The goal is to maximize SNR $= \frac{1}{2}\mathbf{i}^{\mathrm{H}}(\mathbf{Z}^{(1)} - \mathbf{Z}^{(0)})^{\mathrm{H}}\mathbf{K}_{\mathbf{v}}^{-1}(\mathbf{Z}^{(1)} - \mathbf{Z}^{(0)})\mathbf{i}$ subject to the power constraint $\mathbf{i}^{\mathrm{H}}\mathrm{Re}(\bar{\mathbf{Z}})\mathbf{i} \leq P$. With the transformation $\mathbf{x} = \mathrm{Re}(\bar{\mathbf{Z}})^{\frac{1}{2}}\mathbf{i}$ the constraint becomes $\|\mathbf{x}\|^2 \leq P$ and the statement follows from basic linear algebra. $\square$

Proposition 3.6. *Let the noise voltages* $\mathbf{v}_{\mathrm{N}}$ *be statistically independent across different bit indices and consider a receive decoder for long blocks of redundantly coded information. If the decoder is preceded by hard bit detection (i.e. 1-bit quantization) with error probability ϵ then the channel capacity (the largest achievable information rate) C_{hard} in bit/s is given by*

$$\frac{C_{\mathrm{hard}}}{f_{\mathrm{b}}} = 1 - \epsilon \log_2\left(\frac{1}{\epsilon}\right) - (1 - \epsilon)\log_2\left(\frac{1}{1 - \epsilon}\right). \tag{3.59}$$

If the decoder instead uses the raw measurements without any quantization then the formula $\frac{C_{\mathrm{soft}}}{f_{\mathrm{b}}} = 1 - \frac{1}{\sqrt{2\pi}}\int_{-\infty}^{\infty} e^{-\frac{1}{2}(y - \sqrt{\mathrm{SNR}})^2}\log_2(1 + e^{-2y\sqrt{\mathrm{SNR}}})dy$ applies for the channel capacity C_{soft} in bit/s. Thereby $0 \leq C_{\mathrm{hard}} \leq C_{\mathrm{soft}} \leq f_{\mathrm{b}} \cdot \min\{1, \frac{1}{2}\log_2(1 + \mathrm{SNR})\}$.

Proof. The formula for C_{hard} is the capacity of the binary symmetric channel, a basic result from information theory [180, Sec. 7.1.4]. The formula for C_{soft} is from [198]. Thereby $C_{\mathrm{hard}} \leq C_{\mathrm{soft}}$ by the data processing inequality, $C_{\mathrm{hard}} \leq f_{\mathrm{b}}$ and $C_{\mathrm{soft}} \leq f_{\mathrm{b}}$ because the source entropy is 1 bit, and furthermore the binary-input capacity C_{soft} must be smaller than the AWGN-channel capacity of $\frac{1}{2}\log_2(1 + \mathrm{SNR})$ bit per channel use (which uses an entropy-maximizing Gaussian input distribution). $\square$

Note that we did not make any assumptions about the internals of the tags and the coupling: the process is entirely based on the variation of impedance matrix $\mathbf{Z}$ between the reader coil array. Yet, a thorough understanding of the internals is certainly required for establishing large $\mathbf{Z}$-variation through load switching.

Finally, we note that the SNR expression (3.58) can be written as SNR $= \frac{1}{2}\mathbf{i}^{\mathrm{H}}\mathbf{Z}_{\mathrm{RT}}^{\mathrm{H}}\boldsymbol{\Delta}^{\mathrm{H}}\mathbf{Z}_{\mathrm{RT}}^{*}\mathbf{K}_{\mathbf{v}}^{-1}\mathbf{Z}_{\mathrm{RT}}\boldsymbol{\Delta}\mathbf{Z}_{\mathrm{RT}}^{\mathrm{T}}\mathbf{i}$ with $\boldsymbol{\Delta} = (\mathbf{Z}_{\mathrm{T}} + \mathrm{diag}_n Z_{\mathrm{L},n}^{(0)})^{-1} - (\mathbf{Z}_{\mathrm{T}} + \mathrm{diag}_n Z_{\mathrm{L},n}^{(1)})^{-1}$. An important aspect is that SNR is proportional to the fourth power of the tag-to-reader coupling $\mathbf{Z}_{\mathrm{RT}}$. This implies a path loss scaling of SNR $\propto r^{-12}$ and a frequency-dependence SNR $\propto f^4$ in the magnetoquasistatic regime. Far-field operation with electrically small coils on the other hand gives SNR $\propto r^{-4}$ but SNR $\propto f^{12}$. Hence, the usable range of low-frequency load modulation is severely limited.

A numerical evaluation of a cooperative load modulation scenario will be given at the end of Cpt. 6, based on the presented theory.

Chapter 4

The Channel Between Randomly Oriented Coils in Free Space

The position and orientation of a wireless sensor device is subject to the deployment strategy and mobility and can be considered random in the context of wireless link design and performance analysis (as motivated in Sec. 1.3.4). A random coil orientation has a particularly significant impact: it causes severe attenuation for the transmitter-to-receiver mutual impedance Z_{RT} and the resulting channel coefficient h. In Sec. 1.3.4 we discussed the associated state of the art, which is mostly focused on the effect of small lateral and angular offsets on efficient power transfer over short-distance links. This chapter in contrary considers the mathematically tractable domain of the dipole model from (3.49) and studies links between coils with fully random orientation: the orientation vectors $\mathbf{o}_T$ and $\mathbf{o}_R$ are assumed to have uniform distributions on the 3D unit sphere (statistically independent). The approach is illustrated in Fig. 4.1 and models situations like a fully ad-hoc scenario or a moving wireless sensor in an underwater or medical in-body application. In particular, the channel coefficient is given by[1]

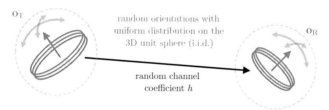

Figure 4.1: Illustration of a SISO link with random coil orientations on both ends, both with a uniform distribution on the 3D unit sphere (all directions are equiprobable).

$h = \mathbf{o}_R^T \mathbf{H}_{3DoF} \mathbf{o}_T$ according to the dipole model (3.49), which applies for mid- and long-range links between coils whose turns are rather flat and have consistent surface orientation. Now h is a random variable because it is a function of the random unit vectors $\mathbf{o}_T, \mathbf{o}_R$.

We shall illustrate the significance of this random channel with a motivating experiment. Fig. 4.2 shows a Monte-Carlo simulation of the statistics of the power attenuation

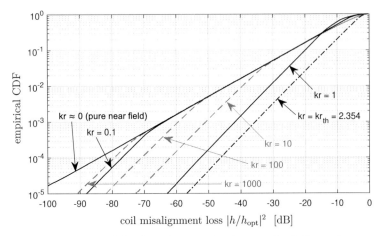

Figure 4.2: Statistics of the channel attenuation due to coil misalignment compared to the absolute-value-optimal channel coefficient (h_{coax} in the near field, h_{copl} in the far field). Severe attenuation occurs with non-negligible probability, especially in the pure near field and pure far field. The transition region benefits from a polarization diversity effect.

$|h/h_{\mathrm{opt}}|^2$ that results from the described distributions of the coil orientations. Thereby the absolute-value-optimal channel coefficient is the coaxial value $h_{\mathrm{opt}} = h_{\mathrm{coax}}$ for near-field distances $kr \leq kr_{\mathrm{th}}$, cf. (2.33), and the coplanar value $h_{\mathrm{opt}} = h_{\mathrm{copl}}$ otherwise.

We observe that severe attenuation occurs with significant probability. In the pure near field the attenuation is worse than $-23.7\,\mathrm{dB}$ in 10% of cases and even worse than $-43.7\,\mathrm{dB}$ in 1% of cases; the pure far field shows very similar behavior.[2] Such losses can certainly cause outage problems in applications with arbitrarily arranged coils. In the transition region between near and far field the attenuation statistics are not as disastrous (yet significant). The reason is the magnetic field polarization in this regime: the magnetic field vector is rotating on an ellipse instead of alternating on a line, as discussed earlier below Proposition 2.3. For the link to be in outage, $\mathbf{o}_{\mathrm{R}}$ must be near-orthogonal to the ellipse, which is less probable than near-orthogonality to a

---

[1]Matrix $\mathbf{H}_{\mathrm{3DoF}} = \alpha((\frac{1}{(kr)^3} + \frac{j}{(kr)^2})\mathbf{F}_{\mathrm{NF}} + \frac{1}{2kr}\mathbf{F}_{\mathrm{FF}})$ from (3.50) with $\mathbf{F}_{\mathrm{NF}}$ from (2.19) and $\mathbf{F}_{\mathrm{FF}}$ from (2.20) is a function of the transmitter-to-receiver direction vector $\mathbf{u}$. Yet we do not have to specify $\mathbf{u}$ since i.i.d. 3D uniform distributions are assumed for $\mathbf{o}_{\mathrm{T}}$ and $\mathbf{o}_{\mathrm{R}}$ and any choice of $\mathbf{u}$ leads to the same channel statistics. Hence one can just set $\mathbf{u} = [1\ 0\ 0]^{\mathrm{T}}$ in this context, associated with $\mathbf{F}_{\mathrm{NF}} = \mathrm{diag}(1, -\frac{1}{2}, -\frac{1}{2})$ and $\mathbf{F}_{\mathrm{FF}} = \mathrm{diag}(0, 1, 1)$.
[2]The results concern an active transmitter with a transmit amplifier, where $|h/h_{\mathrm{opt}}|^2 \approx J_{\mathrm{NF}}^2$ in the near field. For load modulation at a passive RFID tag, the relation becomes $|h/h_{\mathrm{opt}}|^2 \approx J_{\mathrm{NF}}^4$ and the misalignment loss dB-values double (cf. the last paragraph of Sec. 3.9).

line. Polarization diversity is well-studied in the radio context [150, 199, 200] but, to the best of our knowledge, has not been considered for magnetic induction despite the striking benefits seen in Fig. 4.2.

We want to understand all the aspects outlined above in detail. We are furthermore interested in mitigating misalignment loss with coil arrays in combination with a spatial diversity scheme. To this effect, this chapter features the following core results.

- For the random channel between randomly oriented coils we derive the channel statistics in the pure near-field and pure far-field cases in closed form. We show that they are associated with diversity order $\frac{1}{2}$.

- We derive the channel statistics in the mid field (near-far field transition) and show that the associated diversity order is 1.

- We conduct an outage analysis in terms of outage power transfer efficiency, outage capacity, and bit error probability.

- We study spatial diversity schemes for 1×3 and 3×3 links involving colocated orthogonal coil arrays. We study the worst-case performance and performance statistics of the schemes.

To formalize our approach we repeat the channel coefficient formula (3.49)

$$h = \alpha \left(\left(\frac{1}{(kr)^3} + \frac{j}{(kr)^2} \right) J_{\mathrm{NF}} + \frac{1}{2kr} J_{\mathrm{FF}} \right) \tag{4.1}$$

where $\alpha = \frac{\mu_0 A_\mathrm{T} \check{N}_\mathrm{T} A_\mathrm{R} \check{N}_\mathrm{R} f k^3}{\sqrt{4 R_\mathrm{T} R_\mathrm{R}}} \, j e^{-jkr}$ from (3.51) summarizes all technical quantities that are subsidiary for this chapter. We recall the definitions (2.24) and (2.25) of the near- and far-field alignment factors

$$J_{\mathrm{NF}} = \mathbf{o}_\mathrm{R}^\mathrm{T} \boldsymbol{\beta}_{\mathrm{NF}}, \qquad J_{\mathrm{NF}} \in [-\beta_{\mathrm{NF}}, \beta_{\mathrm{NF}}] \subseteq [-1, 1], \tag{4.2}$$

$$J_{\mathrm{FF}} = \mathbf{o}_\mathrm{R}^\mathrm{T} \boldsymbol{\beta}_{\mathrm{FF}}, \qquad J_{\mathrm{FF}} \in [-\beta_{\mathrm{FF}}, \beta_{\mathrm{FF}}] \subseteq [-1, 1], \tag{4.3}$$

and the definitions (2.19) and (2.20) of the scaled near field and the scaled far field

$$\boldsymbol{\beta}_{\mathrm{NF}} = \frac{1}{2}(3\mathbf{u}\mathbf{u}^\mathrm{T} - \mathbf{I}_3)\mathbf{o}_\mathrm{T}, \qquad \beta_{\mathrm{NF}} = \|\boldsymbol{\beta}_{\mathrm{NF}}\| = \frac{1}{2}\sqrt{1 + 3(\mathbf{u}^\mathrm{T}\mathbf{o}_\mathrm{T})^2} \in [\tfrac{1}{2}, 1], \tag{4.4}$$

$$\boldsymbol{\beta}_{\mathrm{FF}} = (\mathbf{I}_3 - \mathbf{u}\mathbf{u}^\mathrm{T})\mathbf{o}_\mathrm{T}, \qquad \beta_{\mathrm{FF}} = \|\boldsymbol{\beta}_{\mathrm{FF}}\| = \sqrt{1 - (\mathbf{u}^\mathrm{T}\mathbf{o}_\mathrm{T})^2} \in [0, 1]. \tag{4.5}$$

Particularly interesting are the specific values of h from (3.54),

$$h_{\text{coax}} = \alpha \left(\frac{1}{(kr)^3} + \frac{j}{(kr)^2} \right) , \qquad h_{\text{copl}} = \frac{\alpha}{2} \left(-\frac{1}{(kr)^3} - \frac{j}{(kr)^2} + \frac{1}{kr} \right) . \qquad (4.6)$$

We note that α and kr do not depend on the coil orientations; this dependency is completely covered by J_{NF} and J_{FF}. When the coil orientations are random (and when they constitute the only randomness in h), then the statistics of h are determined by the joint PDF of J_{NF} and J_{FF}.

Before we embark on this general case, we first study the marginal distributions of J_{NF} and J_{FF}. Those are particularly important because they determine the statistics of h at very small and very large distances, respectively. The formal reason is

$$h \approx h_{\text{coax}} J_{\text{NF}} \qquad \text{if } kr \ll 1 \quad (\text{pure near field}^3) , \qquad (4.7)$$

$$h \approx h_{\text{copl}} J_{\text{FF}} \qquad \text{if } kr \gg 1 \quad (\text{pure far field}) \qquad (4.8)$$

whereby $h_{\text{copl}} \approx \frac{\alpha}{2kr}$ in the latter case.

To obtain more intuition about misalignment loss we consider the power attenuation $|h/h_{\text{coax}}|^2 = J_{\text{NF}}^2$ of the pure near-field case and decompose its dB-value into two effects:

$$10 \log_{10}(J_{\text{NF}}^2) = \underbrace{20 \log_{10}(\beta_{\text{NF}})}_{\text{field magnitude loss}} + \underbrace{20 \log_{10} |\cos \theta_{\text{R}}|}_{\substack{\text{misalignment between} \\ \text{coil axis and field vector}}} , \qquad \cos \theta_{\text{R}} = \mathbf{o}_{\text{R}}^{\text{T}} \frac{\boldsymbol{\beta}_{\text{NF}}}{\beta_{\text{NF}}} . \qquad (4.9)$$

The worst-case field magnitude loss is $-6\,\text{dB}$ when $\beta_{\text{NF}} = \frac{1}{2}$; a deep fade $J_{\text{NF}} \approx 0$ is thus always caused by near-orthogonality between receive-coil orientation $\mathbf{o}_{\text{R}}$ and field vector (i.e. $\cos \theta_{\text{R}} \approx 0$). However this does not imply that $\mathbf{o}_{\text{R}}$ is a more critical parameter than $\mathbf{o}_{\text{T}}$: counterarguments are the symmetry of $J_{\text{NF}} = \mathbf{o}_{\text{R}}^{\text{T}} \mathbf{F}_{\text{NF}} \mathbf{o}_{\text{T}} = \mathbf{o}_{\text{T}}^{\text{T}} \mathbf{F}_{\text{NF}} \mathbf{o}_{\text{R}}$ and the fact that a deep fade can always be prevented by adjusting the direction of $\boldsymbol{\beta}_{\text{NF}}$ through $\mathbf{o}_{\text{T}}$. A similar statement is possible for J_{FF}, in this case however the field magnitude loss is unbounded due to the zero of the radiation pattern on the loop antenna axis (where $\beta_{\text{FF}} = 0$). This can cause a deep fade by itself, although we will find that this aspect is statistically insignificant for SISO links.

---

[3]Of course one can argue that $\frac{1}{(kr)^3} \gg \frac{1}{(kr)^2}$ for $kr \ll 1$, hence discard the $\frac{1}{(kr)^2}$ summand and use $h \approx \frac{\alpha}{(kr)^3} J_{\text{NF}}$ in the pure near field (equivalent to a magnetoquasistatic assumption), but this would harm the physical accuracy of our exposition without yielding a useful mathematical advantage.

4.1 SISO Channel Statistics

This rather technical section derives the channel statistics for a weakly-coupled SISO link between randomly oriented coils (uniform distribution in 3D) and shall provide the mathematical foundation of the outage and diversity analysis that follows.

4.1.1 Near- and Far-Field Marginal Distributions

Lemma 4.1. *Let* $\mathbf{o} \in \mathbb{R}^3$ *be a random vector with uniform distribution* $\mathbf{o} \sim \mathcal{U}(\mathcal{S})$ *on the 3D unit sphere* $\mathcal{S} = \{\mathbf{x} \in \mathbb{R}^3 \mid \|\mathbf{x}\|_2 = 1\}$. *The projection* $\mathbf{o}^\mathrm{T}\mathbf{a}$ *onto any non-random non-zero vector* $\mathbf{a} \in \mathbb{R}^3$ *with magnitude* $a = \|\mathbf{a}\|_2$ *has uniform distribution*

$$\mathbf{o}^\mathrm{T}\mathbf{a} \sim \mathcal{U}(-a, a). \tag{4.10}$$

This special property of 3D space follows as a corollary to well-documented related statements[4], yet there appears to be no documentation of the specific result. Thus we state a short proof which boils down to the basic geometry formula for the lateral surface area of a sphere cap, which is proportional to its height.

Proof. We consider a unit vector $\mathbf{a}$; the general statement follows with a simple scaling. Due to the symmetry of distribution $\mathbf{o} \sim \mathcal{U}(\mathcal{S})$, $\mathbf{o}^\mathrm{T}\mathbf{a}$ has the same distribution for any unit vector $\mathbf{a}$. Hence we can prove the statement by showing that $o_1 = [1\,0\,0]^\mathrm{T}\mathbf{o} \sim \mathcal{U}(-1, 1)$. We consider the cumulative distribution function (CDF) $F(x) = \Pr(o_1 \le x)$. The uniform distribution on $\mathcal{S}$ allows to write the CDF as ratio $F(x) = \mathrm{area}(\mathcal{C}_x) / \mathrm{area}(\mathcal{S})$ where $\mathcal{C}_x = \{\mathbf{q} \in \mathcal{S} \mid q_1 \le x\}$ is a sphere cap of height $1 + x$. By established formulas [204, Sec. 2.5], $\mathrm{area}(\mathcal{S}) = 4\pi$ and $\mathrm{area}(\mathcal{C}_x) = 2\pi(1 + x)$. Hence $F(x) = \frac{1}{2}(1 + x)$, which increases linearly from $F(-1) = 0$ to $F(1) = 1$ and thus amounts to a uniform distribution on $[-1, 1]$. $\qquad\square$

---

[4]An equivalent statement to Lemma 4.1 is Archimedes' hat-box theorem about area-preserving projections from a unit sphere onto a surrounding cylinder [201, 202]. A related and more general statement is: if $\mathbf{o}$ has uniform distribution on the K-dimensional unit sphere, then its first $K - 2$ coordinates have uniform distribution inside the $(K-2)$-dimensional unit ball [203, Corollary 4]. The distribution is also a special case of the Von Mises-Fisher distribution for zero concentration.

Lemma 4.2. *The cumulative distribution function (CDF) F and probability density function (PDF) f of the marginal distributions of the scaled field magnitudes are*

$$F_{\beta_{\mathrm{NF}}}(\beta_{\mathrm{NF}}) = \frac{\sqrt{4\beta_{\mathrm{NF}}^2 - 1}}{\sqrt{3}}, \quad f_{\beta_{\mathrm{NF}}}(\beta_{\mathrm{NF}}) = \frac{4}{\sqrt{3}}\frac{\beta_{\mathrm{NF}}}{\sqrt{4\beta_{\mathrm{NF}}^2 - 1}}, \quad \beta_{\mathrm{NF}} \in [\tfrac{1}{2}, 1], \quad (4.11)$$

$$F_{\beta_{\mathrm{FF}}}(\beta_{\mathrm{FF}}) = 1 - \sqrt{1 - \beta_{\mathrm{FF}}^2}, \quad f_{\beta_{\mathrm{FF}}}(\beta_{\mathrm{FF}}) = \frac{\beta_{\mathrm{FF}}}{\sqrt{1 - \beta_{\mathrm{FF}}^2}}, \quad \beta_{\mathrm{FF}} \in [0, 1]. \quad (4.12)$$

The PDFs are illustrated in Fig. 4.3a and 4.3b. The following statistical moments hold:
$\mathbb{E}[\beta_{\mathrm{NF}}^2] = \tfrac{1}{2}$, $\mathbb{E}[\beta_{\mathrm{NF}}^4] = \tfrac{3}{10}$, $\mathbb{E}[\beta_{\mathrm{FF}}^2] = \tfrac{2}{3}$, $\mathbb{E}[\beta_{\mathrm{FF}}^4] = \tfrac{8}{15}$.

Proof. Let $X := \mathbf{o}_{\mathrm{T}}^{\mathsf{T}}\mathbf{u}$ and note that $X \sim \mathcal{U}(-1, 1)$ by Lemma 4.1. We use the magnitude formulas (2.21),(2.22) to write the CDFs as $F_{\beta_{\mathrm{NF}}}(\beta_{\mathrm{NF}}) = \mathrm{P}[\tfrac{1}{2}\sqrt{1 + 3X^2} \leq \beta_{\mathrm{NF}}] = \mathrm{P}[|X| \leq \sqrt{\frac{4\beta_{\mathrm{NF}}^2-1}{3}}]$ and $F_{\beta_{\mathrm{FF}}}(\beta_{\mathrm{FF}}) = \mathrm{P}[\sqrt{1 - X^2} \leq \beta_{\mathrm{FF}}] = \mathrm{P}[|X| \geq \sqrt{1 - \beta_{\mathrm{FF}}^2}]$. Since $|X| \sim \mathcal{U}(0, 1)$, the property $\mathrm{P}[|X| \leq a] = a$ as well as $\mathrm{P}[|X| \geq a] = 1 - a$ holds for $a \in [0, 1]$, which yields the CDF results. The PDFs follow from differentiation and the statistical moments from integration, e.g. $\mathbb{E}[\beta_{\mathrm{NF}}^2] = \int_{\mathbb{R}} \beta_{\mathrm{NF}}^2 f_{\beta_{\mathrm{NF}}} d\beta_{\mathrm{NF}} = \frac{4}{\sqrt{3}}\int_{\frac{1}{2}}^{1} \frac{\beta_{\mathrm{NF}}^3 d\beta_{\mathrm{NF}}}{\sqrt{4\beta_{\mathrm{NF}}^2-1}} = \frac{2\beta_{\mathrm{NF}}^2+1}{6}\sqrt{\frac{4\beta_{\mathrm{NF}}^2-1}{3}} \Big|_{\frac{1}{2}}^{1} = \frac{1}{2}$. $\square$

Now we bring the random receiver orientation $\mathbf{o}_{\mathrm{R}} \sim \mathcal{U}(\mathcal{S})$ into the picture and formulate the alignment factors $J_{\mathrm{NF}} = \mathbf{o}_{\mathrm{R}}^{\mathsf{T}}\boldsymbol{\beta}_{\mathrm{NF}}$ and $J_{\mathrm{FF}} = \mathbf{o}_{\mathrm{R}}^{\mathsf{T}}\boldsymbol{\beta}_{\mathrm{FF}}$ in terms of inner products. From Lemma 4.1 we immediately obtain the uniform conditional distributions

$$J_{\mathrm{NF}} \,|\, \beta_{\mathrm{NF}} \sim \mathcal{U}(-\beta_{\mathrm{NF}}, \beta_{\mathrm{NF}}), \quad (4.13)$$

$$J_{\mathrm{FF}} \,|\, \beta_{\mathrm{FF}} \sim \mathcal{U}(-\beta_{\mathrm{FF}}, \beta_{\mathrm{FF}}) \quad (4.14)$$

given the field magnitudes β_{NF} and β_{FF}. The distribution $J_{\mathrm{NF}} \,|\, \beta_{\mathrm{NF}}$ is illustrated by the dashed lines in Fig. 4.3c for different values of β_{NF}.

With random transmitter orientation, β_{NF} and β_{FF} are random as well and their distributions are described by Lemma 4.2. This allows for the calculation of the resulting distributions of J_{NF} and J_{FF}.

Proposition 4.1. *When the transmit and receive coil orientations have uniform distributions* $\mathbf{o}_\mathrm{T}, \mathbf{o}_\mathrm{R} \overset{\text{i.i.d.}}{\sim} \mathcal{U}(\mathcal{S})$ *on the 3D unit sphere* $\mathcal{S}$, *then the near-field alignment factor is distributed according to the PDF in Fig. 4.3c which is given by*

$$f_{J_\mathrm{NF}}(J_\mathrm{NF}) = \frac{1}{2\bar{\beta}_\mathrm{NF}} \cdot \begin{cases} 1 & |J_\mathrm{NF}| \leq \frac{1}{2} \\ 1 - \frac{\operatorname{arcosh}(2|J_\mathrm{NF}|)}{\operatorname{arcosh}(2)} & \frac{1}{2} < |J_\mathrm{NF}| < 1 \\ 0 & 1 \leq |J_\mathrm{NF}| \end{cases} . \tag{4.15}$$

For smaller arguments $|J_\mathrm{NF}| \leq \frac{1}{2}$ *the PDF* f_{J_NF} *is equal to the uniform* $f_{J_\mathrm{NF}|\beta_\mathrm{NF}}$ *of* (4.13) *conditioned on* $\beta_\mathrm{NF} = \bar{\beta}_\mathrm{NF}$, *whereby* $\bar{\beta}_\mathrm{NF}$ *is the equivalent field magnitude value*

$$\bar{\beta}_\mathrm{NF} = \frac{\sqrt{3}}{2\operatorname{arcosh}(2)} = 0.6576 \,\hat{=}\, -3.64\,\mathrm{dB}. \tag{4.16}$$

The distribution of the far-field alignment factor has the PDF in Fig. 4.3d, given by

$$f_{J_\mathrm{FF}}(J_\mathrm{FF}) = \frac{1}{2}\left(\frac{\pi}{2} - \arcsin|J_\mathrm{FF}|\right), \qquad J_\mathrm{FF} \in [-1,1]. \tag{4.17}$$

The distributions have zero mean, variances $\mathbb{E}[J_\mathrm{NF}^2] = \frac{1}{6}$ *and* $\mathbb{E}[J_\mathrm{FF}^2] = \frac{2}{9}$, *as well as fourth moments* $\mathbb{E}[J_\mathrm{NF}^4] = \frac{3}{50}$ *and* $\mathbb{E}[J_\mathrm{FF}^4] = \frac{8}{75}$.

Proof. We denote J_*, β_* in equations that hold for both $J_\mathrm{NF}, \beta_\mathrm{NF}$ and $J_\mathrm{FF}, \beta_\mathrm{FF}$. The uniform conditional $J_* \,|\, \beta_* \sim \mathcal{U}(-\beta_*, \beta_*)$ from (4.13) has PDF $f_{J_*|\beta_*} = \frac{1}{2\beta_*}\mathbb{1}_{[-\beta_*,\beta_*]}(J_*)$. For $|J_*| \leq 1$ we find $f_{J_*}(J_*) = \int_\mathbb{R} f_{J_*|\beta_*} f_{\beta_*} \, d\beta_* = \int_0^1 \frac{\mathbb{1}_{[-\beta_*,\beta_*]}(J_*)}{2\beta_*} f_{\beta_*} \, d\beta_* = \frac{1}{2}\int_{|J_*|}^1 \frac{1}{\beta_*} f_{\beta_*} \, d\beta_*$ with marginalization. For the specific PDF f_{β_FF} from (4.12), the antiderivative $\int \frac{1}{\beta_\mathrm{FF}} f_{\beta_\mathrm{FF}} d\beta_\mathrm{FF} = \int \frac{d\beta_\mathrm{FF}}{\sqrt{1-\beta_\mathrm{FF}^2}} = \arcsin(\beta_\mathrm{FF})$ immediately yields f_{J_FF}. For the other specific PDF f_{β_NF} from (4.11), $\frac{1}{2}\int \frac{1}{\beta_\mathrm{NF}} f_{\beta_\mathrm{NF}} d\beta_\mathrm{NF} = \frac{2}{\sqrt{3}}\int \frac{d\beta_\mathrm{NF}}{\sqrt{4\beta_\mathrm{NF}^2 - 1}} = \frac{1}{\sqrt{3}}\operatorname{arcosh}(2\beta_\mathrm{NF})$ but we note that $\operatorname{supp} f_{\beta_\mathrm{NF}} = [\frac{1}{2}, 1]$ can be a subset of the integration domain $\beta_\mathrm{NF} \in [|J_\mathrm{NF}|, 1]$ depending on the value J_NF at which f_{J_NF} is evaluated. Hence, the integration domain must be restricted to the cut set $\beta_\mathrm{NF} \in [|J_\mathrm{NF}|, 1] \cap [\frac{1}{2}, 1]$, i.e. to $\beta_\mathrm{NF} \in [\frac{1}{2}, 1]$ if $|J_\mathrm{NF}| \leq \frac{1}{2}$ or to $\beta_\mathrm{NF} \in [|J_\mathrm{NF}|, 1]$ if $\frac{1}{2} < |J_\mathrm{NF}| < 1$. Applying the above antiderivative to those cases and using $\operatorname{arcosh}(1) = 0$ concludes the derivation.

The moments are $\mathbb{E}[J_*^N] = \mathbb{E}_{\beta_*}[\mathbb{E}_{J_*|\beta_*}[J_*^N]]$ with $J_*|\beta_* \sim \mathcal{U}(-\beta_*, \beta_*)$ by (4.13). Thus $\mathbb{E}_{J_*|\beta_*}[J_*^N] = \frac{1}{2\beta_*}\int_{-\beta_*}^{+\beta_*} J_*^N dJ_* = \frac{\beta_*^N}{N+1}$ for even N, so $\mathbb{E}[J_*^N] = \frac{1}{N+1}\mathbb{E}_{\beta_*}[\beta_*^N]$ and the statistical moments $\mathbb{E}_{\beta_*}[\beta_*^N]$ are given by Lemma 4.2 for $N = 2, 4$. $\qquad\square$

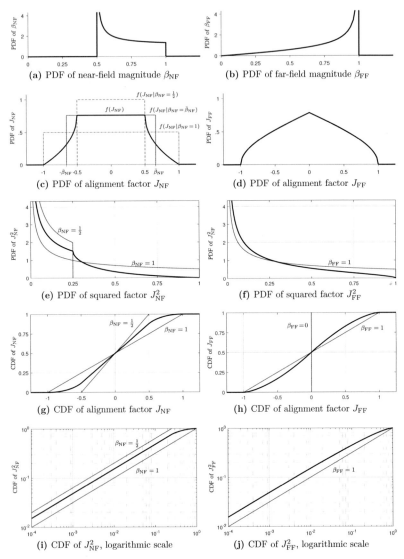

Figure 4.3: Marginal distributions of relevant quantities for the free-space SISO case with random coil orientations on both ends, described by uniformly distributed unit vectors $\mathbf{o}_T, \mathbf{o}_R \overset{\text{i.i.d.}}{\sim} \mathcal{U}(\mathcal{S})$ in 3D.

The squared alignment factors $J_{\mathrm{NF}}^2, J_{\mathrm{FF}}^2$ apply to received power and thus to the power transfer efficiency (PTE) η and the signal-to-noise ratio (SNR). Hence, the statistics of $J_{\mathrm{NF}}^2, J_{\mathrm{FF}}^2$ are crucial for the study of outage. Their PDFs follow from (4.15) and (4.17) and the change-of-variables rule $f_Y(y) = \frac{f_X(\sqrt{y})}{\sqrt{y}}$ which holds for $Y = X^2$ and an even PDF f_X ($f_{J_{\mathrm{NF}}}$ and $f_{J_{\mathrm{FF}}}$ are both even). Therefrom the CDFs

$$
F_{J_{\mathrm{NF}}^2}(s) = \begin{cases} \sqrt{s}/\bar{\beta}_{\mathrm{NF}} & 0 \leq s \leq \frac{1}{4} \\ f_{J_{\mathrm{NF}}}(\sqrt{s})\sqrt{4s} + \sqrt{\frac{4s-1}{3}} & \frac{1}{4} < s \leq 1 \\ 1 & 1 < s \end{cases} , \tag{4.18}
$$

$$
F_{J_{\mathrm{FF}}^2}(s) = 1 - \sqrt{1-s} + \sqrt{s}\left(\frac{\pi}{2} - \arcsin\sqrt{s}\right) \tag{4.19}
$$

can be obtained by integration. Their behavior for smaller arguments is described by

$$
F_{J_{\mathrm{NF}}^2}(s) = \frac{1}{\bar{\beta}_{\mathrm{NF}}}\sqrt{s} = 1.5207 \cdot \sqrt{s}\,, \tag{4.20}
$$

$$
F_{J_{\mathrm{FF}}^2}(s) \approx \frac{\pi}{2}\sqrt{s} = 1.5708 \cdot \sqrt{s}\,. \tag{4.21}
$$

These formulas will prove particularly important for the study of outage performance.

4.1.2 General Case: Near-Far Field Transition

In the transition region between near and far field, neither $kr \ll 1$ nor $kr \gg 1$ holds and the approximations (4.7) and (4.8) are both inadequate. By (4.1), the channel coefficient is the superposition $h = \alpha\left(\frac{1}{(kr)^3} + \frac{j}{(kr)^2}\right) J_{\mathrm{NF}} + \alpha\frac{1}{2kr}J_{\mathrm{FF}}$ of a near- and a far-field channel. These two summands cannot cancel each other because the term $\frac{j}{(kr)^2}$ asserts that they have different phase. This is the essential mechanism of the introduced polarization diversity aspect: h can only fade if both J_{NF} and J_{FF} fade. In the following, we investigate this phenomenon analytically.

In order to describe the channel statistics of this general case, we prepare the necessary set of mathematical tools in the following.

Lemma 4.3. *A random unit vector* $\mathbf{o}$ *with uniform distribution on the 3D unit sphere has covariance matrix* $\mathbf{C} = \mathbb{E}[\mathbf{o}\mathbf{o}^{\mathrm{T}}] = \frac{1}{3}\mathbf{I}_3$.

Proof. The covariance matrix $\mathbf{C}$ is given by $\mathbb{E}[\mathbf{o}\mathbf{o}^{\mathrm{T}}]$ because $\mathbf{o}$ has zero mean. The diagonal elements are $C_{n,n} = \mathbb{E}[o_n^2] = \frac{1}{3}$ as a consequence of $o_n \sim \mathcal{U}(-1,1)$. The off-diagonal elements $C_{m,n} = \mathbb{E}[o_m o_n] = \frac{1}{4\pi}\oiint_{\mathcal{S}} o_m o_n \mathrm{d}\mathbf{s} = \frac{1-1}{4\pi}\iint_{\mathcal{S},o_m \geq 0} o_m o_n \mathrm{d}\mathbf{s} = 0$. $\quad\square$

Lemma 4.4. *Consider a random unit vector* $\mathbf{o}$ *with uniform distribution on the 3D unit sphere and a set of non-random vectors* $\mathbf{A} = [\mathbf{a}_1 \ldots \mathbf{a}_N] \in \mathbb{C}^{3 \times N}$. *The projections* $\mathbf{a}_o = \mathbf{A}^H \mathbf{o} \in \mathbb{C}^N$ *are random with zero mean and covariance matrix* $\mathbb{E}[\mathbf{a}_o \mathbf{a}_o^H] = \frac{1}{3} \mathbf{A}^H \mathbf{A}$.

Proof. $\mathbb{E}[\mathbf{a}_o \mathbf{a}_o^H] = \mathbf{A}^H \mathbb{E}[\mathbf{o}\mathbf{o}^T] \mathbf{A}$ with $\mathbb{E}[\mathbf{o}\mathbf{o}^T] = \frac{1}{3} \mathbf{I}_3$ by Lemma 4.3. □

Lemma 4.5. *Consider a random unit vector* $\mathbf{o}$ *with uniform distribution on the 3D unit sphere* $\mathcal{S}$ *and two orthonormal vectors* $\mathbf{m}, \mathbf{n} \in \mathcal{S}$. *The projections* $m_o = \mathbf{m}^T \mathbf{o}$ *and* $n_o = \mathbf{n}^T \mathbf{o}$ *are random with joint PDF*

$$f(m_o, n_o) = \psi\left(m_o^2 + n_o^2\right), \qquad \psi(x) := \frac{\mathbb{1}_{[0,1]}(x)}{2\pi\sqrt{1-x}}. \qquad (4.22)$$

For $\epsilon \in [0,1]$ *the probability* $p_\epsilon = \mathrm{P}[m_o^2 + n_o^2 \leq \epsilon] = 1 - \sqrt{1-\epsilon}$, *whereby* $\frac{\epsilon}{2} \leq p_\epsilon \leq \epsilon$.

Proof. Due to the symmetry of $\mathcal{S}$ we can prove the statement by deriving the joint PDF of o_1 and o_2, i.e. of the projections of $\mathbf{o}$ onto $\mathbf{m} = [1\,0\,0]^T$ and $\mathbf{n} = [0\,1\,0]^T$. We use polar coordinates $(o_1, o_2) = (R\cos\phi, R\sin\phi)$ and note that $R = \sqrt{o_1^2 + o_2^2} = \sqrt{1 - o_3^2}$ is determined by o_3 since $\|\mathbf{o}\| = 1$. The marginal PDF $f(o_3) = \frac{1}{2}$ leads to PDF $f_R(R) = \frac{R}{\sqrt{1-R^2}}$ for radius $R \in [0,1]$. On the other hand, conditioned on o_3, the pair (o_1, o_2) has uniform distribution on a circle of radius R; hence $f_{\phi|R} = f_\phi = \frac{1}{2\pi}$ and the joint PDF $f_{R,\phi} = f_R \cdot f_{\phi|R} = \frac{R}{2\pi\sqrt{1-R^2}}$. The joint PDF f_{o_1,o_2} now follows from a change of variables from r, ϕ to o_1, o_2 which is a bijective map $[0,1] \times [0, 2\pi] \to [-1,1]^2$. With the appropriate Jacobian determinant, $f_{o_1,o_2} = \frac{1}{R} f_{R,\phi} = \frac{1}{2\pi\sqrt{1-R^2}}$ with $R^2 = o_1^2 + o_2^2$.

The event $o_1^2 + o_2^2 \leq \epsilon \iff 1 - o_3^2 \leq \epsilon \iff |o_3| \geq \sqrt{1-\epsilon}$ has probability $1 - \sqrt{1-\epsilon}$ because $|o_3| \sim \mathcal{U}(0,1)$ by Lemma 4.1. Basic rearrangements prove the bounds. □

Proposition 4.2. *Consider a random unit vector* $\mathbf{o}$ *with uniform distribution on the 3D unit sphere and two linearly independent vectors* $\mathbf{a}, \mathbf{b} \in \mathbb{R}^3$ *with magnitudes* a, b *and correlation coefficient* $\rho = \frac{\mathbf{a}^T \mathbf{b}}{ab}$. *The joint PDF of the projections* $a_o = \mathbf{a}^T \mathbf{o}$ *and* $b_o = \mathbf{b}^T \mathbf{o}$ *is*

$$f_{a_o,b_o}(a_o, b_o) = \frac{1}{ab\sqrt{1-\rho^2}} \, \psi\left(\left\| \mathbf{E}^{-1}\begin{bmatrix} a_o \\ b_o \end{bmatrix} \right\|^2\right), \qquad \mathbf{E} := \begin{bmatrix} a & 0 \\ b\rho & b\sqrt{1-\rho^2} \end{bmatrix} \qquad (4.23)$$

where function ψ *is as in Lemma 4.5.[5] The covariance matrix of* $[a_o \, b_o]^T$ *is* $\frac{1}{3}[\mathbf{a}\,\mathbf{b}]^T[\mathbf{a}\,\mathbf{b}]$ *by Lemma 4.4.*

---

[5]The PDF (4.23) in expanded form is $f_{a_o,b_o}(a_o, b_o) = \frac{1}{2\pi ab}(1 - \rho^2 - (\frac{a_o}{a})^2 - (\frac{b_o}{b})^2 + 2\rho\frac{a_o b_o}{ab})^{-\frac{1}{2}}$ for

Proof. Using the Gram-Schmidt process we construct orthonormal vectors $\mathbf{m} = \frac{\mathbf{a}}{a}$ and $\mathbf{n} = \frac{(\mathbf{I}_3 - \mathbf{m}\mathbf{m}^{\mathrm{T}})\mathbf{b}}{\|(\mathbf{I}_3 - \mathbf{m}\mathbf{m}^{\mathrm{T}})\mathbf{b}\|}$ to express $\mathbf{a} = a\mathbf{m}$ and $\mathbf{b} = b(\rho\,\mathbf{m} + \sqrt{1-\rho^2}\,\mathbf{n})$ and, equivalently, $[\mathbf{a}\ \mathbf{b}] = [\mathbf{m}\ \mathbf{n}]\mathbf{E}^{\mathrm{T}}$. The projections $[a_o\ a_o] = \mathbf{o}^{\mathrm{T}}[\mathbf{a}\ \mathbf{b}] = \mathbf{o}^{\mathrm{T}}[\mathbf{m}\ \mathbf{n}]\mathbf{E}^{\mathrm{T}}$ are related to the projections m_o, n_o onto the orthonormal base by the linear map $[a_o\ b_o]^{\mathrm{T}} = \mathbf{E}[m_o\ n_o]^{\mathrm{T}}$. The joint PDF f_{m_o,n_o} is given by Lemma 4.5 and leads to f_{a_o,b_o} with the following general change-of-variables argument. For a random vector $[m_o\ n_o]^{\mathrm{T}}$ with PDF f_{m_o,n_o} and an invertible linear map $\mathbf{E}$, the vector $[a_o\ b_o]^{\mathrm{T}} = \mathbf{E}\,[m_o\ n_o]^{\mathrm{T}}$ has PDF

$$f_{a_o,b_o}\left(\begin{bmatrix} a_o \\ b_o \end{bmatrix} \right) = \frac{1}{\det(\mathbf{E})} f_{m_o,n_o}\left(\mathbf{E}^{-1} \begin{bmatrix} a_o \\ b_o \end{bmatrix} \right). \qquad\qquad \square$$

The SISO channel coefficient (4.1) is an inner product $h = \mathbf{v}^{\mathrm{T}}\mathbf{o}_{\mathrm{R}}$ of the receiver orientation $\mathbf{o}_{\mathrm{R}}$ and a complex field vector $\mathbf{v} = \alpha((\frac{1}{(kr)^3} + \frac{j}{(kr)^2})\boldsymbol{\beta}_{\mathrm{NF}} + \frac{1}{2kr}\boldsymbol{\beta}_{\mathrm{FF}}) \in \mathbb{C}^3$. Of particular importance are the real and imaginary parts $\mathbf{v} = \mathbf{a} + j\mathbf{b}$ as they give rise to polarization diversity if linearly independent. This can be seen by the term $|h|^2 = (\mathbf{a}^{\mathrm{T}}\mathbf{o}_{\mathrm{R}})^2 + (\mathbf{b}^{\mathrm{T}}\mathbf{o}_{\mathrm{R}})^2$ which vanishes only when $\mathbf{o}_{\mathrm{R}}$ is orthogonal to both $\mathbf{a}$ and $\mathbf{b}$. A precise analysis of this circumstance is enabled by Proposition 4.2 which provides a full description of the statistics of h given a complex field vector $\mathbf{v} = \mathbf{a} + j\mathbf{b}$.

Proposition 4.3. *Consider a channel coefficient of the form $h = \mathbf{v}^{\mathrm{T}}\mathbf{o}_{\mathrm{R}}$ with a non-random vector $\mathbf{v} \in \mathbb{C}^3$ and random receiver orientation $\mathbf{o}_{\mathrm{R}}$ with uniform distribution on the 3D unit sphere. If $\mathbf{a} = \mathrm{Re}(\mathbf{v})$ and $\mathbf{b} = \mathrm{Im}(\mathbf{v})$ are linearly independent, then the statistics of h are described by Proposition 4.2 and the CDF $F_{|h|^2}(s) = \mathrm{P}[|h|^2 \leq s]$ is within the bounds*

$$\frac{s}{2ab\sqrt{1-\rho^2}} \leq F_{|h|^2}(s) \leq \frac{s}{2ab\sqrt{1-\rho^2}} \left(1 - \frac{s}{s_0} \right)^{-\frac{1}{2}} \tag{4.24}$$

under the condition $s < s_0$. Thereby, $a = \|\mathbf{a}\|$, $b = \|\mathbf{b}\|$, $\rho = \frac{\mathbf{a}^{\mathrm{T}}\mathbf{b}}{ab}$, and

$$s_0 = \frac{a^2+b^2}{2} - \sqrt{\left(\frac{a^2+b^2}{2} \right)^2 - a^2b^2(1-\rho^2)}. \tag{4.25}$$

$(\frac{a_o}{a})^2 + (\frac{b_o}{b})^2 - 2\rho\frac{a_o b_o}{ab} \leq 1 - \rho^2$ and $f_{a_o,b_o}(a_o, b_o) = 0$ otherwise. Remarkably, this complicated joint distribution has uniform marginal distributions $a_o \sim \mathcal{U}(-a,a)$ and $b_o \sim \mathcal{U}(-b,b)$ by Lemma 4.1, with $\mathrm{cov}(a_o, b_o) = \frac{\mathbf{a}^{\mathrm{T}}\mathbf{b}}{3}$ by Lemma 4.4. It is important to note that orthogonal $\mathbf{a}, \mathbf{b}$ and thus $\mathrm{cov}(a_o, b_o) = 0$ do not imply statistical independence between a_o and b_o, which would amount to a uniform $f(a_o, b_o)$ over the box $[-a,a] \times [-b,b]$. The actual $f(a_o, b_o)$ for orthogonal $\mathbf{a}, \mathbf{b}$ is supported only on an ellipsis in standard form that is enclosed by the aforementioned box. The statistical dependence is evident by the implications $a_o = \pm a \Rightarrow b_o = 0$ and $b_o = \pm b \Rightarrow a_o = 0$.

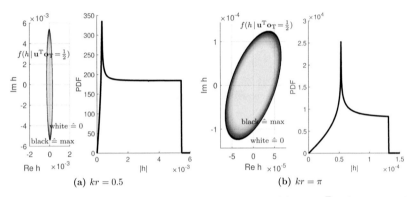

(a) $kr = 0.5$ (b) $kr = \pi$

Figure 4.4: Conditional PDFs of $h \in \mathbb{C}$ and of the resulting $|h|$ given $\mathbf{u}^T \mathbf{o_T}$ (the cosine of the angle of departure). The PDF of h is according to Proposition 4.2; therefrom the PDF of $|h|$ was obtained by numerical integration. In the color plots the probability density is zero in the white areas and infinite on the black boundary ellipse.

Proof. Proposition 4.2 describes the joint statistics of projections $a_o = \mathbf{a}^T \mathbf{o_R}$ and $b_o = \mathbf{b}^T \mathbf{o_R}$. The linear map $\mathbf{E}$ from (4.23) maps points from the 2D unit ball onto the ellipse that is the support of f_{a_o,b_o}. The threshold s_0 is the smaller eigenvalue of $\mathbf{E}^T \mathbf{E}$ and the associated formula (4.25) is found by solving for the roots of the characteristic polynomial $\det(s_i \mathbf{I}_3 - \mathbf{E}^T \mathbf{E})$. Therefore, $a_o^2 + b_o^2 < s_0$ guarantees that (a_o, b_o) is in the interior of supp f_{a_o,b_o}, where $f_{a_o,b_o} < \infty$. In particular, $f_{a_o,b_o} = \frac{\psi(\|\mathbf{E}^{-1}[a_o\ b_o]^T\|^2)}{ab\sqrt{1-\rho^2}} \leq \frac{\psi(s_0)}{ab\sqrt{1-\rho^2}}$ and $P[|h|^2 \leq s] = \iint_{a_0^2 + b_0^2 \leq s} f_{a_o,b_o} da_o db_o \leq \frac{\psi(s_0)}{ab\sqrt{1-\rho^2}} \iint_{a_0^2 + b_0^2 \leq s} da_o db_o = \frac{\psi(s_0)}{ab\sqrt{1-\rho^2}} \pi s$ which, by furthermore using the definition of ψ in (4.22), proves the upper bound. Analogously, the lower bound follows from $f_{a_o,b_o} \geq \frac{\psi(0)}{ab\sqrt{1-\rho^2}} = \frac{1}{2\pi ab\sqrt{1-\rho^2}}$ for $(a_o, b_o) \in$ supp f_{a_o,b_o} which is guaranteed for $a_o^2 + b_o^2 \leq s_0$. □

By investigating Proposition 4.3 we find that the upper bound approaches the lower bound for $s \ll s_0$ because then $(1 - \frac{s}{s_0})^{-\frac{1}{2}} \approx 1$. In this case $F_{|h|^2}(s) \approx \frac{s}{2ab\sqrt{1-\rho^2}}$ in very good approximation, hence $F_{|h|^2}(s) \propto s$ for small s.

The general-case PDF of $h \in \mathbb{C}$ for random coil orientations on both ends and some fixed kr is given by $f_h = \frac{1}{2} \int_{-1}^{+1} f_{h|\mathbf{u}^T \mathbf{o_T} = x} dx$, i.e. by the marginalization of all ellipses. Unfortunately we are unable to evaluate this integral in closed form. Numerical evaluations are shown in Fig. 4.5 and 4.6, whereby a very specific shape can be observed. In particular, the support of f_h (a subset of $\mathbb{C}$ illustrated by Fig. 4.6c) is the rhombus $\{x \cdot h_{\text{coax}} + y \cdot h_{\text{copl}} \,\big|\, x, y \in \mathbb{R}, |x| + |y| \leq 1\}$. The corner points $\pm h_{\text{coax}}$ are reached

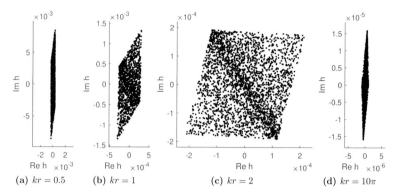

Figure 4.5: Scatter plots of h for coil orientations with 3D uniform distributions $\mathbf{o}_T, \mathbf{o}_R \overset{\text{i.i.d.}}{\sim} \mathcal{U}(\mathcal{S})$ plotted for different values kr (i.e. for distances of $\frac{kr}{2\pi}$ wavelengths). For $kr \ll 1$ or $kr \gg 1$, the set of possible realizations degenerates to a line.

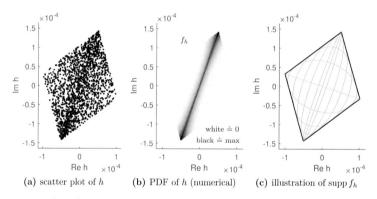

Figure 4.6: Distribution of h for $kr = \pi$, i.e. when the distance r is half a wavelength. Fig. 4.6a shows a Monte Carlo simulation and Fig. 4.6b shows the PDF of h, obtained by computing $\frac{1}{2}\int_{-1}^{+1} f_{h|\mathbf{u}^T\mathbf{o}_T=x}dx$ numerically for every point h with $f_{h|\mathbf{u}^T\mathbf{o}_T}$ from Proposition 4.2. Fig. 4.6c illustrates how the rhombus-shaped support of f_h arises: for a fixed $\mathbf{u}^T\mathbf{o}_T$, $\operatorname{supp} f_{h|\mathbf{u}^T\mathbf{o}_T}$ is an ellipse and $\operatorname{supp} f_h$ is the union of all such ellipses for $\mathbf{u}^T\mathbf{o}_T \in [-1,1]$. The illustration shows that this results in the rhombus. The plot shows ellipses for $\mathbf{u}^T\mathbf{o}_T \in \{0, 0.1, 0.3, 0.5, 0.7, 0.9, 1\}$, which degenerate to a line for $\mathbf{u}^T\mathbf{o}_T \in \{0, 1\}$. The ellipses for $\mathbf{u}^T\mathbf{o}_T \approx 0$ cause a significant concentration of probability mass, corresponding to the red line in Fig. 4.6b.

by a degenerated ellipsis in the cases $\mathbf{u}^\mathsf{T}\mathbf{o}_\mathrm{T} = \pm 1$ and the corner points $\pm h_\mathrm{copl}$ by a degenerated ellipsis in the case $\mathbf{u}^\mathsf{T}\mathbf{o}_\mathrm{T} = 0$. Most probability mass is near the line that connects $\pm h_\mathrm{copl}$ because, loosely speaking, the 3D coil orientations $\mathbf{o}_\mathrm{T}$ and $\mathbf{o}_\mathrm{R}$ are more likely to be near-coplanar than near-coaxial due to an extra degree of freedom.

4.2 SISO Channel: Performance and Outage

4.2.1 Power Transfer: Loss and Outage

We first study the statistics of the power transfer efficiency (PTE) $\eta = |h|^2$ over the considered SISO link. Due to the random coil orientations, η is a random fraction of its maximum value $|h_\mathrm{opt}|^2$ for the given distance and technical parameters in α. Consider a required PTE η_req, e.g., the necessary PTE to sustain the operation of a sensor. An outage event occurs when $\eta < \eta_\mathrm{req}$. The outage probability

$$\epsilon = \mathrm{P}\left[\eta < \eta_\mathrm{req}\right] = \mathrm{P}\left[|h|^2 < \eta_\mathrm{req}\right] = F_{|h|^2}(\eta_\mathrm{req}) \qquad (4.26)$$

is characterized in its behavior by the CDF of $|h|^2$. It makes sense to refer to the η-value which corresponds to a given outage probability ϵ as the outage PTE

$$\eta_\epsilon = F_{|h|^2}^{-1}(\epsilon). \qquad (4.27)$$

We investigate these quantities in the pure near-field case (4.7) with $\eta = |h_\mathrm{coax}|^2 J_\mathrm{NF}^2$ and the pure far-field case (4.8) with $\eta = |h_\mathrm{copl}|^2 J_\mathrm{FF}^2$, giving

$$\epsilon \approx F_{J_\mathrm{NF}^2}\left(\frac{\eta_\mathrm{req}}{|h_\mathrm{coax}|^2}\right) = \frac{\sqrt{\eta_\mathrm{req}}}{\bar{\beta}_\mathrm{NF}\,|h_\mathrm{coax}|} \approx \frac{(kr)^3\sqrt{\eta_\mathrm{req}}}{\bar{\beta}_\mathrm{NF}\,|\alpha|} \qquad \text{for } kr \ll 1 \qquad (4.28)$$

$$\epsilon \approx F_{J_\mathrm{FF}^2}\left(\frac{\eta_\mathrm{req}}{|h_\mathrm{copl}|^2}\right) \approx \frac{\pi\sqrt{\eta_\mathrm{req}}}{2\,|h_\mathrm{copl}|} \approx \frac{\pi kr\sqrt{\eta_\mathrm{req}}}{|\alpha|} \qquad \text{for } kr \gg 1 \qquad (4.29)$$

whereby the more detailed expressions hold for $\frac{\eta_\mathrm{req}}{|h_\mathrm{coax}|^2} \leq \frac{1}{4}$ and $\frac{\eta_\mathrm{req}}{|h_\mathrm{copl}|^2} \ll 1$, respectively. The scaling behavior of these expressions reveals the drastic effect of random coil misalignment: to decrease ϵ by a factor of 10 one must lower the target η_req by a factor of 100, i.e. tolerate $-20\,\mathrm{dB}$ received power. The associated outage PTE values are

$$\eta_\epsilon = F_{J_{\mathrm{NF}}^2}^{-1}(\epsilon) \cdot |h_{\mathrm{coax}}|^2 = \bar{\beta}_{\mathrm{NF}}^2 \epsilon^2 \, |h_{\mathrm{coax}}|^2 = 0.4324 \cdot \epsilon^2 \, |h_{\mathrm{coax}}|^2 \qquad \text{for } kr \ll 1 \qquad (4.30)$$

$$\eta_\epsilon = F_{J_{\mathrm{FF}}^2}^{-1}(\epsilon) \cdot |h_{\mathrm{copl}}|^2 \approx \frac{4}{\pi^2} \epsilon^2 |h_{\mathrm{copl}}|^2 = 0.4053 \cdot \epsilon^2 \, |h_{\mathrm{copl}}|^2 \qquad \text{for } kr \gg 1 \qquad (4.31)$$

whereby the detailed expressions holds for $\epsilon \leq \frac{\mathrm{arcosh}(2)}{\sqrt{3}} = 0.7603$ and $\epsilon \ll 1$, respectively. The drastic robustness problems of these setups are highlighted by the dependence $\eta_\epsilon \propto \epsilon^2$. For example, the near-field PTE $\eta_{\epsilon=0.01}$ that can be supported with 99% reliability is $-43.6\,\mathrm{dB}$ below $|h_{\mathrm{coax}}|^2$, making robust power transfer extremely inefficient. This problem would not be solved by fixing the transmitter orientation: this case shows the same scaling behavior $\eta_\epsilon = \bar{\beta}_{\mathrm{NF}}^2 \epsilon^2 \, |h_{\mathrm{coax}}|^2$. Because of channel reciprocity the same holds true if only the receiver is fixed.

Remarkably, the pure far-field case exhibits the same scaling behavior as the pure near-field case (apart from the distance dependence) even though β_{FF} can fade to zero while β_{NF} is lower-bounded by $\frac{1}{2}$. We infer that this effect is overshadowed by the impact of misalignment between receive-coil orientation and field vector.

Interesting related results are the expected values $\mathbb{E}[\eta] = \mathbb{E}[J_{\mathrm{NF}}^2] \cdot |h_{\mathrm{coax}}|^2 = \frac{1}{6} \, |h_{\mathrm{coax}}|^2$ (i.e. $-7.8\,\mathrm{dB}$) in the pure near field and $\mathbb{E}[\eta] = \mathbb{E}[J_{\mathrm{FF}}^2] \cdot |h_{\mathrm{copl}}|^2 = \frac{2}{9} \, |h_{\mathrm{copl}}|^2$ (i.e. $-6.5\,\mathrm{dB}$) in the pure far field, following from the statistical moments in Proposition 4.1.

We extend the analysis to the general case where both near- and far-field propagation are considered. We begin with the setting where $\mathbf{o}_{\mathrm{T}}$ is fixed and the channel coefficient $h = \mathbf{o}_{\mathrm{R}}^{\mathrm{T}} \mathbf{v}$ is characterized by a given field vector $\mathbf{v}$. The PDF $f_{h|\mathbf{u}^{\mathrm{T}}\mathbf{o}_{\mathrm{T}}}$ is supported on an ellipse in $\mathbb{C}$. Proposition 4.3 states that if $\mathbf{v} = \mathbf{a} + j\mathbf{b}$ has linearly independent real and imaginary parts $\mathbf{a}$ and $\mathbf{b}$ then $F_{|h|^2}(s) \approx \frac{1}{c} s$ with $c = 2ab\sqrt{1 - \rho^2}$ and $\rho = \frac{\mathbf{a}^{\mathrm{T}}\mathbf{b}}{ab}$. In this case the outage probability $\epsilon = \frac{1}{c}\eta_{\mathrm{req}}$ is linear in η_{req} and the outage PTE $\eta_\epsilon = c \cdot \epsilon$ is linear in ϵ. We notice a clear advantage over the pure near- and far-field cases due to the polarization diversity, reminiscent of the behavior witnessed earlier in Fig. 4.2. This implies that the described behavior transfers to the case where both $\mathbf{o}_{\mathrm{T}}$ and $\mathbf{o}_{\mathrm{R}}$ are random, with the PDF f_h shown in Fig. 4.5 and 4.6. This is intuitive because linear dependence occurs probability zero: $\mathrm{Re}(\mathbf{v})$ and $\mathrm{Im}(\mathbf{v})$ can not be linearly dependent unless $\mathbf{u}^{\mathrm{T}}\mathbf{o}_{\mathrm{T}} = 0$ or $\mathbf{u}^{\mathrm{T}}\mathbf{o}_{\mathrm{T}} = \pm 1$ (the simple proof is omitted). A rigorous argument for the scaling behavior is however unavailable because we can not evaluate the marginalization integral $f_h = \frac{1}{2} \int_{-1}^{+1} f_{h|\mathbf{u}^{\mathrm{T}}\mathbf{o}_{\mathrm{T}}=x} \, dx$ in closed form.

It is worthwhile to compare the results to the well-studied Rayleigh fading model $h \sim \mathcal{CN}(0,1)$ for radio channels with rich multipath propagation. The relevant descriptions are $f_{|h|}(x) = 2x \cdot e^{-x^2}$ and $f_{|h|^2}(s) = e^{-s}$ as well as $F_{|h|^2}(s) = 1 - e^{-s}$. For small arguments $F_{|h|^2}(s) \approx s$, associated with diversity order 1. Again, $\mathrm{Re}(h)$ and $\mathrm{Im}(h)$ are non-zero with probability 1. If the distribution would degenerate to a line in $\mathbb{C}$, e.g. with the purely real-valued distribution $h \sim \mathcal{N}(0,1)$, then the diversity order would also drop to $\frac{1}{2}$ because, loosely speaking, one half of the propagation mechanisms were lost. An interesting parallel is that we can describe Rayleigh fading as $h = \mathbf{o}_{\mathrm{R}}^{\mathrm{T}} \mathbf{v}$ with $\mathbf{v} \sim \mathcal{CN}(0, \mathbf{I}_3)$; the distribution of $\mathbf{o}_{\mathrm{R}}$ is irrelevant because of the symmetry of $\mathcal{CN}(0, \mathbf{I}_3)$. Again, $\mathrm{Re}(\mathbf{v})$ and $\mathrm{Im}(\mathbf{v})$ are linearly independent with probability 1.

4.2.2 Outage Capacity

We shift our focus to the communication performance of this random channel for reception in additive white Gaussian noise (AWGN) with variance σ^2. The signal-to-noise ratio $\mathrm{SNR} = \eta P_{\mathrm{T}}/\sigma^2$ is a random variable because it is multiplied by the random PTE η. Hence all the statements made above for $\eta_{\mathrm{req}}, \eta_{\mathrm{coax}}, \eta_{\mathrm{copl}}, \eta_{\epsilon}$ hold analogously for $\mathrm{SNR}_{\mathrm{req}}, \mathrm{SNR}_{\mathrm{coax}}, \mathrm{SNR}_{\mathrm{copl}}, \mathrm{SNR}_{\epsilon}$. In this context a well-established quantity is the (narrowband) outage capacity [90, Eq. 5.57]

$$C_{\epsilon} = \Delta_f \log_2 \left(1 + \frac{F_{|h|^2}^{-1}(\epsilon) \cdot P_{\mathrm{T}}}{\sigma^2} \right) = \Delta_f \log_2 \left(1 + \frac{\eta_{\epsilon} P_{\mathrm{T}}}{\sigma^2} \right), \qquad (4.32)$$

the largest data rate that can be supported with an outage probability that does not exceed a fixed ϵ. With log-linearization we obtain $C_{\epsilon} \leq \frac{\Delta_f}{\log(2)} \frac{\eta_{\epsilon} P_{\mathrm{T}}}{\sigma^2}$ which is tight in the power-limited regime $\frac{\eta_{\epsilon} P_{\mathrm{T}}}{\sigma^2} \ll 1$ (we already established that η_{ϵ} is small often). In the pure near-field case this translates to $C_{\epsilon} \leq \frac{\Delta_f}{\log(2)} \frac{|h_{\mathrm{coax}}|^2 P_{\mathrm{T}}}{\sigma^2} \cdot 0.4324 \cdot \epsilon^2$. Again we find that $C_{\epsilon} \propto \epsilon^2$ leads to terrible performance whenever some level of robustness (low ϵ) is required. In contrary, in the bandwidth-limited regime we can identify an absolute data rate penalty of $\Delta_f \log_2(0.4324 \cdot \epsilon^2)$ due to misalignment. In Fig. 4.7a we illustrate the behavior of C_{ϵ} as a function of $\mathrm{SNR}_{\mathrm{coax}} = \frac{|h_{\mathrm{coax}}|^2 P_{\mathrm{T}}}{\sigma^2}$ and in comparison to the narrowband capacity of a coaxial near-field link $C_{\mathrm{coax}} = \Delta_f \log_2(1 + \mathrm{SNR}_{\mathrm{coax}})$.

4.2.3 Bit Error Probability and Diversity Order

Another popular characteristic of a random channel is the mean bit error probability of antipodal signaling (BPSK modulation) in AWGN, given by $p_{\mathrm{e}} = \mathbb{E}[Q(\sqrt{2\,\mathrm{SNR}}\,)]$ where SNR is considered as a random variable affected by the channel fluctuations. We consider the pure near- and far-field cases, first with fixed transmitter orientation

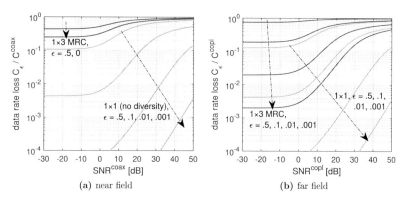

(a) near field   (b) far field

Figure 4.7: Outage capacity $C_\epsilon = \Delta_f \log_2(1 + F_{J^2}^{-1}(\epsilon)\mathrm{SNR}_*)$ in comparison to the channel capacity $\Delta_f \log_2(1 + \mathrm{SNR}_*)$ under ideal coil alignment.

(only $\mathbf{o}_\mathrm{R}$ is random). Here $\mathrm{SNR} \approx J_\mathrm{NF}^2 \mathrm{SNR}_\mathrm{coax}$ and $\mathrm{SNR} \approx J_\mathrm{FF}^2 \mathrm{SNR}_\mathrm{copl}$, respectively, and by (4.13) and (4.14) this case is associated with uniform conditional distributions $J_*|\beta_* \sim \mathcal{U}(-\beta_*, +\beta_*)$ given the field magnitude β_* (the placeholder $_*$ represents $_\mathrm{NF}$ or $_\mathrm{FF}$ and $_\mathrm{coax}$ or $_\mathrm{copl}$). For this simple distribution we can evaluate the expected value in closed form through integration by parts (the steps are omitted), giving

$$p_\mathrm{e} = \frac{1}{2\beta_*} \int_{-\beta_*}^{+\beta_*} Q\big(\sqrt{2J_*^2\,\mathrm{SNR}_*}\,\big)\,dJ_* = Q\big(\sqrt{2\beta_*^2\,\mathrm{SNR}_*}\,\big) + \frac{1 - e^{-\beta_*^2\,\mathrm{SNR}_*}}{\sqrt{4\pi\beta_*^2\,\mathrm{SNR}_*}}. \qquad (4.33)$$

With $Q(x) < \frac{1}{x}\frac{1}{\sqrt{2\pi}}e^{-x^2/2}$ we find the upper bound $p_\mathrm{e} \leq \frac{1}{\beta_*\sqrt{4\pi}}\mathrm{SNR}_*^{-\frac{1}{2}}$ which is tight at large SNR. This expression has the standard form for fading channels $p_\mathrm{e} = c \cdot \mathrm{SNR}^{-L}$ from [90, Eq. 3.158]; by comparison we deduce that the diversity order $L = \frac{1}{2}$.

An important insight is that this upper bound also applies to the fully random SISO-case (where the coils on both ends have random orientation) if we set β_* according to $\frac{1}{2\beta_*} = f_{J_*}(0)$. This establishes an appropriate description of the fading behavior and holds because both f_{J_NF} and f_{J_FF} are even with maximum value $f_{J_*}(0)$ that is non-zero and finite. In particular we obtain the bounds

pure near field: $\qquad p_\mathrm{e} \leq \dfrac{\mathrm{arcosh}(2)}{\sqrt{3\pi}}\,\mathrm{SNR}_\mathrm{coax}^{-1/2} = 0.4290 \cdot \mathrm{SNR}_\mathrm{coax}^{-1/2}\,,$ \qquad (4.34)

pure far field: $\qquad p_\mathrm{e} \leq \dfrac{\sqrt{\pi}}{4}\,\mathrm{SNR}_\mathrm{copl}^{-1/2} = 0.4431 \cdot \mathrm{SNR}_\mathrm{copl}^{-1/2}$ \qquad (4.35) which are again tight for large SNR_*. Therefore, the distributions of J_{NF} and J_{FF} are both associated with a diversity order of just $\frac{1}{2}$. The 1×1-curves in Fig. 4.8 show p_e as a function of SNR_*, whereby the above high-SNR descriptions are clearly observable.

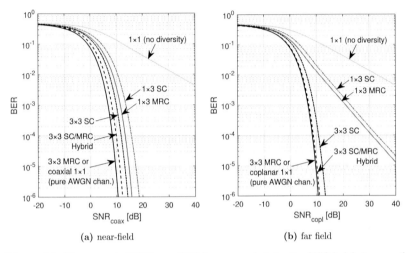

(a) near-field   (b) far field

Figure 4.8: Bit error rate (BER) over SNR for a magneto-inductive SISO link between coils with random orientations (all 3D directions are equiprobable) as well as randomly oriented SIMO (or MISO) and MIMO links with orthogonal arrays of three coils and the use of different spatial diversity schemes (the topic of Sec. 4.3).

The performance is far from the pure AWGN channel associated with a perfectly aligned SISO link (i.e. without fading). The near-field case has a small advantage of 0.281 dB over the far-field case (the same effect as a coding gain).

4.3 Spatial Diversity Schemes

In this section we study the use of orthogonal coil arrays in combination with a spatial diversity scheme for misalignment mitigation, motivated by the terrible outage performance of the randomly arranged SISO link. We exclusively consider the pure near- and far-field cases and characterize any scheme in terms of equivalent alignment factors

$\bar{J}_{\text{NF}}$ and $\bar{J}_{\text{FF}}$ of an equivalent SISO link in the sense that

$$\eta = \bar{J}_{\text{NF}}^2 |h_{\text{coax}}|^2 \qquad \text{for } kr \ll 1 \qquad (4.36)$$

$$\eta = \bar{J}_{\text{FF}}^2 |h_{\text{copl}}|^2 \qquad \text{for } kr \gg 1 \qquad (4.37)$$

describe the actual PTE η that is achieved with the spatial diversity scheme under investigation. Thereby h_{coax} and h_{copl} concern a SISO link with the same link distance and technical parameters as the SIMO, MISO or MIMO scheme at hand.

Ideally a diversity scheme should establish a strictly positive worst-case value $\min \bar{J}_*^2 > 0$ in order to assert $\eta \geq |h_*|^2 \cdot \min \bar{J}_*^2 > 0$ and thereby prevent deep fading. Furthermore $\bar{J}_*^2$ should take on large values with high probability. For simplicity we define that the realized $\bar{J}_*$ is always positive. Fig. 4.9 shows the statistics of $\bar{J}_{\text{NF}}$ and $\bar{J}_{\text{FF}}$ for all considered diversity schemes, which shall provide guidance.

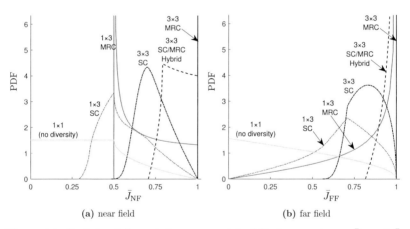

(a) near field

(b) far field

Figure 4.9: Statistics of the equivalent near- and far-field alignment factors $\bar{J}_{\text{NF}}$ and $\bar{J}_{\text{FF}}$ for magneto-inductive links between nodes with random orientations (all 3D directions are equiprobable) in SISO, SIMO (or MISO), and MIMO configuration and the use of different spatial diversity schemes. Any array consists of three colocated, orthogonal coils. The transmitter and receiver orientations are statistically independent.

4.3.1 SIMO Maximum-Ratio Combining

Consider a magneto-inductive SIMO[6] setup with one transmit coil and a colocated receive array of $N_R = 3$ coils with orthogonal orientations $\mathbf{O}_R = [\mathbf{o}_{R,1}, \ \mathbf{o}_{R,2}, \ \mathbf{o}_{R,3}]$. According to (3.49) the channel vector is given by the projections $\mathbf{h} = \mathbf{O}_R^T \mathbf{v}$ of a field vector $\mathbf{v} = \alpha \left(\left(\frac{1}{(kr)^3} + \frac{j}{(kr)^2} \right) \boldsymbol{\beta}_{\mathrm{NF}} + \frac{1}{2kr} \boldsymbol{\beta}_{\mathrm{FF}} \right)$. In (3.28) we established that maximum-ratio combining (MRC) leads to $\eta = \|\mathbf{h}\|^2$ or rather $\eta = \|\mathbf{O}_R^T \mathbf{v}\|^2 = \|\mathbf{v}\|^2$ because $\mathbf{O}_R$ is an orthogonal matrix. Hence an equivalent SISO channel coefficient is given by $\|\mathbf{v}\|$ and, more specifically, by $h_{\mathrm{coax}} \beta_{\mathrm{NF}}$ in the pure near-field case and $h_{\mathrm{copl}} \beta_{\mathrm{FF}}$ in the pure far-field case. Thus the scaled field magnitudes constitute $\bar{J}_{\mathrm{NF}} = \beta_{\mathrm{NF}}$ and $\bar{J}_{\mathrm{FF}} = \beta_{\mathrm{FF}}$ according to (4.37). We know that $\beta_{\mathrm{NF}} \in [\frac{1}{2}, 1]$ and $\beta_{\mathrm{FF}} \in [0,1]$ from (4.4) and (4.5). Therefore, when using MRC in the discussed SIMO-setup, deep fades are prevented in the near field but can still occur in the pure far field. The intuitive reason is: if the receive array is located in the radiation pattern zero of the transmit coil then all receive coils are faded, which can not be fixed by a receive diversity scheme. The statistical description is readily available in Lemma 4.2: the PDF $f_{\bar{J}_{\mathrm{NF}}} = f_{\beta_{\mathrm{NF}}}$ and $f_{\bar{J}_{\mathrm{FF}}} = f_{\beta_{\mathrm{FF}}}$.

An important observation in this context is that $f_{\beta_{\mathrm{FF}}}(\beta_{\mathrm{FF}}) \approx \beta_{\mathrm{FF}}$ for small arguments, which follows from a first-order Taylor series at zero and can be seen in Fig. 4.3b. This behavior is associated with a CDF $F_{\beta_{\mathrm{FF}}^2}(\beta_{\mathrm{FF}}^2) \approx \beta_{\mathrm{FF}}^2$ and thus $F_{\bar{J}_{\mathrm{FF}}^2}(\bar{J}_{\mathrm{FF}}^2) \approx \bar{J}_{\mathrm{FF}}^2$ for small arguments. Hence SIMO MRC in the pure far-field has a diversity order of just 1 due to the radiation pattern zero of the single-coil end.

4.3.2 SIMO Selection Combining

We consider the same setup as in Sec. 4.3.1 but now with selection combining as spatial diversity scheme, which just uses the receive coil with the best channel. In this case the equivalent alignment factors $\bar{J}_{\mathrm{NF}} = \max_{m=1,2,3} |J_{\mathrm{NF}}|$ and $\bar{J}_{\mathrm{FF}} = \max_{m=1,2,3} |J_{\mathrm{FF}}|$. The worst cases are characterized by $\bar{J}_{\mathrm{NF}} = \frac{\min \beta_{\mathrm{NF}}}{\sqrt{3}} = \frac{1}{2\sqrt{3}}$ and $\bar{J}_{\mathrm{FF}} = \frac{\min \beta_{\mathrm{FF}}}{\sqrt{3}} = 0$ whereby the factor $\frac{1}{\sqrt{3}}$ occurs when all receive-coil orientations $\mathbf{O}_R = [\mathbf{o}_{R,1}, \ \mathbf{o}_{R,2}, \ \mathbf{o}_{R,3}]$ are at the same angle to the impinging field. To see this, consider $\mathbf{O}_R = \mathbf{I}_3$ and $\boldsymbol{\beta}_{\mathrm{NF}} = \beta_{\mathrm{NF}} \frac{\mathbf{a}}{\|\mathbf{a}\|}$ with vector $\mathbf{a} = [1 \ 1 \ 1]^T$, resulting in $\mathbf{O}_R^T \boldsymbol{\beta}_{\mathrm{NF}} = \frac{\beta_{\mathrm{NF}}}{\sqrt{3}} [1 \ 1 \ 1]^T$.

---

[6]For every evaluation of a 1×3 SIMO setup in this section, the equivalent spatial diversity scheme applied to a 3×1 MISO setup leads to the same results. This will not be pointed out repeatedly. In our exposition we prefer the SIMO setup because it is more suitable from a didactic perspective.

Proposition 4.4. *For SIMO selection combining the conditional PDF*

$$f_{\bar{J}_*|\beta_*}(\bar{J}_*|\beta_*) = \begin{cases} \frac{3}{\beta_*}\left(1 - \frac{4}{\pi}\arccos\frac{|\bar{J}_*|}{\sqrt{\beta_*^2 - \bar{J}_*^2}}\right) & \frac{\beta_*}{\sqrt{3}} \leq \bar{J}_* < \frac{\beta_*}{\sqrt{2}} \\ \frac{3}{\beta_*} & \frac{\beta_*}{\sqrt{2}} \leq \bar{J}_* \leq \beta_* \\ 0 & \text{otherwise} \end{cases} \qquad (4.38)$$

holds; the PDFs of $\bar{J}_{\text{NF}}$ and $\bar{J}_{\text{FF}}$ are obtained by computing $f_{\bar{J}_*} = \int_0^1 f_{\bar{J}_*|\beta_*} f_{\beta_*}\, d\beta_*.$

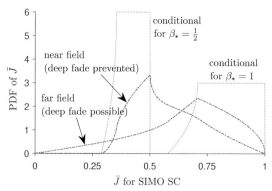

Figure 4.10: Performance statistics of SIMO selection combining (SC) and illustration of the mathematical background.

Proof. We set $\beta_* = 1$ and discard the symbol (the general result follows from simple scaling). Let sel_1 denote the event where coil 1 gets selected because $J_1^2 \geq \max\{J_2^2, J_3^2\}$. Due to the uniformly distributed array orientation $f(\bar{J}) = f(|J_1| \,|\, \text{sel}_1)$, so we derive this PDF instead. With $\beta_* = 1$ in the SIMO case, $\mathbf{j}_* = \mathbf{O}_R^T \boldsymbol{\beta}_*$ is a unit vector, so $J_1^2 + J_2^2 + J_3^2 = 1$. If $J_1^2 > \frac{1}{2}$ then J_1^2 is obviously largest, so $\text{P}[\text{sel}_1|J_1^2 > \frac{1}{2}] = 1$, but $\text{P}[\text{sel}_1|J_1^2 < \frac{1}{3}] = 0$ in contrary. The remaining transitional case $\frac{1}{3} \leq J_1^2 \leq \frac{1}{2}$ is non-trivial: with J_1 fixed, $\mathbf{j}_*$ has uniform distribution on a circle of radius $R = \sqrt{1 - J_1^2}$ and the largest element depends on the particular position on that circle. We use the parameterization $\mathbf{j}_* = [J_1 \ R\cos(\phi) \ R\sin(\phi)]^T$ with $\phi \sim \mathcal{U}(0, 2\pi)$. Depending on ϕ,

91

$J_2^2 > J_1^2$ or $J_3^2 > J_1^2$ may hold (but not both because $\frac{1}{3} \le J_1^2$). Thus,

$$\mathrm{P}\left[\mathrm{sel}_1 \,\Big|\, \tfrac{1}{3} \le J_1^2 \le \tfrac{1}{2}\right] = 1 - 2 \cdot \mathrm{P}\left[J_2^2 > J_1^2 \,\Big|\, \tfrac{1}{3} \le J_1^2 \le \tfrac{1}{2}\right] = 1 - 2 \cdot \mathrm{P}\left[\cos^2(\phi) > \frac{J_1^2}{R^2}\right]$$

$$= 1 - 2 \cdot \mathrm{P}\left[\frac{\phi}{2\pi} < \frac{1}{2\pi}\arccos\left(\frac{|J_1|}{R}\right)\right] = 1 - \frac{4}{\pi}\arccos\left(\frac{|J_1|}{R}\right).$$

We note that $\mathrm{P}[\mathrm{sel}_1] = \frac{1}{3}$ due to symmetry and $f_{|J_1|} = 1$ because $J_1 \sim \mathcal{U}(-1,1)$. With Bayes' rule we determine the PDF $f_{|J_1| \,|\, \mathrm{sel}_1} = \frac{1}{\mathrm{P}[\mathrm{sel}_1]}\mathrm{P}[\mathrm{sel}_1 \,|\, |J_1|] \cdot f_{|J_1|} = 3 \cdot \mathrm{P}[\mathrm{sel}_1 \,|\, |J_1|]$. Collecting the results for $\mathrm{P}[\mathrm{sel}_1 \,|\, |J_1|]$ for the three different cases finishes the proof. $\square$

Also for SIMO SC we find that the PDF of $\bar{J}_{\mathrm{FF}}$ is linear for small arguments, i.e. $f_{\bar{J}_{\mathrm{FF}}} \approx c \cdot \bar{J}_{\mathrm{FF}}$ (we deduce $\sqrt{2} \le c \le \sqrt{3}$ with basic estimates and $c \approx 1.5152$ numerically). As discussed in Sec. 4.3.1 this is associated with diversity order 1.

4.3.3 MIMO Maximum-Ratio Combining

We consider a 3×3 link between orthogonal arrays with orientations $\mathbf{O}_\mathrm{T}$ and $\mathbf{O}_\mathrm{R}$; the setup studied earlier in Sec. 3.7.5. The channel matrix $\mathbf{H} = \mathbf{O}_\mathrm{R}^\mathrm{T}\mathbf{H}_{\mathrm{3DoF}}\mathbf{O}_\mathrm{T}$ according to (3.53). With maximum-ratio combining $\eta = \lambda_{\max}(\mathbf{H}^\mathrm{H}\mathbf{H})$ according to (3.28). We find that the array orientations are irrelevant, hence $\eta = \lambda_{\max}(\mathbf{H}_{\mathrm{3DoF}}^\mathrm{H}\mathbf{H}_{\mathrm{3DoF}})$. The eigenvalues of $\mathbf{H}_{\mathrm{3DoF}}$ are h_{coax} and h_{copl} and we obtain $\eta = |h_{\mathrm{coax}}|^2$ for $kr \le kr_{\mathrm{th}}$ and $\eta = |h_{\mathrm{copl}}|^2$ for $kr > kr_{\mathrm{th}}$. This corresponds to $\bar{J}_{\mathrm{NF}} = 1$ and $\bar{J}_{\mathrm{FF}} = 1$, i.e. no misalignment loss.

4.3.4 MIMO Selection Combining

We consider the same 3×3 link as above and consider selection combining on both ends, yielding $\eta = \max_{m,n}|(H)_{m,n}|$. We do not attempt the complicated task of deriving the statistics of the associated $\bar{J}_{\mathrm{NF}}$ and $\bar{J}_{\mathrm{FF}}$ and instead focus on their worst-case values.

Proposition 4.5. *Selection combining on both ends limits the misalignment loss of the considered 3×3 MIMO link according to*

$$0.47978 \le \bar{J}_{\mathrm{NF}} \le 1\,, \tag{4.39}$$

$$\sqrt{\frac{3 + \sqrt{2}}{14}} = 0.56152 \le \bar{J}_{\mathrm{FF}} \le 1 \tag{4.40}$$

whereby the value 0.47978 is the greatest real root of $(24x^3 - 8x^2 - x + 1)^2 - 16x^3$.

Proof. We denote $\mathcal{J}$ for the smallest positive $\bar{J}_{\mathrm{NF}}$ that occurs by 3×3 selection combining. It is the solution of $\mathcal{J} = \min_{\mathbf{O}_{\mathrm{T}}, \mathbf{O}_{\mathrm{R}}} \max_{m,n} |(\mathbf{J}_{\mathrm{NF}})_{m,n}|$ over all orthogonal array orientations $\mathbf{O}_{\mathrm{R}}, \mathbf{O}_{\mathrm{T}}$ whereby $\mathbf{J}_{\mathrm{NF}} = \mathbf{O}_{\mathrm{R}}^{\mathrm{T}} \mathrm{diag}(1, -\frac{1}{2}, -\frac{1}{2}) \, \mathbf{O}_{\mathrm{T}}$. By numerical simulation (a rigorous argument is unavailable at this moment) we found that any $\mathcal{J}$-achieving matrix $\mathbf{J}_{\mathrm{NF}}$ has six elements with absolute value $\mathcal{J}$ and two with the same absolute value a. Furthermore, by flipping coil orientations and indices, which does not affect optimality, any $\mathcal{J}$-attaining $\mathbf{J}_{\mathrm{NF}}$ can be transformed to

$$\mathbf{J}_{\mathrm{NF}} = \begin{bmatrix} \mathcal{J} & \mathcal{J} & \mathcal{J} \\ \mathcal{J} & \mathcal{J} & a \\ \mathcal{J} & a & b \end{bmatrix}, \quad \lambda_1(\mathbf{J}_{\mathrm{NF}}) = 1, \quad \lambda_2(\mathbf{J}_{\mathrm{NF}}) = \frac{1}{2}, \quad \lambda_3(\mathbf{J}_{\mathrm{NF}}) = -\frac{1}{2} \quad (4.41)$$

with the specific stated eigenvalues. They give $\mathrm{tr}(\mathbf{J}_{\mathrm{NF}}) = 1$ and $\det(\mathbf{J}_{\mathrm{NF}}) = -\frac{1}{4}$ as well as $\mathrm{tr}(\mathbf{J}_{\mathrm{NF}}^2) = \frac{3}{2}$. However trace and determinant can also be expressed in terms of $\mathcal{J}, a, b$ from the given structure of $\mathbf{J}_{\mathrm{NF}}$. We equate the terms to obtain the system of equations $2\mathcal{J} + b = 1$, $\mathcal{J}(\mathcal{J} - a)^2 = \frac{1}{4}$, $6\mathcal{J}^2 + 2a^2 + b^2 = \frac{3}{2}$ with the numerical solution $\mathcal{J} = 0.479788$, $a = -0.242059$, $b = 0.040423$. By substitution we find that $\mathcal{J}$ is a real root of $(24\mathcal{J}^3 - 8\mathcal{J}^2 - \mathcal{J} + 1)^2 - 16\mathcal{J}^3$; it turns out $\mathcal{J}$ is the greatest real root.

We now consider the far-field quantity and denote Γ for the smallest occurring positive $\bar{J}_{\mathrm{FF}}$. By the same arguments as above we find a Γ-attaining setup with

$$\mathbf{J}_{\mathrm{FF}} = \begin{bmatrix} \Gamma & \Gamma & \Gamma \\ \Gamma & -\Gamma & c \\ \Gamma & c & 0 \end{bmatrix}, \quad \lambda_1(\mathbf{J}_{\mathrm{FF}}) = 1, \quad \lambda_2(\mathbf{J}_{\mathrm{FF}}) = -1, \quad \lambda_3(\mathbf{J}_{\mathrm{FF}}) = 0. \quad (4.42)$$

Equating $\det(\mathbf{J}_{\mathrm{FF}})$ due to the structure and due to the eigenvalues yields the equation $2c\Gamma^2 + \Gamma^3 - c^2\Gamma = 0$ which we rearrange to $4c^2\Gamma^2 - (c^2 - \Gamma^2)^2 = 0$. Likewise, equating $\mathrm{tr}(\mathbf{J}_{\mathrm{FF}}^2)$ due to structure and eigenvalues yields $6\Gamma^2 + 2c^2 = 2$, hence $c^2 = 1 - 3\Gamma^2$. We substitute this into the first equation and after some calculation obtain $\Gamma^4 - \frac{3}{7}\Gamma^2 + \frac{1}{28} = 0$, a quadratic equation in Γ^2 with solutions $\Gamma^2 = \frac{3 \pm \sqrt{2}}{14}$. A comparison to numerical experiments suggests the positive solution of $\Gamma^2 = \frac{3 + \sqrt{2}}{14}$ (and $c = -\sqrt{\frac{5 - 3\sqrt{2}}{14}}$). $\qquad\square$

4.3.5 MIMO MRC/SC Hybrid

Again we consider a magneto-inductive 3×3 link between orthogonal coil arrays, now with maximum-ratio combining on one end and selection combining on the other end.

Proposition 4.6. *The worst-case misalignment loss of the MIMO MRC/SC hybrid scheme is characterized by* $\bar{J}_{NF} \geq \frac{1}{\sqrt{2}} = 0.7071$ *and* $\bar{J}_{FF} \geq \sqrt{\frac{2}{3}} = 0.8165$.

Proof. As seen in Sec. 4.3.1, receive-side MRC captures the entire field magnitude β_*, hence transmit-side SC corresponds to $\bar{J}_* = \max_n \beta_{*,n}$ for transmit coils $n = 1, 2, 3$. For the near field $\beta_{NF,n} = \frac{1}{2}\sqrt{1 + 3(\mathbf{u}^T\mathbf{o}_{T,n})^2}$. With the argument of Sec. 4.3.2 we find that the minimum value of $\bar{J}_{NF}$ is attained when all $\beta_{NF,n}$ are equal because of $\mathbf{u}^T\mathbf{o}_{T,n} = \frac{1}{\sqrt{3}} \forall n$, hence $\min \bar{J}_{NF} = \frac{1}{2}\sqrt{1 + 3\frac{1}{3}} = \frac{1}{\sqrt{2}}$. With $\beta_{FF,n} = \sqrt{1 - (\mathbf{u}^T\mathbf{o}_{T,n})^2}$ an analogous argument yields $\min \bar{J}_{NF} = \sqrt{1 - \frac{1}{3}} = \sqrt{\frac{2}{3}}$. $\square$

4.4 Further Stochastic Results

As outlined in Cpt. 1 and Sec. 3.9, an important use of magnetic induction is data transmission from passive tags with the use of load modulation, primarily in RFID technology. By writing the rather general SNR-result from Sec. 3.9 for a SISO link (single tag, single-coil reader) we obtain the proportionality SNR $\propto |Z_{RT}|^4$ where Z_{RT} is the coil mutual impedance. In the pure near field (the typical regime for an RFID system) this implies SNR $\propto J_{NF}^4$, associated with even more severe misalignment losses than the previously considered setup with an active transmitter: the dB-loss-values double and the diversity order drops from $\frac{1}{2}$ to an extremely poor $\frac{1}{4}$. In Fig. 4.11 we show the behavior of the bit error probability for this random channel.

Next up, we apply the developed theory of random coil orientations to magneto-inductive massive MIMO links, in particular in the context of Sec. 3.7.4.

Proposition 4.7. *Consider a weakly-coupled MIMO link between two massive arrays, both colocated (or near-colocated) and uncoupled by assumption. All coils have i.i.d. random orientation with uniform distribution on the 3D unit sphere and full CSI is available. The reception is subject to AWGN with covariance matrix $\sigma^2 \mathbf{I}_{N_R}$. Then the channel capacity is characterized by the equivalent 3×3 noise-whitened channel matrix*

$$\bar{\mathbf{H}}' = \frac{1}{\sigma}\sqrt{\frac{N_T}{3}}\sqrt{\frac{N_R}{3}}\,\mathbf{H}_{3DoF} \qquad (4.43)$$

which is equivalent to a 3×3 MIMO link between orthogonal arrays, cf. (3.55), but with an array gain of $\frac{N_T}{3} \cdot \frac{N_R}{3}$ in terms of SNR.

Proof. The $N_R \times N_T$ link has noise-whitened channel matrix $\bar{\mathbf{H}} = \frac{1}{\sigma}\mathbf{O}_R^T\mathbf{H}_{3DoF}\mathbf{O}_T$ as a result of (3.52) and uncoupled arrays. With the singular value decompositions

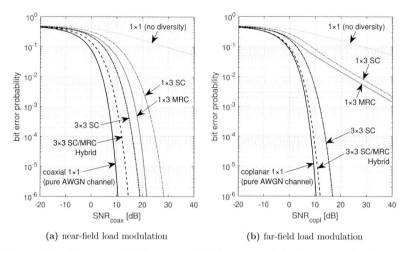

(a) near-field load modulation   (b) far-field load modulation

Figure 4.11: Bit error probability p_e over SNR$_*$ for magneto-inductive load modulation over SISO links between coils with random orientation (all 3D directions are equiprobable) as well as SIMO and MIMO links between randomly oriented orthogonal arrays with different spatial diversity schemes. Coil misalignment has a particularly drastic effect for load modulation because of the backscatter nature of the approach (the wireless channel applies twice).

of $\mathbf{O}_R$ and $\mathbf{O}_T$ we obtain $\bar{\mathbf{H}} = \frac{1}{\sigma}\mathbf{V}_R^T\mathbf{\Sigma}_R^T\mathbf{U}_R\mathbf{H}_{3\text{DoF}}\mathbf{U}_T\mathbf{\Sigma}_T\mathbf{V}_T^T$. The unitary $\mathbf{V}_R$ and $\mathbf{V}_T$ can be compensated with signal processing without affecting the capacity. Furthermore $\mathbf{\Sigma}_R \approx \sqrt{\frac{N_R}{3}}\mathbf{I}_{3\times N_R}$ in the large-N_R limit and and $\mathbf{\Sigma}_T \approx \sqrt{\frac{N_T}{3}}\mathbf{I}_{3\times N_T}$ in the large-N_T limit because $\frac{1}{N_R}\mathbf{O}_R\mathbf{O}_R^T \approx \frac{1}{3}\mathbf{I}_3$ and $\frac{1}{N_T}\mathbf{O}_T\mathbf{O}_T^T \approx \frac{1}{3}\mathbf{I}_3$ by Lemma 4.3. Thus $\frac{1}{\sigma}\sqrt{\frac{N_R}{3}}\sqrt{\frac{N_T}{3}}\,\mathbf{I}_{N_R\times 3}\mathbf{U}_R\mathbf{H}_{3\text{DoF}}\mathbf{U}_T\mathbf{I}_{3\times N_T}$ is an equivalent channel matrix. We note that all elements apart from the upper-left 3×3 block are zero and thus can not be used for communication, hence this block $\frac{1}{\sigma}\sqrt{\frac{N_R}{3}}\sqrt{\frac{N_T}{3}}\mathbf{U}_R\mathbf{H}_{3\text{DoF}}\mathbf{U}_T$ constitutes an equivalent channel matrix. Again, the unitary matrices do not affect capacity and can be discarded, leaving only $\frac{1}{\sigma}\sqrt{\frac{N_R}{3}}\sqrt{\frac{N_T}{3}}\mathbf{H}_{3\text{DoF}}$. $\qquad\square$

Another interesting application of the developed stochastic theory is the study of spatial correlation. In particular we are interested in the covariance between the channel coefficients between different fixed-orientation transmit coils and a single-coil receiver with random orientation. The setup is illustrated in Fig. 4.12 and the result will be used for the design of localization algorithms in Cpt. 7.

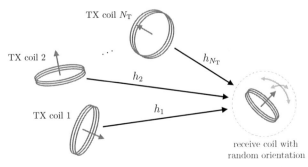

Figure 4.12: MISO link between fixed-orientation transmit coils and a receive coil with random orientation (uniform distribution in 3D).

Proposition 4.8. *Consider a MISO link between a (possibly distributed) array of N_T fixed-orientation coils and a receive coil whose random orientation $\mathbf{o}_T$ has uniform distribution on the 3D unit sphere. We assume a receive coil that is small compared to the link distances and has consistent turn surface orientation such that the link coefficients are scalar products $h_n = \mathbf{o}_R^T \mathbf{v}_n$ (the field vector $\mathbf{v}_n \in \mathbb{C}^3$ is due to the n-th transmitter). Then the channel coefficients have covariance*

$$\mathrm{cov}(h_m, h_n) = \mathbb{E}[h_m h_n^*] = \frac{1}{3}\mathbf{v}_m^T \mathbf{v}_n^* . \tag{4.44}$$

Hence the covariance matrix of the channel vector is $\frac{1}{3}\mathbf{V}^H\mathbf{V}$ with $\mathbf{V} = [\mathbf{v}_1^* \dots \mathbf{v}_{N_T}^*]$.

Proof. $\mathrm{cov}(h_m, h_n)$ is given by $\mathbb{E}[h_m h_n^*]$ because $\mathbb{E}[h_n] = 0 \ \forall n$. We expand the term $\mathbb{E}[h_m h_n^*] = \mathbf{v}_m^T \mathbb{E}[\mathbf{o}_R \mathbf{o}_R^T]\mathbf{v}_n^* = \mathbf{v}_m^T(\frac{1}{3}\mathbf{I}_3)\mathbf{v}_n = \frac{1}{3}\mathbf{v}_m^T\mathbf{v}_n^*$ with the use of Lemma 4.3. $\qquad\square$

Chapter 5

Randomly Placed Passive Relays: Effects and Utilization

In Cpt. 1 we discussed passive relaying as an interesting opportunity for magneto-inductive wireless systems. The idea is to place resonant passive relay coils in the vicinity of the transmitter and/or receiver. The magnetic field generated by the transmitter induces a current in the relay coils which subsequently generates a secondary magnetic field. The effect of this additional magnetic field at the receiver position can improve the link, e.g., for range extension or higher data rates. The corresponding state of the art was detailed in Sec. 1.3.5, where we argued that existing studies of magneto-inductive passive relaying almost exclusively consider well-arranged scenarios such as contiguous waveguides of coplanar or coaxial relays, allowing simplified assumptions on the mutual coil couplings. Such well-defined arrangements are however meaningless for mobile or ad-hoc sensor applications, which demand a study of magneto-inductive passive relaying in arbitrary (or random) arrangement. The effects and performance statistics of this context are currently unclear.

In this regards, this chapter contains the following specific contributions.

- We present an analysis of magneto-inductive passive relaying in arbitrary arrangements, with a focus on the case of passive relays in close vicinity of one link end.

- We show that the resulting channel has characteristics similar to multipath fading: the channel gain is governed by a noncoherent sum of phasors, resulting in increased frequency selectivity. We decompose the relaying gain into two major effects: the gain from the transmitter-receiver mutual impedance change and the loss from increased encountered coil resistance (due to coupling with lossy relays).

- For better utilization of the relaying channel we propose an optimization scheme based on adaption of the passive relay loads via load switching. We demonstrate reliable and significant performance gains and thus establish the scheme as a powerful opportunity for magneto-inductive wireless applications.

- We characterize the random frequency-selective fading channel caused by a random cluster of passive relays, in terms of coherence bandwidth and affected bandwidth versus relay density. We find that this channel offers significant frequency diversity which, when utilized with a channel-aware transmission scheme such as waterfilling, yields great data rate improvements even without any relay optimization.

As a preparatory step for the intended study we integrate the effect of N_Y passive relays into the system model of Cpt. 3. The only required change is the adaptation of the $(N_T + N_R) \times (N_T + N_R)$ impedance matrix $\mathbf{Z}_C$ in (3.1) that holds all self- and mutual impedances of and between all transmit and receive coils. The adapted impedance matrix (comprising the action of loaded passive relays) is denoted $\tilde{\mathbf{Z}}_C$. To do so we first express the impedance matrix between all transmit, receive, and relay coils

$$
\begin{bmatrix} \mathbf{Z}_{C:T} & \mathbf{Z}_{C:TR} & \mathbf{Z}_{C:TY} \\ \mathbf{Z}_{C:RT} & \mathbf{Z}_{C:R} & \mathbf{Z}_{C:RY} \\ \mathbf{Z}_{C:YT} & \mathbf{Z}_{C:YR} & \mathbf{Z}_{C:Y} \end{bmatrix} = \begin{bmatrix} \mathbf{Z}_C & & \mathbf{Z}_{C:TY} \\ & & \mathbf{Z}_{C:RY} \\ \mathbf{Z}_{C:YT} & \mathbf{Z}_{C:YR} & \mathbf{Z}_{C:Y} \end{bmatrix} \in \mathbb{C}^{N\times N} \cdot \Omega \qquad (5.1)
$$

where $N = N_T + N_R + N_Y$ is the total number of coils. We now terminate the relay ports with passive loads with impedance matrix $\mathbf{Z}_L \in \mathbb{C}^{N_Y \times N_Y} \cdot \Omega$. The concept is illustrated in Fig. 5.1; the depicted relay loads are resonance capacitors and thus the impedance matrix $\mathbf{Z}_L = \mathrm{diag}_{l=1\ldots N_Y} \frac{1}{j\omega C_l}$. With the use of (2.45) we directly obtain the adapted impedance matrix between all $N_T + N_R$ transmit and receive coils,

$$
\tilde{\mathbf{Z}}_C = \begin{bmatrix} \tilde{\mathbf{Z}}_{C:T} & \tilde{\mathbf{Z}}_{C:TR} \\ \tilde{\mathbf{Z}}_{C:RT} & \tilde{\mathbf{Z}}_{C:R} \end{bmatrix} = \mathbf{Z}_C - \begin{bmatrix} \mathbf{Z}_{C:TY} \\ \mathbf{Z}_{C:RY} \end{bmatrix} \left(\mathbf{Z}_{C:Y} + \mathbf{Z}_L \right)^{-1} \begin{bmatrix} \mathbf{Z}_{C:YT} & \mathbf{Z}_{C:YR} \end{bmatrix} . \qquad (5.2)
$$

This notion has been employed previously by [26, 87, 100, 103]. The charm of this approach is that through just replacing $\mathbf{Z}_C$ by $\tilde{\mathbf{Z}}_C$ the MIMO signal and noise model presented in Cpt. 3 applies without further ado and yields the channel matrix $\mathbf{H}$ and noise covariance matrix $\mathbf{K}$.[1]

---

[1]This simple adaptation even accounts for changes in the thermal noise statistics due to passive relays near the receive coils: $\mathbf{Z}_C^{\mathrm{out}}$ changes to $\tilde{\mathbf{Z}}_C^{\mathrm{out}}$ and $\mathbf{Z}_R^{\mathrm{out}}$ to $\tilde{\mathbf{Z}}_R^{\mathrm{out}}$ and this alters the noise covariance matrix $\mathbf{K}$. Recall from (3.19) that the thermal-noise portion of $\mathbf{K}$ (which is given by $R_{\mathrm{ref}}\mathbf{K}_i$) is proportional to $\mathrm{Re}(\tilde{\mathbf{Z}}_R^{\mathrm{out}})$.

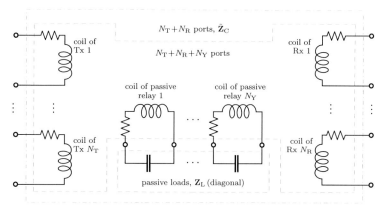

Figure 5.1: Incorporating the effect of N_Y passive relays into the $(N_T + N_R)$-port network between all transmit and receive coils, using the formula (5.2) from Fig. 5.1.

5.1 Effects and General Properties

The effect of passive relays can be described most compactly with the channel matrix expression $\mathbf{H} = \frac{1}{2}\operatorname{Re}(\tilde{\mathbf{Z}}_{C:R})^{-\frac{1}{2}}\tilde{\mathbf{Z}}_{C:RT}\operatorname{Re}(\tilde{\mathbf{Z}}_{C:T})^{-\frac{1}{2}}$ from (3.47) which concerns a MIMO link with ideal power matching on both ends and weak transmitter-receiver coupling. The resulting SISO-case channel power gain (i.e. the power transfer efficiency) is

$$\eta = |h|^2 = \frac{|\tilde{Z}_{C:RT}|^2}{4 \cdot \operatorname{Re}(\tilde{Z}_{C:R}) \cdot \operatorname{Re}(\tilde{Z}_{C:T})} \tag{5.3}$$

which conforms with the weak-coupling limit of the η formula for simultaneous power matching in (3.44) (e.g., the formula (5.3) would be invalid for the dense coaxial relaying arrangement in Fig. 5.3). Due to the crucial role of the mutual impedance $\tilde{Z}_{C:RT}$ in (5.3), a main goal of passive relaying is to increase $|\tilde{Z}_{C:RT}|$. However also $\operatorname{Re}(\tilde{Z}_{C:T})$ or $\operatorname{Re}(\tilde{Z}_{C:R})$ of the denominator can increase significantly if passive relays are near the transmitter or the receiver, respectively, as discussed earlier below Proposition 2.6. This effect is obviously detrimental to the goal of achieving large $|h|^2$. The problem stems from the fact that the presence of passive relays near the transmitter increases the transmit power $P_T = |i_T|^2\operatorname{Re}(\tilde{Z}_{C:T})$ necessary for running a target current i_T through the transmit coils. The analogous effect may occur at the receive side, where the necessary induced voltage for inducing a target current increases.

An interesting question is the specification of the relay loads. A natural choice is one resonance capacitor per relay coil [26, 70, 86] as indicated in Fig. 5.1, i.e. $\mathbf{Z}_L$ becomes a diagonal matrix with $(\mathbf{Z}_L)_{l,l} = -j\mathrm{Im}(\mathbf{Z}_{C:Y})_{l,l}$ for $l = 1 \ldots N_Y$. In the magnetoquasistatic case this results in the structure

$$\mathbf{Z}_{C:Y} + \mathbf{Z}_L = \operatorname*{diag}_{l=1\ldots N_Y} \left(R_{Y,l} + \frac{1}{j\omega C_l} \right) + j\omega \mathbf{M} \tag{5.4}$$

where $\mathbf{M} \in \mathbb{R}^{N_Y \times N_Y}$ holds the mutual inductances between the relay coils in its off-diagonal elements and the self-inductances $L_{Y,l}$ on the diagonal. With a resonant design at some ω_{res} the diagonal elements of (5.4) attain the real value $(\mathbf{Z}_{C:Y} + \mathbf{Z}_L)_{l,l} = R_{Y,l}$ for $\omega = \omega_{\mathrm{res}}$ while the off-diagonal values are purely imaginary. We will see that this method of specifying the loads is not necessarily the best choice. A more sophisticated (but technically more costly) method choosing the loads such that some performance metric is maximized. Analogous load-optimization approach have been studied for magneto-inductive waveguides in [83] and for the utilization of passive radio antennas in compact arrays in [100].

The inverse $(\mathbf{Z}_{C:Y} + \mathbf{Z}_L)^{-1}$, which plays a crucial role in (5.2), exhibits a vastly involved dependence on ω and on all other setup parameters. This complicates analytic or intuitive attempts of understanding passive relaying in general arrangements. Nevertheless we attempt just that in the following. For this purpose we consider only the off-diagonal element $\tilde{Z}_{C:RT}$ of $\tilde{\mathbf{Z}}_C$, i.e. the transmitter-to-receiver mutual impedance, which can be rearranged as follows:

$$\tilde{Z}_{C:RT} = Z_{C:RT} - \mathbf{z}_{C:RY}^T \left(\mathbf{Z}_{C:Y} + \mathbf{Z}_L \right)^{-1} \mathbf{z}_{C:YT} \tag{5.5}$$

$$\overset{\mathrm{MQS}}{=} j\omega M_{RT} + \omega^2 \mathbf{m}_{RY}^T \left(\mathbf{Z}_{C:Y} + \mathbf{Z}_L \right)^{-1} \mathbf{m}_{YT} \tag{5.6}$$

$$= j\omega M_{RT} + \omega^2 \sum_{l=1}^{N_Y} \sum_{i=1}^{N_Y} M_{RY,l} \left((\mathbf{Z}_{C:Y} + \mathbf{Z}_L)^{-1} \right)_{l,i} M_{YT,i} \tag{5.7}$$

$$\overset{\substack{\text{uncoupled} \\ \text{relays}}}{=} j\omega M_{RT} + \omega^2 \sum_{l=1}^{N_Y} \frac{M_{RY,l} M_{YT,l}}{R_{Y,l} + j\omega L_{Y,l} + Z_{L,l}} \tag{5.8}$$

$$\overset{\substack{\text{resonant} \\ \text{loading}}}{=} j\omega_{\mathrm{res}} M_{RT} + \omega_{\mathrm{res}}^2 \sum_{l=1}^{N_Y} \frac{M_{RY,l} M_{YT,l}}{R_{Y,l}} \tag{5.9}$$

$$= \omega_{\mathrm{res}} \sqrt{L_R L_T} \left(j\kappa_{RT} + \sum_{l=1}^{N_Y} \kappa_{RY,l}\, \kappa_{YT,l}\, Q_{Y,l} \right). \tag{5.10}$$

Thereby $\kappa_{RT} = M_{RT} / \sqrt{L_T L_R} \in [-1, +1]$ is the coupling coefficient between transmitter and receiver and $Q_{Y,l} = \omega_{\mathrm{res}} L_{Y,l} / R_{Y,l}$ is the Q-factor of the l-th relay coil.

Before discussing the implications of the above mathematics we shall study two introductory examples in terms of equivalent circuits. A first example in Fig. 5.2a shows of a SISO link affected by one resonant passive relay ($N_Y = 1$, $\omega = \omega_{res}$). From

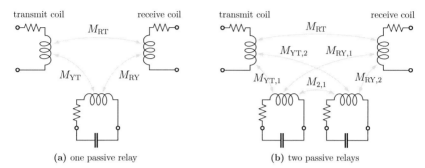

(a) one passive relay   (b) two passive relays

Figure 5.2: Introductory examples of passive relaying in terms of equivalent circuits, describing the two-port network between transmit coil and receive coil in the presence of one or two resonant passive relay coils. The illustration indicates the possible propagation paths from transmitter to receiver, via the direct path or via the passive relay(s).

(5.9) we obtain the MQS transmitter-to-receiver mutual impedance at the resonance frequency,

$$\tilde{Z}_{\text{C:RT}} = \underbrace{j\omega_{\text{res}}M_{\text{RT}}}_{\substack{\text{direct path,}\\ \text{imaginary}}} + \underbrace{\frac{\omega_{\text{res}}^2 M_{\text{RY}} M_{\text{YT}}}{R_Y}}_{\substack{\text{via relay,}\\ \text{real valued}}} . \tag{5.11}$$

We note that $|\tilde{Z}_{\text{C:RT}}| \geq |Z_{\text{C:RT}}| = |\omega_{\text{res}}M_{\text{RT}}|$ because the above summands are the real and imaginary parts of $\tilde{Z}_{\text{C:RT}}$. In words, the presence of a single resonant relay can only increase the (absolute value of the) transmitter-to-receiver mutual impedance. This is an exclusive property of the magnetoquasistatic case (where $Z_{\text{C:RT}}, Z_{\text{C:RY}}, Z_{\text{C:YT}}$ are all purely imaginary, and where the relay introduces 90° phase shift between induced current and generated field) at the resonant design frequency and does not hold in general (which will be seen later in Fig. 5.6b).

Fig. 5.2b shows the same experiment for the presence of two passive relay coils ($N_Y = 2$), again evaluated at $\omega = \omega_{res}$. The two relay coils are assumed to have the same resistance R_Y. By considering (5.4) for this case and computing its inverse in closed form we obtain

$$(\mathbf{Z}_{\text{C:Y}} + \mathbf{Z}_{\text{L}})^{-1} = \begin{bmatrix} R_Y & j\omega_{\text{res}} M_{1,2} \\ j\omega_{\text{res}} M_{2,1} & R_Y \end{bmatrix}^{-1} = \frac{1}{R_Y^2 + \omega_{\text{res}}^2 M_{2,1}^2} \begin{bmatrix} R_Y & -j\omega_{\text{res}} M_{1,2} \\ -j\omega_{\text{res}} M_{2,1} & R_Y \end{bmatrix}$$

$$(5.12)$$

The matrix is symmetric due to reciprocity but not written that way for didactic purposes. According to (5.6) this results in a SISO-case mutual impedance

$$\tilde{Z}_{\text{C:RT}} = \overbrace{j\omega_{\text{res}} M_{\text{RT}}}^{\text{direct path}} + \frac{\omega_{\text{res}}^2}{R_Y^2 + \omega_{\text{res}}^2 M_{2,1}^2} \Big(\overbrace{M_{\text{RY},1} R_Y M_{\text{YT},1}}^{\text{via relay 1}} + \overbrace{M_{\text{RY},2} R_Y M_{\text{YT},2}}^{\text{via relay 2}}$$

$$- j\omega_{\text{res}} \underbrace{M_{\text{RY},1} M_{1,2} M_{\text{YT},2}}_{\text{via relay 1 after 2}} - j\omega_{\text{res}} \underbrace{M_{\text{RY},2} M_{2,1} M_{\text{YT},1}}_{\text{via relay 2 after 1}} \Big) \quad (5.13)$$

The expression is involved as a result of the inter-relay coupling ($M_{1,2}$, $M_{2,1}$). The five conceivable propagation paths observable in Fig. 5.2b also emerge mathematically: the direct path (90° phase shift), two paths via a single relay each (180°), and two paths via both relays (270°). For $\omega \neq \omega_{\text{res}}$ the factor $\frac{\omega_{\text{res}}^2}{R_Y^2 + \omega_{\text{res}}^2 M_{2,1}^2}$ is replaced by the involved complex-valued expression $\frac{\omega^2}{(R_Y + j\omega L_Y + \frac{1}{j\omega C_Y})^2 + \omega^2 M_{2,1}^2}$, affecting gain and phase shift.

After these specific examples we shall now discuss the implications of the more general statements (5.5)-(5.10). A particularly interesting expression is (5.7); it shows that the mutual impedance $\tilde{Z}_{\text{C:RT}}$ is the sum of $1 + N_Y^2$ complex numbers. For an arbitrary relay arrangement the elements $(\mathbf{Z}_{\text{C:Y}} + \mathbf{Z}_{\text{L}})^{-1})_{l,i}$ have general phase, hence the sum is noncoherent. In consequence $|\tilde{Z}_{\text{C:RT}}|$ can be very large if the complex numbers happen to add up constructively, or close to zero if destructive addition occurs by virtue of the arrangement realization. Furthermore, $\tilde{Z}_{\text{C:RT}}$ clearly depends on ω in a complicated way. We will later find that this gives rise to frequency-selective fading. The circumstance is comparable to Rayleigh fading of radio channels with rich multipath propagation, which also results from the non-coherent sum of many complex path amplitudes.

The dependencies are less convoluted if the coupling between relays is zero, as seen at (5.8) where the number of summands reduces to $1 + N_Y$ (one from the direct path, one per passive relay). When furthermore the considered frequency is the resonant design frequency then the direct path determines $\text{Im}(\tilde{Z}_{\text{C:RT}})$ and the passive relays determine $\text{Re}(\tilde{Z}_{\text{C:RT}})$, see (5.9) and (5.10).

An interesting question are the conditions for a passive relay to have a significant effect on the transmitter-to-receiver link. From expression (5.10) for the case of a single passive relay, i.e. for $N_Y = 1$, we find the criterion

$$|\kappa_{RY}\,\kappa_{YT}|\,Q_Y \nll |\kappa_{RT}| \quad \Longrightarrow \quad \text{passive relay has an appreciable effect} \quad (5.14)$$

where $\nll$ denotes "not much smaller than". This criterion can be fulfilled by different circumstances. Firstly, if transmitter and receiver are misaligned so that the direct link is in a deep fade $\kappa_{RT} \approx 0$, then the propagation path via the passive relays becomes significant (assuming of course that neither κ_{YT} nor κ_{RY} are faded). This is analogous to multipath propagation allowing for a radio link in a non-line-of-sight situation. Another way for a passive relay to have an appreciable effect is strong coupling to either the transmitter or the receiver. To see this, assume a setup where the passive relay is close to the transmitter and $|\kappa_{RT}| \approx |\kappa_{RY}|$ (as they relate to similar distances), which yields the criterion $|\kappa_{YT}|\,Q_Y \nll 1$. The corresponding criterion for proximity to the receiver is $|\kappa_{RY}|\,Q_Y \nll 1$. Note that a small coupling coefficient can be compensated by a large coil quality factor, e.g., strong coupling could occur for a distant relay made of superconducting material. Lastly we note that (5.14) is a sufficient but not necessary criterion: e.g., if $\kappa_{RY} \approx 0$ due to misalignment then the link can still be heavily affected by the relay if κ_{YT} is large. In this case the presence of the passive relay affects $\tilde{Z}_{C:T}$ by detuning the transmit coil.

In summary, we state the following conditions on passive relays to improve $|h|^2$ of the transmitter-receiver link: (i) $|\tilde{Z}_{C:RT}|$ is appreciably larger than $|Z_{C:RT}|$, i.e. at least one passive relay has an appreciable effect and no destructive phasor addition occurs, and ii) power dissipation by the relay circuits does not outweigh the increase in $|\tilde{Z}_{C:RT}|$. While these insights explain important aspects of magneto-inductive passive relaying, the behavior of arbitrarily arranged networks is still obscured by the analytical intractability of $(\mathbf{Z}_{C:Y} + \mathbf{Z}_L)^{-1}$ in the general case and the geometric dependencies of the mutual coil couplings. Therefore we will now shift the focus to simulation results; the parameters assumed throughout the chapter are listed in Table 5.1.

As first simulation example we consider a magneto-inductive waveguide between transmitter and receiver, constituted by equidistantly spaced passive relays, whereby all coils are in coaxial arrangement. Here the idea is to achieve a large link gain by establishing a contiguous paths of strongly coupled relays from the transmitter to the receiver. This approah is followed by many studies on magneto-inductive passive relaying in the literature [64–68]. At high relay density the channel power gain $|h|^2$ approaches

number of turns $\overset{\circ}{N}$	10		
coil diameter D_c	$10\,\mathrm{mm}$		
coil length (height) l_c	$10\,\mathrm{mm}$		
coil wire diameter D_w	$0.5\,\mathrm{mm}$		
design frequency f_res	$50\,\mathrm{MHz}$	wavelength λ	$6\,\mathrm{m}$
temperature T	$300\,\mathrm{K}$	wavenumber k	$1.047\,\frac{1}{\mathrm{m}}$
antenna temperature T_A	$300\,\mathrm{K}$	coil Q-factor at f_res	≈ 474
LNA noise variance σ_i^2/Δ_f	$2\cdot 10^{-22}\,\mathrm{A}^2/\mathrm{Hz}$	coil 3 dB bandwidth	$\approx 106\,\mathrm{kHz}$
LNA noise resistance R_N	$40\,\Omega$	coil self-resonance f	$\approx 258\,\mathrm{MHz}$
LNA noise corr. coeff. ρ	$0.5 + 0.7j$	coil wire length l_w	$0.326\,\mathrm{m}$
iid noise variance $\sigma_\mathrm{iid}^2/\Delta_f$	$10^{-19}\,\mathrm{V}^2/\mathrm{Hz}$	electrical size l_w/λ	$0.054\,(\ll 1)$

(a) specified parameters   (b) resulting parameters

Table 5.1: The simulation parameters used throughout this chapter. The electrical size of the coil is sufficiently small for AC circuit theory to apply but also sufficiently large for radiation to occur appreciably. The effect of the chosen noise parameters has order of magnitude $\sigma_i^2 R_\mathrm{ref} = -170\,\mathrm{dB\text{-}mW/Hz}$, $R_\mathrm{N}^2 \sigma_i^2 / R_\mathrm{ref} \approx -172\,\mathrm{dB\text{-}mW/Hz}$, and $\sigma_\mathrm{iid}^2 / R_\mathrm{ref} \approx -177\,\mathrm{dB\text{-}mW/Hz}$, which compares to $k_\mathrm{B} T \approx -174\,\mathrm{dB\text{-}mW/Hz}$ of thermal noise. For the antenna temperature we choose a small value of just $300\,\mathrm{K}$ because we assume a shielded environment, otherwise it could be much larger for low-frequency operation [178, Fig. 2].

0 dB because then the arrangement essentially forms a cable between transmitter and receiver. In this case the channel bandwidth widens significantly due to the strong interaction of the many resonant modes (such resonant mode splitting is described by coupled mode theory [205] and specifically for magnetic induction in [69, 206, 207]).

5.2 One Passive Relay Near the Receiver

Placing many passive relays between transmitter and receiver in order to form a dense and contiguous waveguide as in Fig. 5.3a is clearly an elusive idea for any sensor application with mobile ad-hoc character. Yet, there is an interesting use case for magneto-inductive passive relays in the sensor context: a passive relay in close proximity to the transmitter or the receiver can yield a significant link gain (although certainly shy from the large gains in Fig. 5.3b) by utilizing the effect of strongly coupled magnetic resonances [69, 70, 206]. The concept is detailed in Fig. 5.4 which shows that, under certain technical conditions, a 3 dB gain is achieved by just placing a resonant passive relay near the receiver.

For ease of exposition we consider the relay near the receiver and never near the transmitter. The receive coil impedance is affected if the relay is in close proximity,

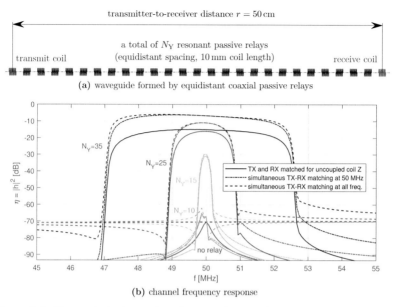

(b) channel frequency response

Figure 5.3: Magneto-inductive passive relaying in the coaxial waveguide arrangement presented in the original publication [64]. The placement is adapted to the number of relays N_Y; the coils are equidistantly spaced for any N_Y. The available space can hold $N_Y \leq 49$ relays. All mutual impedances are calculated by evaluating the double line integral (2.16) numerically.

hence the evaluation in Fig. 5.4 assumes adaptive power matching as in [190]. We assume such adaptive matching throughout the chapter in order to focus on the potential of passive relaying rather than matching. For the relay load capacitance C_Y we consider two different cases: (i) C_Y is set such that the relay is resonant at $f_{res} = 50$ MHz and (ii) C_Y is set to the value which maximizes $|h|^2$ (adaptively for any relay placement). This is realized by solving a one-dimensional optimization problem.

The evaluation studies the passive relaying gain, which is the effect of the relay on $|h|^2$ from (5.3). This gain is decomposed into the two key influences: the gain from the increased mutual impedance $|\tilde{Z}_{C:RT}|$ and the loss from the increased real part $\mathrm{Re}(\tilde{Z}_{C:R})$ of the encountered receive coil impedance in (5.3). The case of a constant resonant relay load hardly yields any gain because the two influences compensate each other. Note that this poor relaying gain holds for adaptive receiver matching; with static matching the passive relay would mostly have a detrimental effect (not shown for brevity). An

105

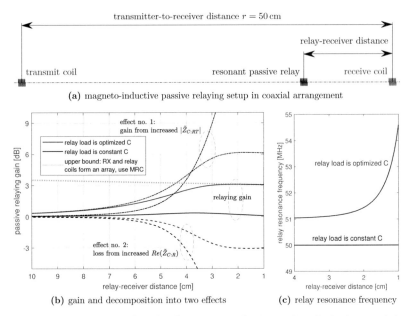

(a) magneto-inductive passive relaying setup in coaxial arrangement

(b) gain and decomposition into two effects

(c) relay resonance frequency

Figure 5.4: Relaying gain achieved with a magneto-inductive passive relay in close proximity to the receive coil, evaluated for a constant relay load capacitance (for resonance at the design frequency) as well as an optimized load capacitance to maximize the link power gain. The receiver uses adaptive power matching. All coils are single-layer solenoids with 1 cm diameter and 1 cm length.

optimized relay load capacitance however gives rise to a significant link gain which approaches 3 dB as well as the baseline of a two-coil receive array (associated with 3 dB array gain). This is because the relay and receiver coils together manage to draw twice the power from the magnetic field, and the strong relay-receiver coupling allows for near-lossless power transport from the relay to the receiver. Thereby the optimized capacitance prevents resonant mode splitting from destroying the link at f_{res} in the case of strong relay-receiver coupling. The optimized system behaves like a receive coil with doubled turn number, associated with $+6$ dB from doubled induced voltage and -3 dB from the doubled resistance due to the longer wire.

The above experiment with coaxial coil arrangement is not representative of sensor applications with mobile ad-hoc character, where the application dictates the node positions and orientations. Following the tone of this thesis, we shall now study the potential of a randomly placed relay with random orientation in a setting with arbitrary transmitter and receiver orientations. In particular, we evaluate the setup of Fig. 5.4a but now the relay is randomly placed near the receiver, as shown in Fig. 5.5, with a certain given relay-receiver distance. The transmitter-receiver distance is now 2 m.

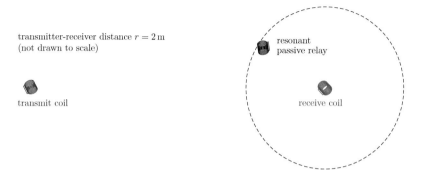

transmitter-receiver distance $r = 2\,\mathrm{m}$
(not drawn to scale)

resonant
passive relay

transmit coil

receive coil

Figure 5.5: Considered scenario with one passive relay ($N_Y = 1$) randomly placed on a sphere, which is centered around the receiver, with uniform distribution. The sphere radius specifies the relay-receiver distance. The transmitter-receiver distance is 2 m. All coils have random orientation whereby all directions in 3D are equiprobable.

The random arrangement renders $|h|^2$ a random variable; its statistics are shown as a function of the relay-receiver distance in Fig. 5.6. We observe that, remarkably, the presence of the passive relay with a constant resonant load hardly affects the statistics of $|h|^2$ even though it does affect the two major influences significantly. Again, optimization of the relay load capacitance is crucial and yields a relaying gain that is often significant and never negative (in any case a severely detuned load can be chosen, such that the relay does not affect the link). While the scheme is capable of improving an already decent link by 3 dB, its most important feat is the prevention of deep fading due to transmitter-receiver misalignment.[2] This is apparent in the scatter plots of h in Fig. 5.7: the presence of a load-optimized passive relay prevents that the channel fades close to zero. With the chosen parameters the passive relay is effective up to a relay-receiver distance of about 6 cm (six times its coil size).

We conclude that a single magneto-inductive passive relay has the potential for significant performance gains even in a setting with random node positions and ori-

---

[2]Note that the considered link distance $r = 2\,\mathrm{m}$ compares to a wavelength $\lambda = 6\,\mathrm{m}$ (and $kr = \frac{2\pi}{3}$) at the chosen 50 MHz, i.e. the receiver is located in the near-far-field transition and benefits from the polarization diversity effect described in Sec. 4.2. In the pure near field the misalignment losses (and the relay's ability to compensate them) would be much more drastic. Yet, the observed misalignment compensation abilities of passive relaying are significant even in the near-far-field transition.

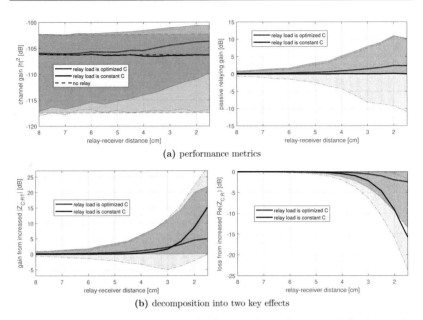

(a) performance metrics

(b) decomposition into two key effects

Figure 5.6: Link performance statistics for the case of a single magneto-inductive passive relay that is placed randomly near the receive coil, whereby all three involved coils have random orientation. In particular, the statistics of $|h|^2$, the passive relaying gain and its two key influences (same as in Fig. 5.4) are shown. The solid lines are the median value, the colored areas span from the 5th to the 95th percentile (i.e. they comprise 90% of realizations). A related evaluation was done in [26, Fig. 4.13].

entations and large transmitter-receiver separation, given that the relay is strongly coupled to either the transmitter or the receiver. In the following section we increase the number of randomly placed passive relays.

5.3 Random Relay Swarm Near the Receiver

For the same problem context as in the previous section, we want to study the potential of a random swarm of many passive relays around the receiver. The considered setup is depicted in Fig. 5.8.

In all following experiments we specify the (volumetric) relative relay density ρ_Y. For example, the value $\rho_Y = 0.03$ means that a random point in the ball hits one of the cylindrical volumes that encloses a passive relay with 3% chance. This density

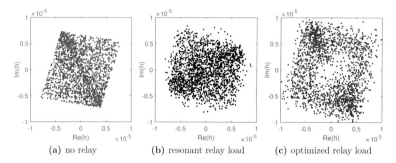

(a) no relay   (b) resonant relay load   (c) optimized relay load

Figure 5.7: Scatter plot of the random channel coefficient h resulting from random orientations of transmitter, receiver and passive relay as well as random relay placement on a sphere around the receiver. The transmitter-receiver distance is $2\,\mathrm{m}$, the relay-receiver distance is $2.5\,\mathrm{cm}$ (comparing to a coil size of $1\,\mathrm{cm}$). The no-relay case is almost equivalent to Fig. 4.5c (there $kr = 2$, here $kr = \frac{4\pi}{6} = 2.094$).

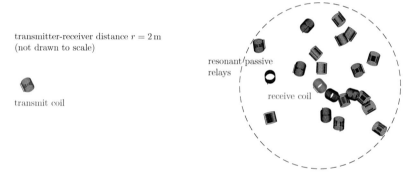

Figure 5.8: Henceforth considered scenario with $N_\mathrm{Y} = 20$ passive relays randomly placed inside a ball around the receiver (with uniform distribution but without collisions). The ball radius follows from the specified the relative relay density. The transmitter-receiver distance is $2\,\mathrm{m}$. All relays have random orientation whereby all directions in 3D are equiprobable.

together with $N_\mathrm{Y} = 20$ determines the radius of the ball (e.g., the radius is $16.3\,\mathrm{cm}$ for $\rho_\mathrm{Y} = 1\%_0$ (0.001) or $7.6\,\mathrm{cm}$ for $\rho_\mathrm{Y} = 10\%_0$). The passive relay positions are sampled from a uniform distribution (iid) in this ball, which is repeated until no collision between (the cylindrical hulls of) the coils occurs. All relay coil orientations are random with all possible directions being equiprobable.

To understand the effect of such a swarm of passive relays around the receiver we look at some example realizations of the channel frequency spectrum in Fig. 5.10. To ensure the comparability to the no-relays case (direct path) we fixed the orientations of the transmit and receive coils; we choose orientations such that the near-field alignment factor equals its RMS value $J_{NF} = 1/\sqrt{6}$ from Proposition 4.1. We observe that the presence of passive relays introduces spectral channel fluctuations, sometimes with a gain and sometimes with a loss. The affected bandwidth is significantly larger than the coil 3 dB-bandwidth, which is consistent with the theory on resonant node splitting that occurs for strongly coupled resonances [205]. Based on these observations it makes sense to characterize the random passive relay channel as a frequency-selective fading channel, consistent with the sum-of-noncoherent-phasors arguments in Sec. 5.1. The intensity of the fluctuation and the affected bandwidth increases with the relay density (somewhat comparable to the variance of Rician fading, which increases when the K-factor decreases [36]).

For the described random arrangement, Fig. 5.10 shows the channel statistics over frequency for different relay density. The implications are analogous to the preceding discussion: the presence of resonant passive relays may yield a significant gain but may also cause a deep fade at the 50 MHz design frequency. Clearly the probability for the noncoherent phasors in (5.7) to add up destructively is significant, but the probability of highly coherent phasors is very low for a random arrangement.

In the following we use available degrees of freedom to enforce constructive addition of coherent phasors: the relay load capacitances, which allow some control over the gain and phase shift induced by a passive relay. In particular we choose the load switching strategy illustrated in Fig. 5.11 because it allows for an effective optimization method and furthermore suggests a low-complexity circuit implementation of the adaptive loading concept. Such load adaptation at passive relays has been studied for sum rate maximization in interference radio channels [100] and for resolving ambiguities in near-field localization [26]. We assume that the passive relay is equipped with some logic for opening and closing switches after receiving corresponding commands. This requirement could be realized with effort in terms of hardware and protocol.

Out of the $(N_s + 1)^{N_Y}$ possible switching states of the passive relay swarm network we want to find the switching state that maximizes $|h|^2$ at the design frequency. This binary optimization problem is intractable for $N_Y = 20$, but genetic algorithms are an efficient means to find decent heuristic solutions [208]. We therefore employ a genetic algorithm for relay load switching: starting from random switching states, we simulate 100 generations whereby in each generation the 30 strongest individuals (i.e. the load switching states of largest $|h|^2$) survive. Per generation, every individual produces a mutated child (i.e. a similar switching state with a few switches flipped at

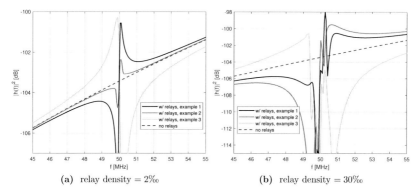

(a) relay density $= 2‰$ **(b)** relay density $= 30‰$

Figure 5.9: Example channel frequency spectra resulting from a random swarm realization around the receive coil, shown for different relay densities.

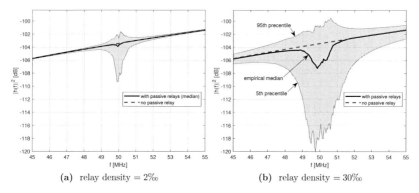

(a) relay density $= 2‰$ **(b)** relay density $= 30‰$

Figure 5.10: Statistics of the channel frequency spectrum resulting from a random swarm around the receive coil, shown for different relay densities.

random, which may or may not improve $|h|^2$) while 10 parent individuals give birth to 45 recombined individuals (two switching states combined with a XOR operation) to alleviate the problem of local maxima. We do not claim to attain global optimality with this approach. Fig. 5.12 shows an example frequency response of the passive relaying channel after load switching optimization with the described genetic algorithm, which establishes a strong peak at the design frequency. The implementation uses $N_s = 100$ whereby the ratio between $C_1, \ldots, C_{N_s}$ is the same as between the first hundred prime numbers. Furthermore C_0 is chosen for resonance at $\frac{1}{2\pi\sqrt{L_Y C_0}} = 55\,\text{MHz}$ (i.e. when

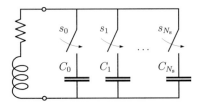

Figure 5.11: Circuit representation of a passive relay with an adaptive load capacitance, realized in the form of $N_{\mathrm{s}} + 1$ capacitances with open-circuit switches in parallel. The net capacitance is $C_{\mathrm{Y}} = \sum_{i=0}^{N_{\mathrm{s}}} s_i C_i$ and the resonance frequency of the relay $f_{\mathrm{res}} = \frac{1}{2\pi\sqrt{L_{\mathrm{Y}}C_{\mathrm{Y}}}}$. The binary switching states $s_i \in \{0,1\}$ allow for adaptation of the capacitive load. The capacitance C_0 establishes resonance near 50 MHz, at 55 MHz. If $s_0 = 0$ then the passive relay is essentially deactivated. The smaller capacitances $C_1, \ldots, C_{N_{\mathrm{s}}}$ allow for fine tuning of the resonance frequency around 50 MHz.

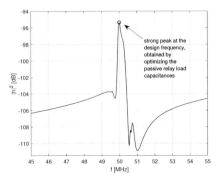

Figure 5.12: Example of a channel frequency response after maximizing the channel gain at the 50 MHz design frequency via load switching and a genetic algorithm. The considered scenario has a random swarm of passive relays around the receive coil. The relative relay density is high (30‰).

only switch 0 is closed) and furthermore $\frac{1}{2\pi\sqrt{L_{\mathrm{Y}}C_{\mathrm{max}}}} = 45$ MHz with $C_{\mathrm{max}} = \sum_{i=0}^{N_{\mathrm{s}}} C_i$ (all switches closed).

We consider two additional and particularly simple protocols for load switching. Firstly, a 1-relay scheme where only the one relay that yields the largest $|h|^2$ is activated (after an exhaustive search over all N_{Y} relays) with the idea of mitigating transmitter-receiver misalignment and possibly achieving a small relay gain beyond that. Secondly, an $N_{\mathrm{Y}} - 1$ relay scheme where only the one relay whose deactivation leads to the largest $|h|^2$ is open-circuited, while all other relays remain resonant at 50 MHz. The idea is to prevent destructive phasor addition by deactivating a detrimental relay and, this way, achieving a decent relay gain.

Another considered scheme is frequency tuning which simply operates at the frequency with the best channel realization $\max_f |h(f)|^2$; the resonant relay design remains unchanged. Besides the simplicity of this one-dimensional search, frequency tuning is attractive because of its low hardware complexity: it requires only tunable filters and voltage-controlled oscillators, but no adjustments to the relays.

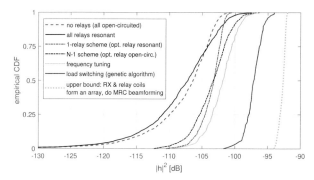

Figure 5.13: Channel power gain statistics for the scenario in Fig. 5.8 with a random swarm of passive relays around the receive coil. We compare various schemes for operating on this channel, foremost optimization of the passive relay loads to maximize the power gain. All coils have random orientation. The relative relay density is 30‰.

In Fig. 5.13 we compare the discussed scheme for the scenario in Fig. 5.8. Now all involved coils have random orientation (transmitter, receiver and all relays) and we consider a high relative relay density of 30‰. The observed distribution for the no-relays case (with deep fading due to misalignment) was described in detail in Cpt. 4. The addition of a passive relay swarm with static resonant loads affects the channel as described earlier: they sometimes yield a gain and sometimes a loss. The simple 1-relay and $N_Y - 1$ relay schemes already bring a significant performance improvement, particularly by preventing deep fading. They are however outperformed by the simple frequency tuning scheme which reliably find reasonably constructive conditions in the frequency domain. The variance of $|h|^2$ is smaller with the 1-relay scheme, consistent with the observation that the intensity of channel fluctuations increases with the relay density. Adaptive load switching, controlled by the described genetic algorithm, yields by far the best performance of the different schemes, with a 4.8 dB gain over frequency tuning in the median and high reliability (i.e. little residual variance). The comparison to frequency tuning shows that load switching achieves much more than just shifting resonance peaks back to the design frequency. The scheme is only 4.5 dB (median) below the highly optimistic baseline which relates to a hypothetical coil array consisting of the receive coil and all 20 relay coils, performing coherent receive beamforming. This shows that the presence of a passive relay swarm with optimized loads allows the receiver to draw significantly more power from the magnetic field (the effect of the passive relays is comparable to a parabolic reflector near a radio antenna).

5.4   Utilizing Spectral Fluctuations

In this final section we conduct a communication-theoretic study of the spectral fluctuations induced by magneto-inductive passive relays with static resonant loads. We are particularly interested in the utilization for high data-rate communication. With the plots in Fig. 5.9 and 5.10 we already demonstrated that randomly deployed magneto-inductive passive relays give rise to a frequency-selective fading channel when strongly coupled to one of the communicating nodes. In order to study this phenomenon systematically we consider the correlation coefficient between the channel at the resonant design frequency $h(f_{\text{res}})$ and the channel $h(f_{\text{res}} + \Delta)$ for a frequency shift Δ, given by

$$\rho(\Delta) = \frac{\mathbb{E}[AB^*]}{\sqrt{\mathbb{E}[|A|^2] \cdot \mathbb{E}[|B|^2]}}, \qquad A = h(f_{\text{res}}) - \mathbb{E}[h(f_{\text{res}})], \qquad (5.15)$$

$$B = h(f_{\text{res}} + \Delta) - \mathbb{E}[h(f_{\text{res}} + \Delta)].$$

Naturally, the correlation coefficient fulfills $|\rho(\Delta)| \leq 1$. We define the channel *coherence bandwidth* [90, Fig. 2.13] as one half of the main lobe width of $\rho(\Delta)$, measured between two points[3] with a decorrelation of $\rho(\Delta) = \frac{1}{\sqrt{2}}$.

In the discussion of Fig. 5.9 and 5.10 we already noted that the channel bandwidth which is significantly affected by passive relays increases with the relay density. We want to capture this effect quantitatively. For that purpose we consider for the channel coefficient the relative RMS deviation

$$\sigma_{\text{rel}}(f) = \frac{\sqrt{\mathbb{E}[|h(f) - h_0(f)|^2]}}{|h_0(f)|} \qquad (5.16)$$

compared to the channel response $h_0(f)$ when no relays are present. We fix the transmitter and receiver orientation the same way as for Fig. 5.9 and 5.10 so that $h_0(f)$

---

[3]At low relay density, side lobes which also reach $\rho(\Delta) = \frac{1}{\sqrt{2}}$ may occur; these are disregarded by our definition of coherence bandwidth.

becomes deterministic. $\sigma_{\text{rel}}(f)$ is the relative RMS deviation of h from h_0 at a frequency of interest f. We define the *affected bandwidth* as the main lobe width of $\rho(\Delta)$, measured between two points which fulfill $\sigma_{\text{rel}}(f) = \frac{1}{10}$.

Fig. 5.14a shows the evolution of $\rho(\Delta)$ over frequency shift Δ for high relay density. We observe that the channel decorrelates significantly even for small Δ, i.e. the coherence bandwidth is small. It decays with increasing relay density and attains a value similar to the coil 3 dB bandwidth. However $\rho(\Delta)$ never decays to zero (no full decorrelation occurs) which we explain as follows. Each relay swarm realization establishes a general trend across most affected frequencies, i.e. by virtue of the realization the channel is either rather weak or rather strong across most frequencies. This explanation is supported by Fig. 5.9 to some extend. The evolution of $\sigma_{\text{rel}}(f)$ in Fig. 5.14b confirms the earlier discussion: the intensity of the channel fluctuations and the affected bandwidth increases with the relay density.

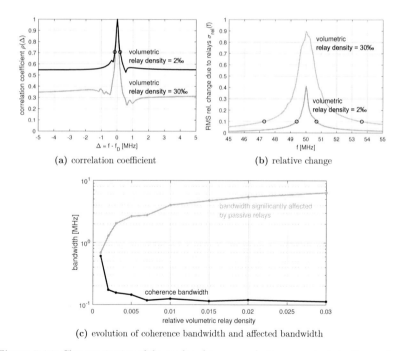

(a) correlation coefficient   (b) relative change

(c) evolution of coherence bandwidth and affected bandwidth

Figure 5.14: Characterization of the random frequency-selective magneto-inductive passive relaying channel that arises from a random swarm of passive relays around the relay coil.

We note that the affected bandwidth is much larger than the coherence bandwidth at high relay density. We conclude the availability of multiple frequency bands whose fading statistics are decorrelated to some extend (note that the channel spectrum is determined by 20 statistically independent relay deployments). This can be utilized by the transmission scheme to achieve frequency diversity over this SISO channel.

In other words, when the channel is faded at the design frequency there is still a good chance to find a good channel at a different frequency: the peak(s) may have shifted due to resonant mode splitting. This can be utilized by adapting the transmit power allocation to the channel spectrum (ideally with the waterfilling principle) when channel state information is available at the transmitter. We investigate the resulting performance in Fig. 5.15 in terms of the achievable data rates with a transmit power of $P_{\mathrm{T}} = 10\,\mu\mathrm{W}$. When the transmit power allocation is adapted to the channel via the waterfilling principle then the presence of passive relays, even without any optimization measures, gives rise to significant and reliable data rate improvements. This notion of improving the communication performance over a fading channel with transmit-side channel state information is well-known in the radio communication context (concerning fluctuations in time, frequency, and space) [90, Sec. 5.4.6]. It is particularly effective at low SNR, where using just the most constructively faded band proves very beneficial. The results in Fig. 5.15 also include the achievable rate with waterfilling after the adaptive load switching scheme (controlled by the genetic algorithm described earlier) has been used to maximize $|h(f_{\mathrm{res}})|^2$. This yields further significant improvements, consistent with the comparison of frequency tuning and load switching in Fig. 5.13.

We showed that passive relaying gives rise to significant and reliable improvements of the communications performance if (i) the transmit signaling is adapted to the channel spectrum realization which results from the specific coil arrangement or (ii) the relay loads are adapted in order to optimize the channel at the operating frequency. In any case the matching of the affected end (e.g. the receiver matching) must be adapted to the coupling conditions at hand. Future work should investigate possible forms of coordinating load adaptation (e.g., load switching) with little to no intelligence and communication capabilities at the passive relays. A possible approach could be a greedy scheme where each passive relay adapts its load capacitance in an attempt to maximize its own induced current.

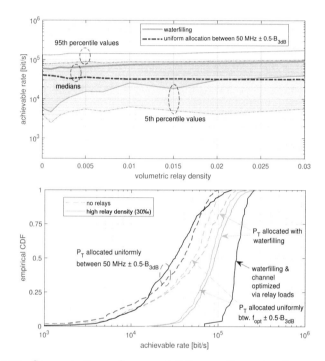

Figure 5.15: Communication performance statistics over the random frequency-selective magneto-inductive passive relaying channel that arises from a random swarm of passive relays around the relay coil. We consider different signaling schemes and different channel states (no relays present, resonant relays present, relays with optimized load capacitances present). The assumed transmit power is $P_T = 10\,\mu\text{W}$. The frequency tuning scheme allocates the transmit power in a band around $f_{\text{opt}} = \arg\max |h(f)|^2$ (bandwidth equal coil 3 dB bandwidth), determined for each realization individually.

117

Chapter 6

A Study of Magnetic Induction for Small-Scale Medical Sensors

With the use of the introduced concepts, this chapter studies the highly relevant technological context of enabling wireless powering and communication for medical in-body microsensors (microrobots). In Cpt. 1 we already discussed that this significant goal of biomedical research is expected to provide untethered diagnostic sensing and treatment in future medical applications. A big problem of the approach is that, on the one hand, the sensors must be sufficiently small for minimally-invasive maneuvering in cavities of the human body but, on the other hand, a very small sensor device can not be equipped with a useful battery. Wireless powering as a potential alternative however also does not allow for an arbitrarily small device size because efficient wireless power reception requires a rather large antenna aperture. This holds especially true in the biomedical context where tissue absorption causes significant signal attenuation.

In the face of these problems, a natural question is the minimum sensor size that still allows for running the device with wireless powering. Related questions concern the achievable data rate of wireless data transmission from the sensor to an external device (uplink) and whether this transmission should be realized in an active or passive fashion, i.e. whether a transmit amplifier or load modulation via circuit switching should be employed at the sensor. In this chapter we consider magnetic induction in this context, because of its capabilities in terms of power transfer and media penetration (see Cpt. 1). Specifically, this chapter makes the following contributions:

- We develop the biomedical sensor problem context from a wireless engineering perspective and discuss key design considerations and trade-offs.

- We evaluate the data uplink from a wireless-powered in-body sensor using active transmission and compare the results to load modulation. We do so for different important cases of the transmission scheme and assumptions on the channel knowledge (full knowledge, sensor location knowledge, no knowledge).

- We show that placing several small-scale passive relay coils near the sensor node yields significant performance gains if (and only if) the relay loads are optimized.

- We find that load modulation is a very promising data transmission scheme for small in-body sensors, however the performance and scaling behavior depends critically on the measurement accuracy of the RFID-reader-type receiver device (foremost on its ability to suppress self-interference).

- We show that sensor cooperation allows for a significant improvement of the data uplink, either with active transmission or load modulation.

6.1 Biomedical Setup and Link Design

In the following we study the performance and feasibility limits of magneto-inductive wireless powering (downlink) and data transmission (uplink) for micro-scale devices in the biomedical in-body problem setup illustrated in Fig. 6.1. This setup is relevant to contemporary applications such as gastrointestinal endoscopy and presumably to future applications of medical microrobots. The setup comprises several micro-scale in-body

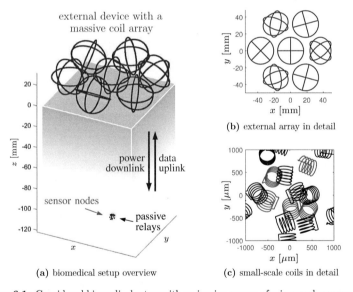

Figure 6.1: Considered biomedical setup with an in-vivo swarm of micro-scale sensor nodes, each equipped with a multi-turn coil, here shown 12 cm beneath the skin. The sensors receive power from and send data to an external device that is equipped with an array of 21 coils. The sensors are accompanied by resonant passive relay coils for potential performance gains.

sensors beneath the skin. An external device, which sits above the skin, serves as the transmitting field source for wireless powering (downlink) as well as the receiving data sink of sensor data (uplink). We assume a downlink transmit power of $P_{\mathrm{T,DL}} = 1\,\mathrm{W}$ for wireless powering, which is deemed sufficiently small to prevent any thermal injuries to the patient. The device features a massive array of 21 coils in order to obtain vast spatial diversity, an array gain, and potentially even a spatial multiplexing gain (see Sec. 3.7.5, Sec. 3.8.2, and Sec. 4.3).

We will later also consider the sensor coil(s) being accompanied by resonant passive relay coils of equal size, as indicated in the illustration. These passive relay coils could be inactive sensor nodes engaging in passive cooperation. We are interested in the passive relaying gains (e.g., as seen in Cpt. 5) available in this biomedical context.

6.1.1   Choosing the Operating Frequency

The operating frequency (or rather: the design frequency of the matching circuits) f_{c} should be as large as possible for strong magnetic induction (cf. the role of ω in (2.23)) but sufficiently low to penetrate tissue and, furthermore, to avoid a performance limitation due to the radiation resistance. In other words, the wavelength should exceed the coil wire length considerably. We note that a wavelength with similar order of magnitude as the link distance would provide the polarization diversity effect of the near-far-field transition described in Cpt. 4. Based on these arguments we consider $f_{\mathrm{c}} = 300\,\mathrm{MHz}$ as an interesting design frequency (associated with $1\,\mathrm{m}$ wavelength).

6.1.2   External Coil Design

After a careful study based on the theory in Sec. 2.2, we choose the following external-side coil geometry in order to achieve a large Q-factor and thus a good link performance at the chosen $300\,\mathrm{MHz}$. We choose circular single-turn coils with $10\,\mathrm{cm}$ circumference, which is about the maximum size where they are still electrically small ($\frac{1}{10}$ of the design wavelength). We choose a large wire diameter of $3\,\mathrm{mm}$ to establish a small ohmic resistance. This coil geometry exhibits $R_{\mathrm{ohm}} \approx 45\,\mathrm{m\Omega}$ and $R_{\mathrm{rad}} \approx 19\,\mathrm{m\Omega}$ as well as a realizable Q-factor of ≈ 1380 at the $300\,\mathrm{MHz}$ design frequency.

6.1.3   Tissue Attenuation and Sensor Depth

We shall now investigate the choice of operating frequency in more detail and under due consideration of the desired sensor depth and the electrical properties of tissue.

We employ the theory from Sec. 2.1.6 to account for tissue attenuation: the penetrated material is modeled in terms of its complex propagation constant

$$\gamma = j\omega\sqrt{\mu\epsilon\left(1 - \frac{j\sigma}{\omega\epsilon}\right)} \tag{6.1}$$

as stated earlier in (2.34). We assume that the penetrated material consist entirely of muscle tissue and evaluate γ as a function of frequency with the use of numerical values for the permittivity ϵ and conductivity σ given by the so-called Debye model for muscle tissue in [209, Fig. 3]. In the process we assume a relative permeability of 1 which is adequate for most organisms (humans, animals, plants) [54]. The effect of tissue on channel coefficients is calculated as follows. Any link distance r of an external-to-internal link is decomposed into a distance r_{free} traveled in free space and a distance r_{tissue} traveled in tissue, i.e. $r = r_{\text{free}} + r_{\text{tissue}}$. We then replace the factor e^{-jkr} that would apply to the free-space channel coefficient with $e^{-jkr_{\text{free}}}e^{-\gamma r_{\text{tissue}}}$, whereby k is the wavenumber. From Sec. 2.1.6 we recall the depth of penetration $1/\text{Re}(\gamma)$, the distance into tissue where an amplitude attenuation of $e^{-1} = -8.7\,\text{dB}$ applies [52].

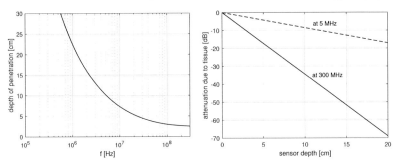

Figure 6.2: Depth of penetration $1/\text{Re}(\gamma)$ and attenuation $20\log_{10}e^{-\text{Re}(\gamma)\cdot r_{\text{tissue}}}$ in muscle tissue, based on the complex propagation constant γ from (6.1). The calculation uses permittivity and conductivity values over frequency given by the Debye model in [209, Fig. 3].

The numerical evaluation in Fig. 6.2 of this tissue attenuation model shows the decrease of penetration depth with increasing frequency. We observe that a sensor depth of 12 cm is associated with severe attenuation if the operating frequency exceeds about 4 MHz. Smaller sensor depths such as 5 cm allow the use of a much larger frequency (and hence stronger induction and more communication bandwidth) with little attenuation. The implications are analogous to those in [209, Fig. 6], where they suggest a frequency slightly above 1 GHz for sensor depths of a few centimeters. For more detailed studies on magnetic fields in tissue we refer to [209–211].

6.1.4 Sensor Coil Design

As a next step towards a detailed study we need to define the geometry of the sensor coils. First we note that micro-scale coils are certainly electrically small at any meaningful operating frequency. We consider the following size constraint on the coil geometry: the coil must fit into a cube of a given size. The sensor coil design objective is to achieve a large Q-factor within this limited volume. This is achieved by choosing a large turn number, coil diameter, and wire diameter without violating the constraint. Using the formulas in Sec. 2.2 we find a suitable design in the form of a single-layer solenoid coil whose height equals its diameter (henceforth called *size*, which is equal to the edge length of the aforementioned cube) as seen in Fig. 6.1c. After careful comparisons we decide to use 5 turns and a turn spacing of 1.5 wire diameters. For the wire material we assume the conductivity of copper.

We note that, in order to obtain a decent Q-factor, the coil geometry must be spread out such that the size constraints are exhausted in all three dimensions. For example, a flat printed coil is a bad design because its small wire cross section would result in large ohmic resistance and thus a small Q-factor. Because of this volume-limitation problem we also decide to use a single coil per sensor device instead of an array of even smaller coils (thereby we sacrifice potential gains from further spatial diversity and spatial multiplexing but keep the complexity small).

Finally we have to specify the sensor-side matching network. Each sensor coil is matched individually with a two-port matching network of two capacitors in L-structure (as in Fig. 3.6b). Thereby we use power matching via the 2×2 version of rule (3.39). This matching network applies to both the down- and uplink. The matching bandwidth is large because the coil is small and the Q-factor decreases with the coil size (e.g., $Q \approx 63.7$ for 0.5 mm size, $Q \approx 23.5$ for 0.2 mm size, and $Q \approx 10$ for 0.1 mm size).

6.1.5 Tissue Attenuation and Sensor Depth Revisited

Now that we have decided on specific coil designs, we shall proceed with a more detailed study of tissue attenuation at the example of the wireless powering downlink. We consider a setup similar to Fig. 6.1 but much simpler, with just a SISO link between a single external coil (one of the central tri-axial cluster) as transmitter and just a single sensor as receiver. We assume that the two coils are in coplanar arrangement. Furthermore, we do not consider any passive relays. Fig. 6.3 shows an evaluation of the power transfer efficiency over frequency for two different cases in terms of sensor size and depth. The results show that the operating frequency must be chosen under

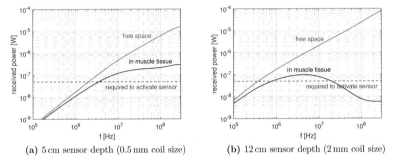

(a) 5 cm sensor depth (0.5 mm coil size)   (b) 12 cm sensor depth (2 mm coil size)

Figure 6.3: A simple evaluation of SISO power transfer efficiency versus frequency. The transmitter is a single external coil (of the central cluster), the receiver a single sensor coil. The two coils are in coplanar arrangement.

due consideration of the desired sensor depth. This is made apparent by the significant differences between the black graphs of the two cases. For the low-frequency regime, where the depth of penetration exceeds the sensor depth, we note that increasing the frequency is always beneficial. At larger frequencies, where tissue causes severe signal attenuation, we observe that a frequency increase can yield limited gains or significant losses, depending on the sensor depth. In all subsequent experiments we will consider the smaller sensor depth of 5 cm and stick with the 300 MHz operating frequency.

The red line in each plot indicates the assumed minimum downlink received power $P_0 = 50$ nW that is required to activate the sensor chip (for comparison, the 2009 paper [212] describes a biomedical sensor implementation that required 450 nW).

6.2   Wireless Powering Downlink

This section studies the performance of the MISO downlink from the 21-coil array to a single sensor, with transmit beamforming used at the array. Before that we have to clarify a few last design aspects though.

6.2.1 Array Transmit Matching and Spacing

First, we specify the placement of the coils that constitute the *external array* which consists of seven tri-axial coil clusters. Their orientations are chosen in a heuristic fashion with the idea of limiting the inter-array coupling. For all subsequent experiments we choose a hull-to-hull (not center-to-center) spacing of $3\,\mathrm{cm}$ between neighboring tri-axial coil clusters. This choice will be dictated by the uplink reception properties of the array and argued in detail later in Sec. 6.3.1.

We assume that the transmit amplifiers of the high-complexity external device are capable of driving any desired currents through the coil array. Our primary interest are thereby the coil currents, which determine the power of the generated electromagnetic field and the resulting exposure for the patient. Potential intrinsic power losses, e.g. due to mismatch or the use of class A amplifiers with high linearity but poor power efficiency, are of secondary nature. To this effect, we employ the assumption of *ideal transmit-side power matching* from (3.47) for the powering downlink because it is the simplest formal way of capturing just the effects of interest.

6.2.2 Downlink Performance

The performance of transmit beamforming depends critically on the employed scheme and the availability of channel information, i.e. whether the channel vector is known to the external device. We study three different cases:

- **Maximum-Ratio Combining:** The transmit signal vector is set $\mathbf{x} = \mathbf{h}^* \sqrt{P_{\mathrm{T,DL}}}$ based on full knowledge of the channel vector $\mathbf{h}$. This way we attain the maximum received-power value $P_{\mathrm{R,DL}} = |\mathbf{h}^{\mathrm{T}}\mathbf{x}|^2 = \|\mathbf{h}\|^2 P_{\mathrm{T,DL}}$ (cf. Sec. 3.4).

- **Using Sensor Location Knowledge:** Since the receiving sensor coil is a single-layer solenoid with small pitch angle, we can use the channel vector description $\mathbf{h} = \mathbf{V}^{\mathrm{H}}\mathbf{o}_{\mathrm{R}}$ in terms of projections of field vectors generated by the external coils onto the receive-coil orientation $\mathbf{o}_{\mathrm{R}}$ (cf. Sec. 3.7.3 and Proposition 4.3). We assume that the sensor location is known and that, consequently, the field vectors in $\mathbf{V}$ can be calculated.[1] The sensor coil orientation $\mathbf{o}_{\mathrm{R}}$ is however assumed unknown and random with a uniform distribution in 3D. For this case, Proposition 4.8 states that the channel vector covariance matrix is $\mathbb{E}[\mathbf{h}\mathbf{h}^{\mathrm{H}}] = \frac{1}{3}\mathbf{V}^{\mathrm{H}}\mathbf{V}$. We utilize this property for beamforming: the transmit vector is set to $\mathbf{x} = \mathbf{q}\sqrt{P_{\mathrm{T,DL}}}$ where $\mathbf{q}$ is a unit-length eigenvector of $\mathbf{V}^{\mathrm{H}}\mathbf{V}$ associated with its largest eigenvalue. This way we maximize $\mathbb{E}_{\mathbf{o}_{\mathrm{R}}}[P_{\mathrm{R,DL}}] = \mathbb{E}_{\mathbf{o}_{\mathrm{R}}}[|\mathbf{h}^{\mathrm{T}}\mathbf{x}|^2]$ and attain the value

$\mathbb{E}_{\mathbf{o}_R}[P_{R,DL}] = \frac{1}{3} \cdot \lambda_{\max}(\mathbf{V}^H\mathbf{V}) \cdot P_{T,DL}$, although without robustness guarantees regarding the actual realizations of $P_{R,DL}$ (i.e. fading may occur).

- **Random Beamforming:** When no channel information is available, we resort to the use of a random transmit vector $\mathbf{x} = \frac{\mathbf{u}}{\|\mathbf{u}\|}\sqrt{P_{T,DL}}$ where $\mathbf{u}$ is sampled from a complex-valued Gaussian distribution $\mathbf{u} \sim \mathcal{CN}(\mathbf{0}, \mathbf{I}_{N_T})$ with $N_T = 21$. Fading will occur whenever the chosen $\mathbf{x}$ is near-orthogonal to the channel vector.

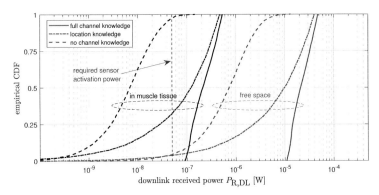

Figure 6.4: Statistics of the received power at a sensor device with random orientation via the MISO downlink, evaluated for different transmit beamforming schemes. The sensor depth is 5 cm, the sensor coil size is 0.5 mm, and the transmitting external 21-coil array uses a sum power of $P_{T,DL} = 1\,\mathrm{W}$. This is a single-frequency evaluation at 300 MHz.

The received power statistics for a randomly oriented sensor are shown in Fig. 6.4. As expected, the availability of channel knowledge and resulting choice of beamforming scheme has a huge effect on the mean performance and robustness. To interpret the results we shall first note that the sensor depth (or, put differently, the operating frequency) is too small to really benefit from the polarization diversity effect in the near-far-field transition.[2] The misalignment losses are thus essentially characterized by the

---

[1]We note that this location-knowledge-based transmit beamforming scheme, which employs Proposition 4.8, is based on the simplified propagation model described in Sec. 6.1.3. The field vector direction may be different (and hard to predict for a given position) in an actual biomedical application, where propagation is affected by various types and shapes of tissue and medium transitions.

[2]At 300 MHz, $kr \approx 0.414$ applies between the external coils of the central triaxial cluster and a 5 cm deep sensor coil. Through Fig. 4.2 we find that, in this regime, polarization diversity only affects misalignment losses that are already worse than $-30\,\mathrm{dB}$. The polarization diversity effect would be most pronounced at a distance of 0.3747 m, which corresponds to the near-far threshold value $kr = 2.3540$ from (2.33).

pure near-field regime. This is consistent with the observed performance of maximum-ratio combining, which varies in a 6 dB window, as predicted by $\bar{J}_{\mathrm{NF}} = \beta_{\mathrm{NF}} \in [\frac{1}{2}, 1]$ in Sec. 4.3.1 (this is just the field magnitude loss, cf. (4.9)). Thereby, devastating orthogonality between field vector and coil axis is prevented.

This mechanism is unavailable to the location-knowledge scheme, which, as a result, frequently results in severe losses. In particular, the location-knowledge scheme just maximizes the field magnitude at the sensor location without regarding the sensor orientation. This provision establishes a large median received power, which makes it an appealing for biomedical applications where the approximate sensor location will often be known. However, the CDF tail for very small values shows that the outage behavior of the location-knowledge scheme is even worse than that of the random beamforming scheme. This is an effect of the dominance of near-field propagation in this setting, in particular because the eigenvector $\mathbf{q}$ associated with $\lambda_{\max}(\mathbf{V}^{\mathrm{H}}\mathbf{V})$, which determines the transmit vector $\mathbf{x} = \mathbf{q}\sqrt{P_{\mathrm{T,DL}}}$, exhibits near-collinearity of $\mathrm{Re}(\mathbf{q})$ and $\mathrm{Im}(\mathbf{q})$ (and because the channel vector $\mathbf{h}$ has the same behavior).[3] In contrast, the random beamforming scheme $\mathbf{x} = \frac{\mathbf{u}}{\|\mathbf{u}\|}\sqrt{P_{\mathrm{T,DL}}}$ with $\mathbf{u} \sim \mathcal{CN}(\mathbf{0}, \mathbf{I}_{21})$ typically exhibits near-orthogonal $\mathrm{Re}(\mathbf{x})$ and $\mathrm{Im}(\mathbf{x})$, resulting in slightly better outage behavior.

6.3 Data Uplink

While supplying power to a sensor is an interesting challenge in its own right, we shall now shift our focus to the data transmission capabilities of micro-scale in-body sensors. We first consider the data uplink via *active transmission*, i.e. the sensor device drives its coil with its own transmit amplifier. The uplink transmit power is assumed to be

$$P_{\mathrm{T,UL}} = \max\left\{0, \ \tfrac{1}{2}(P_{\mathrm{R,DL}} - P_0)\right\} \tag{6.2}$$

or in words, $P_{\mathrm{T,UL}}$ is half of the received downlink power in excess of the assumed chip activation power $P_0 = 50\,\mathrm{nW}$. For simplicity we make the idealistic assumption that the power downlink and data uplink operate simultaneously in a full-duplex fashion. An investigation of appropriate practical duplexing schemes is out of scope.

---

[3]This suggests a possible adaptation to the location-knowledge beamforming scheme: set the transmit signal according to $\mathbf{x} = \frac{1}{\sqrt{2}}(\mathbf{q}_1 + j\mathbf{q}_2)\sqrt{P_{\mathrm{T,DL}}}$ where $\mathbf{q}_1$ and $\mathbf{q}_2$ are eigenvectors of $\mathbf{V}^{\mathrm{H}}\mathbf{V}$ corresponding to the largest and second-largest eigenvalue, respectively. This adaptation improves the outage behavior but sacrifices the discussed optimality property regarding the maximization of $\mathbb{E}_{\mathbf{o}_{\mathrm{R}}}[P_{\mathrm{R,DL}}]$.

6.3.1 Array Receive Matching and Spacing Revisited

The assumptions on the receive matching network of the external coil array require careful consideration in order to obtain practically relevant results for the uplink performance. A reactive noise matching network fulfilling (3.43) would be SNR-optimal at the design frequency and could theoretically be realized with the approach in Fig. 3.7. However, such a network between the $2 \cdot 21 = 42$ ports at hand would exhibit a very small matching bandwidth and can not realistically be implemented (see the discussion in Sec. 3.6). Hence we discard this approach. Instead we choose to match the array with *an individual two-port network per coil-load pair*, each consisting of three reactive lumped elements in T-pad structure (i.e. we use a total of 21 two-port networks; the concept is described by Fig. 3.5 together with Fig. 3.6a).

Tuning the component value of these two-port networks is crucial and challenging if the spacing between the seven tri-axial coil clusters is small, because then strong inter-array coupling occurs between these high-Q coils (we demonstrated the associated problems in Fig. 3.7). Choosing a small spacing is however incentivized by the desire of maintaining short distances between outer clusters and sensors. We shall demonstrate and study this trade-off for the data uplink for different array matching paradigms:

1. Full 21-coil array with 21 individual T-pad two-port networks for noise matching each coil individually for its uncoupled impedance.

2. Same as above but the two-port networks are iteratively adapted to the inter-array coupling conditions in a round robin fashion and according to the two-port noise matching rule, using the encountered coil impedance.

3. Same as above but numerically optimized for maximum average SNR at the design frequency, averaged over the random sensor orientation and optimized with a single run of an iterative gradient-search algorithm.

4. Baseline: ideal multiport noise-matching network (implementation unrealistic).

5. Baseline: using just the 3-coil array constituted by the central tri-axial sub-array.

We observe that using more than one tri-axial cluster is beneficial (owing to a receive array gain) if and only if an appropriate matching strategy is employed. Approaching the performance of the theoretically ideal multiport strategy with a practical strategy requires significant optimization effort; simplistic practical matching strategies are clearly associated with significant losses.

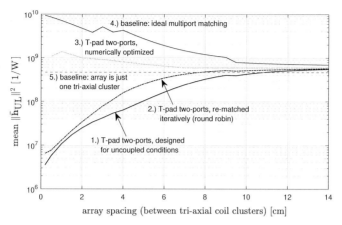

array spacing (between tri-axial coil clusters) [cm]

Figure 6.5: Data uplink performance versus external array spacing. The plotted performance measure is the squared norm $\|\bar{\mathbf{h}}_{UL}\|^2 = \mathbf{h}_{UL}^H \mathbf{K}^{-1} \mathbf{h}_{UL}$ of the noise-whitened uplink channel vector $\bar{\mathbf{h}}_{UL}$. In particular we evaluate its mean value; the randomness stems from the assumed random orientation of the sensor coil. The considered noise bandwidth is 100 kHz. The evaluation considers muscle tissue, 300 MHz carrier frequency, 5 cm sensor depth, and 0.5 mm sensor size.

Based on the observed behavior we decide on a 3 cm array spacing between tri-axial coil clusters. We use matching strategy number 3, i.e. individual T-pad two-port networks whose component values are tuned with an iterative gradient search that maximizes the mean $\|\bar{\mathbf{h}}_{UL}\|^2$, without claiming global optimality. In particular, we aim for a larger matching bandwidth by maximizing the sum of mean $\|\bar{\mathbf{h}}_{UL}\|^2$ values in dB at the frequency points 296, 298, 300, 302, 304 MHz (the 300 MHz summand has double weight). This shall allow for large uplink data rates despite the small 3-dB bandwidth of the external coils ($\approx$ 215 kHz). The outcome is seen in Fig. 6.6c.

6.3.2 Achievable Rates with Active Transmission

We will now study the system from a broadband perspective, i.e. we carefully consider and utilize spectral channel fluctuations. The frequency spectra of various key quantities are depicted in Fig. 6.6. We note that Fig. 6.6a yields very useful insights for wireless powering downlink: the PTE spectrum exhibits a significant resonance peak at 304 MHz, a consequence of the strong inter-array coupling.[4] We choose to use these 304 MHz instead of 300 MHz for the downlink and this way obtain a spectacular

129

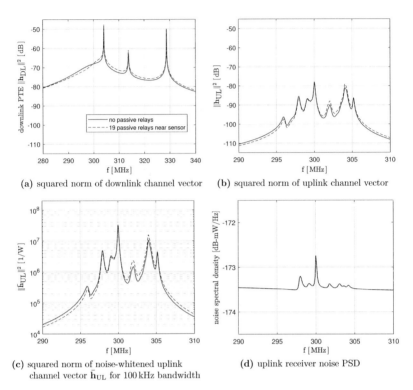

(a) squared norm of downlink channel vector

(b) squared norm of uplink channel vector

(c) squared norm of noise-whitened uplink channel vector $\bar{\mathbf{h}}_{\mathrm{UL}}$ for 100 kHz bandwidth

(d) uplink receiver noise PSD

Figure 6.6: Frequency spectra of the key quantities that determine uplink communication performance. The evaluation considers a sensor of 0.5 mm size which is located 5 cm deep in muscle tissue. The transmitting sensor uses an L-pad two-port matching network designed at 300 MHz, the receiving external array is matched as described in Sec. 6.3.1. The passive relays near the sensors (which are of equal size) only have a minor effect because their Q-factor is small due to the small coil size. The shown noise PSD is of the horizontal coil of the central cluster.

19.5 dB gain for wireless powering.

---

[4]Future work should study the precise conditions for such resonance peaks to arise and the feasibility of realizing them in practice. At this point we can only state that they are caused by strong inter-array coupling. One potential issue of the evaluation at hand is that the array as a whole, whose outer circumference of $\approx$ 48.9 cm is about one half of the 1 m wavelength, may not qualify as an electrically small circuit (although its constituent coils are indeed electrically small) and thus may not be eligible for AC circuit analysis. Also, capacitive coupling and feed wires might have a significant effect. These aspects should be clarified by comparing to full-wave simulation and/or experiments.

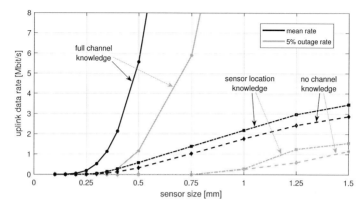

Figure 6.7: Achievable rate of the data uplink whereby the transmit power is determined by the downlink received power. The wireless-powered sensor is 5 cm deep into muscle tissue, the required sensor activation power is 50 nW. The external array is the receiving end.

Given the available uplink transmit power $P_{\mathrm{T,UL}}$ we are interested in the achievable uplink data rate (in the information-theoretic sense as in Sec. 3.5) for reception in additive Gaussian noise[5] according to the model in Sec. 3.2. We consider the same three cases of channel knowledge as earlier in Fig. 6.4. We conduct receive beamforming at the external array, which formally transforms the broadband SIMO channel into a broadband SISO channel. For the case of full channel knowledge we assume receive-side maximum-ratio combining, for the other cases we assume selection combining.

Fig. 6.7 shows the evolution of achievable uplink data rate over sensor coil size. The data rate is random due to the random sensor orientation; the plot shows the mean value as well as the 5th-percentile value. With full transmit-side channel knowledge, we achieve channel capacity by using waterfilling as in (3.33) for spectral power allocation, based on the norm of the noise-whitened channel vector. In the other cases the power is allocated uniformly over the 3-dB bandwidth of the external array.

We observe that a sensor can be activated and transmit data to outside the body if its coil size is larger than about 0.35 mm. Such size would be sufficiently small for many medical target applications [23]. With increasing size, the data rate grows rapidly as the sensor coil Q-factor and the mutual impedances to the external coils increase. We

---

[5]This experiment (and all subsequent active-uplink considerations) use the noise parameters in Table 5.1 of Cpt. 5. To describe the spatial correlation matrix $\mathbf{\Phi}$ of extrinsic noise, which occurs at the external array and impairs reception in the uplink, we account for spatial correlation with the model $\Phi_{mn} = J_0(kr_{mn})\, \mathbf{o}_m^{\mathrm{T}} \mathbf{o}_n$, which uses the center-to-center coil distance r_{mn}, the orientation vectors $\mathbf{o}_m$ and $\mathbf{o}_n$ of the involved coils, and the Bessel function J_0 (cf. [103, 213]).

observe that the required minimum sensor size depends heavily on the availability of channel information and the desired outage rate.

6.3.3   Small-Scale Swarms of Passive Relays

We are interested in potential performance gains from *passive relays*, whose capabilities have been demonstrated in Cpt. 5, in the biomedical context. We consider a randomly arranged swarm of 19 passive resonant relay coils around the sensor node (and with the same coil geometry, now with an assumed 0.35 mm size). They could be placed in hopes of a performance gain or just represent nearby idle sensor nodes. Strong coupling to a relay causes resonance splitting and induced frequency shifts of the resonance peaks on the order of the sensor-coil 3 dB bandwidth, which is $\approx$ 7 MHz for the considered 0.35 mm coil size. Likewise, dense and arbitrarily arranged swarms of passive relays cause f-selective fading as described in Cpt. 5. We are interested in the implications for our application. All coil orientations are random with uniform distribution in 3D, see Fig. 6.1c, and the passive relay locations are sampled per coordinate from a Gaussian distribution about the sensor location (the standard deviation is two times the coil size and we re-sample until no coils collide in terms of their cylindrical hulls). All relay coil orientations are random whereby all possible directions being equiprobable. The sensor matching is adapted to the relay coupling conditions. We use the dominant peak at 304 MHz in the downlink and waterfilling in the uplink.

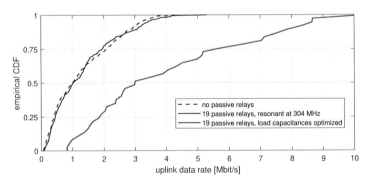

Figure 6.8: Data uplink performance of a single actively transmitting sensor in close vicinity of a dense cluster of passive relays. The coils are 0.35 mm in size and located 5 cm deep in tissue. The scheme employs adaptive sensor matching, frequency-tuning in the powering downlink and waterfilling in the uplink. Optimization of relay load capacitances is done via the genetic-algorithm-based switching scheme from Sec. 5.3.

Fig. 6.8 shows the resulting uplink rates for many realizations of the random swarm geometry. We observe that, without further ado, the presence of passive relays hardly affects the performance statistics. This is in contrary to the results of the previous chapter, in particular to Sec. 5.4, where the utilization of passive-relay-induced spectral channel fluctuations together with transmit-side channel information (waterfilling) led to significant data rate improvements. This discrepancy can be accredited to the vast asymmetry of the link: the coil 3-dB bandwidth is 32 times larger on the sensor side compared to the external side. Because of this, a large portion of the relay-induced channel fluctuation falls outside the essentially usable frequency band(s).

Still, a large performance gain can be realized when relay load capacitances are optimized in a controlled fashion. Specifically, we employ the load switching scheme from Sec. 5.3, which uses a genetic algorithm, to maximize $\|\mathbf{h}_{\mathrm{DL}}\|^2$ at 304 MHz. Out-of-band effects are not an issue with this controlled approach. We observe vast resulting performance improvements, especially to the lower percentile of data rates (i.e. to cases with a severely misaligned sensor orientation). In this context note that passive relaying here improves both the downlink and the uplink.

In conclusion, we can state that this form of passive cooperation poses an interesting and powerful technique also in this biomedical context. How such an optimization scheme could be coordinated between passive nodes, equipped with minimal technical capabilities, is left as an open problem.

6.3.4 Achievable Rates with Load Modulation

Finally we want to investigate the data uplink via *load modulation at a passive sensor* and decoding of the transmitted bit at the external array, which now acts like an RFID reader. Here we do not assume a minimum required power to activate the device (e.g., it might gather sufficient power from energy harvesting to run the load switching process). The evaluation follows the theory in Sec. 3.9, i.e. the data rate is evaluated with the capacity formula (3.59) for the binary symmetric channel, expressed in bit per channel use. The data rate is random because the sensor orientation is random; again we evaluate the mean rate and 5th-percentile value. We do so for two cases: (i) based on full knowledge of the channel, the current vector at the external array is set to its SNR-optimal value under a 1 W sum-power constraint based on Proposition 3.5 and (ii) without channel knowledge we use a random current vector with the appropriate sum power. We assume that the noise bandwidth equals the channel access rate f_{b} (the number of channel uses per second). The maximum channel access rate is determined

133

by transients signal at the high-quality external coils; we choose $f_{\mathrm{b}} = 2 \cdot 10^5$ which is very close to the ratio $f_{\mathrm{c}}/Q_{\mathrm{ext}}$. We compare two different noise models: thermal noise only and the more pessimistic model described below (3.57), which comprises error from limited fidelity and residual self-interference.

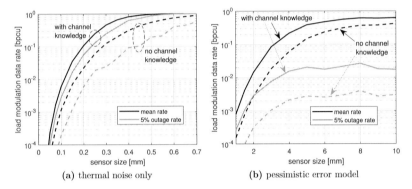

(a) thermal noise only   (b) pessimistic error model

Figure 6.9: Uplink data rate with load modulation at a passive sensor, 5 cm deep into muscle tissue. The results are expressed in terms of channel capacity (3.59) in bit per channel use. The access rate is $2 \cdot 10^5$ channel uses per second, associated with 200 kHz noise bandwidth. The external array, which acts like an RFID reader in this case, uses 1 W for field generation. The data rate is random by virtue of the random sensor coil orientation.

The results in Fig. 6.9 show that load modulation can be a powerful low-complexity alternative to active transmission: if thermal noise constitutes the only limitation, then decent data rates can be achieved reliably with a sensor smaller than 1 mm. However, the performance depends critically on the measurement accuracy and fidelity of the receiving array. In particular, when the external array suffers from significant self-induced errors, i.e. it drives an inaccurate current and/or fails to cancel the self-induced voltage (a problem described in detail by [63]), then a sensor size larger than 5 mm (which is incompatible with most in-body applications) is required for somewhat useful performance. Future work should study these practical limits, especially feasible noise levels in between the discussed optimistic and pessimistic model, in greater detail.

6.4 Cooperative Data Uplink

We consider a swarm of small-scale in-body sensors as illustrated in 6.1c. All sensor devices are equipped with transmit amplifiers and are supplied with power wirelessly as described in 6.2. Assume that one of the sensors has the acute need to transmit data to the external device while all other sensors are idle. Instead of remaining idle, they assist in the uplink data transmission via physical layer cooperation (distributed beamforming). This requires that the initiating sensor broadcasts its data to the other sensors, but because of the short distances between the sensors we assume that this first hop does not constitute a limitation. Instead we focus on the final hop: the MIMO link from the distributed sensor coils to the external coil array. We assume that the sensors can establish phase synchronization which is feasible because of the sub-GHz operating frequency. Furthermore, full channel knowledge is assumed for both ends.

We also consider a scenario where the sensor swarm is intertwined with a swarm of 15 passive relays. Those shall propel spectral and spatial channel variations, which can be exploited by the transmit signaling. The positions and orientations of the sensor and relay coils are randomly sampled like in Sec. 6.3.3. In any scenario, the sensor matching networks are adapted to the swarm coupling conditions.

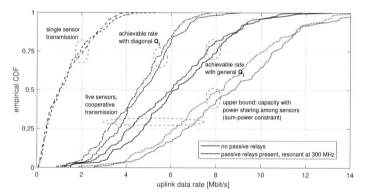

Figure 6.10: Uplink data rates from five cooperating in-body sensors, with and without nearby passive resonant relay coils in random arrangement, to an external 21-coil array. Either case considers 20 in-body coils of 0.35 mm size, 5 cm deep into muscle tissue. The external device uses 1 W to supply power wirelessly. The results are shown as cumulative distribution function (CDF). The data rate is random because of the random orientation of the sensor coils and passive relays.

The wireless powering downlink uses maximum-ratio transmit combining at the external array to maximize the received sum power over this frequency-selective MIMO channel (which results in a concentration of transmit power into the 304 MHz frequency described in Sec. 6.3.2). This way we feed vast power to sensors with a good channel, with the idea that those sensors should also see a good channel in the uplink (which depends on array matching and noise statistics though). The beamforming is constrained to activate the initiating sensor (i.e. assert received power $\geq P_0$) whenever possible.

The resulting uplink transmit power $P_{\mathrm{T},n}$ of the n-th sensor ($n = 1 \ldots 5$) is according to (6.2). Therewith, we operate on the frequency-selective MIMO uplink channel as follows. We allocate the transmit power $P_{\mathrm{T},n} = \sum_k P_{k,n}$ available at the n-th sensor to frequency bands k with a heuristic approach: via waterfilling over the hypothetical parallel SISO channels that would arise when all other sensors are silent (but present) and maximum-ratio combining is done at the receiving external array in each band k. The achievable uplink rate is $D_{\mathrm{UL}} = \sum_k \Delta_f \bar{D}_k$ whereby $\bar{D}_k$ is the achievable MIMO rate in the narrow band k. For each k, we evaluate $\bar{D}_k$ in various different way. First, the suboptimal but easy-to-determine achievable rate (3.37), resulting from a diagonal transmit covariance matrix. Second, the channel capacity under per-transmitter power constraints given by $P_{k,n}$ for the coupled transmitters $n = 1 \ldots 5$, given by (3.36), which is calculated by numerically solving the optimization problem. For comparison we include the channel capacity under a sum-power constraint (3.32), which could be achieved if the sensors were able to share their available power. We also compare to the case without cooperation (i.e. to Sec. 6.3.2).

Fig. 6.10 shows that physical layer cooperation yields a significant increase of uplink data rate (while the same power is fed into the system). This is the result of utilizing spatial signal fluctuations with distributed cooperation. We furthermore observe that the introduction of passive relays, in combination with the spectral and spatial awareness of the chosen transmit signaling, gives rise to appreciable performance gains. The effect is limited though, because of the rather small Q-factor of the small-scale relay coils (as discussed in Sec. 6.3.3).

To complete the picture we also want to evaluate *node cooperation for load modulation*. We employ the ideas and assumptions of Sec. 3.9: all sensors simultaneously transmit the same bit by switching to the appropriate load for the bit duration. This shall achieve a stronger effect at the receiver (comparable to an array gain) and spatial diversity. Perfect synchronization and data exchange are assumed. On the one hand we study the use of canonical switching loads: a capacitance for resonance at $300\,\mathrm{MHz}$ for bit 0, an open circuit for bit 1. On the other hand we also study optimized switching loads, whereby the optimization was conducted with an iterative gradient search for maximization of the SNR in (3.58).

The results in Fig. 6.11 show that passive relays are not helpful here; they are rather slightly harmful. We accredit this to their power consumption outweighing their benefits for this scheme. More importantly, we observe that cooperation yields a

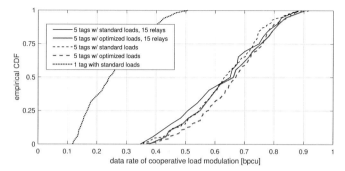

Figure 6.11: Uplink data rate with load modulation of five cooperating passive sensor tags, with and without optimized switching loads, with and without passive relays present nearby. The results are expressed in terms of channel capacity (3.59) in bit per channel use (bpcu). The access rate is $2 \cdot 10^5$ channel uses per second. The coils are 0.25 mm large and 5 mm deep into muscle tissue. Full channel knowledge is assumed. All other assumptions are analogous to Sec. 6.3.4.

large performance gain over load modulation of a single sensor. This promises a large performance potential of cooperative load modulation between a massive number of sensors, even when their load terminations are left unoptimized. The coordination of the distributed scheme remains as an open problem.

Chapter 7

Position and Orientation Estimation of an Active Coil in 3D

In Cpt. 1 we emphasized the technological need for localizing wireless devices, i.e. determining their position and (if desired) their orientation. Besides acoustic or camera-based localization techniques, radio localization is a natural approach for devices that are already equipped with wireless technology. Thereby one evaluates the geometrical information contained in signals received via the wireless channel, in particular the signal phase or time of arrival, the angle of arrival, or the received signal strength [214]. Radio localization however faces severe challenges from radio channel distortions in dense propagation environments [51, 120, 121]. In particular, the effects of line-of-sight blockage and multipath propagation drastically limit its use inside of buildings, in the underground or underwater, or in medical applications. In contrast, low-frequency magnetic near-fields are hardly affected by the environment as long as no major conducting objects are nearby [56–61] (see item 2 of Sec. 1.2). Thus, the magnetic near-field at some position relative to the source (a driven coil or a permanent magnet) can be predicted accurately with a free-space model. This allows to localize an agent[1] relative to a stationary setup of coils with known locations (anchor coils) [58–61, 98]. In particular, position and orientation estimates can be obtained by fitting a free-space channel model (e.g., a dipole model) to measurements of induced voltage or of a related quantity [59–61, 98].[2]

While these circumstances present the prospect of highly accurate localization via magneto-inductive signal measurements, no such system appears to be in widespread use. We shall review published accuracy information for magneto-inductive localization system implementations. An average error of 11 cm is reported by [57] for a $(7\,\mathrm{m})^2$ setup with a tri-axial magnetometer and a two-coil anchor (4 W power, 270 turns, 2071 Hz) in a magnetic laboratory. They mention calibration and interference as possi-

---

[1]We exclusively consider localization of an active agent. This is in contrary to localizing a passive resonant coil such as a near-field RFID tag, which has been studied, e.g., by [26]. This latter case has a severely limited range due to SNR $\propto r^{-12}$ for larger r in the pure near field (cf. Sec. 3.9).

[2]An alternative approach would be location fingerprinting which had some success in the context of magnetic fields [215]. We refrain from this technique because of the effort in acquiring a database of tuples $(\mathbf{h}_{\mathrm{meas}}, \mathbf{p}_{\mathrm{ag}}, \mathbf{o}_{\mathrm{ag}})$ that adequately cover the five-dimensional sample space of $\mathbf{p}_{\mathrm{ag}}$ and $\mathbf{o}_{\mathrm{ag}}$.

ble imperfections. Pasku et al. [58] achieve 30 cm average error (2D) over a 15 m × 12 m office space using coplanar coils (20 turns with 7 cm radius, 0.14 W, 24.4 kHz). A 50-turn coil wound around a football, driven with 0.56 W at 360 kHz, is localized in a $(27\,\mathrm{m})^2$ area with 77 cm mean error in [60]. They use large receivers and techniques to mitigate self-interference from induced signals in the long cables. Abrudan et al. [59] report 30 cm mean accuracy in undistorted environments and 80 cm otherwise over about 10 m distance. They use tri-axial 80-turn coils at 2.5 kHz.

From the listed performance data we observe that the relative error (the ratio of typical position error to setup size) of the referenced system implementations does not beat $\approx 2\%$, although no work identifies the error source responsible for this apparent performance bottleneck. The identification of this bottleneck is, in our opinion, currently the most important research question in magneto-inductive localization. We suspect the following error sources as potential bottlenecks:

- The employed localization algorithm.

- Noise, interference, and quantization.

- Channel model inadequacies such as weak coupling, the dipole assumption, poorly calibrated model parameters, or unconsidered propagation effects such as radiation[3] (direct path or multipath) or field distortion due to eddy currents induced in nearby conductors.[4]

Regarding these shortcomings, this chapter makes the following contributions.

- In Sec. 7.1, for the purpose of localization, we employ the free-space dipole model (3.49), a complex-valued model that comprises mid- and far-field propagation modes and phase shifts. A slight adaptation yields a great measurement fit.

- Sec. 7.2 formalizes joint estimation of the position $\mathbf{p}_{ag}$ and orientation $\mathbf{o}_{ag}$ of an agent coil from measurements of magneto-inductive channel coefficients $h_{meas,n}$ to anchor coils $n = 1 \ldots N$. We state the likelihood function and Fisher information matrix and consequently derive and discuss the Cramér-Rao lower bound on the position error (called the position error bound [126]) and orientation error.

---

[3]The authors of [60] note that radiative propagation should be considered by future solutions.

[4]When such distortions are due to a large conducting ground (e.g. a building floor with reinforcing bars) then they can be captured by an image source model. This led to appreciable but limited improvements of the localization accuracy in [60,61].

- In Sec. 7.3 the position error bound is used as a tool for studying the potential accuracy on the indoor scale. This way we determine a suitable operating point.

- After demonstrating the poor convergence behavior of attempted likelihood function maximization by iterative gradient search, we design and investigate various alternative algorithms in Sec. 7.4. We propose two algorithms with robust real-time capabilities. Firstly, a weighted least squares algorithm termed WLS3D where the orientation parameter is effectively eliminated and the cost function relaxed by a position-dependent weighting. Secondly, an algorithm termed Magnetic Gauss based on the random misalignment theory developed in Cpt. 4.

- In Sec. 7.5 we present a system implementation which uses flat spiderweb coils tuned to $500\,\mathrm{kHz}$ for $N = 8$ anchors. We evaluate the achievable accuracy in an office setting after thorough calibration. To allow for a rigorous evaluation, all measurements are made with a multiport network analyzer, i.e. the agent is tethered and furthermore mounted on a controlled positioner device.

- In Sec. 7.6 we investigate the different error sources and conjecture that field distortions due to reinforcement bars cause the accuracy bottleneck for our system. We conclude with accuracy projections for more ideal circumstances, based on the position error bound.

7.1 Problem Formulation and Channel Modeling

We consider a single-coil agent[5] with center position $\mathbf{p}_{\mathrm{ag}} \in \mathbb{R}^3$ and coil axis orientation $\mathbf{o}_{\mathrm{ag}} \in \mathbb{R}^3$ (unit vector). We furthermore consider the presence of N single-coil anchors with center positions $\mathbf{p}_n \in \mathbb{R}^3$ and coil axis orientations $\mathbf{o}_n \in \mathbb{R}^3$ (unit vectors) for $n = 1 \ldots N$. An exemplary setup is illustrated in Fig. 7.1. If we assume for the moment a transmitting agent (the transmission direction is irrelevant to our formalism) then we encounter a SIMO channel from the agent coil to the anchor coils. All coils are assumed to be power matched at the same design frequency; we consider the narrowband channel at the design frequency. We assume that the agent is never very close to an anchor, such that all anchor-agent couplings are weak at all times in the sense of Sec. 3.7.2. This allows to use the simple model $\mathbf{h} = \frac{1}{2} \operatorname{Re}(\mathbf{Z}_{\mathrm{C:R}})^{-\frac{1}{2}} \mathbf{z}_{\mathrm{C:RT}} \operatorname{Re}(Z_{\mathrm{C:T}})^{-\frac{1}{2}}$ from (3.47).

---

[5]Various related work considers orthogonal tri-axial coil arrays, e.g. [59, 61, 216, 217], however we deem the form factor and hardware complexity of such arrays undesired for many applications. We, in contrary, assume an unobtrusive setup consisting of planar coils, allowing for an integrated agent coil and anchor coils which could be flush-mounted on walls without obstructing any activities in between.

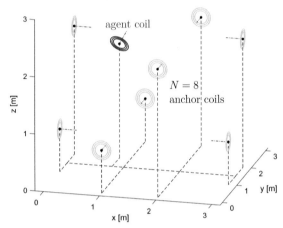

Figure 7.1: Three-dimensional localization setup with $N = 8$ single-coil anchors surrounding a cubic space with $3\,\mathrm{m}$ side length. The single-coil agent has position $\mathbf{p}_{\mathrm{ag}} \in \mathbb{R}^3$ somewhere in this volume and arbitrary orientation $\mathbf{o}_{\mathrm{ag}} \in \mathbb{R}^3$ (unit vector). The coils are considered flat such that the anchors could be flush-mounted on room walls or furniture and the agent could be an integrated printed coil. The coils are not drawn to scale.

The distributed array constituted by the anchor coils is just weakly coupled, i.e. the impedance matrix between the anchor coils $\mathbf{Z}_{\mathrm{C:R}}$ is approximately diagonal and thus $\mathrm{Re}(\mathbf{Z}_{\mathrm{C:R}}) \approx \mathrm{diag}(R_1, \ldots, R_N)$, because the inter-anchor distances are much larger than the coil diameters. Likewise we denote $\mathrm{Re}(Z_{\mathrm{C:T}}) = R_{\mathrm{ag}}$. We obtain a componentwise description $h_n = \frac{(z_{\mathrm{C:RT}})_n}{\sqrt{4 R_n R_{\mathrm{ag}}}}$ and furthermore note that an individual two-port matching network per anchor coil (as in Fig. 3.5) suffices for power matching of the array.

We consider that a measurement of the channel vector $\mathbf{h}_{\mathrm{meas}} \in \mathbb{C}^N$ is available, i.e. channel estimation has been performed. The channel vector is furthermore described by a deterministic model $\mathbf{h}_{\mathrm{model}} \in \mathbb{C}^N$ which is considered as a function $\mathbf{h}_{\mathrm{model}}(\mathbf{p}, \mathbf{o})$ of a position hypothesis $\mathbf{p}$ and an orientation hypothesis $\mathbf{o}$, given full knowledge of the anchor topology and all other technical parameters. Any channel model will exhibit a model error

$$\boldsymbol{\epsilon} = \mathbf{h}_{\mathrm{meas}} - \mathbf{h}_{\mathrm{model}}(\mathbf{p}_{\mathrm{ag}}, \mathbf{o}_{\mathrm{ag}}), \qquad\qquad \boldsymbol{\epsilon} \in \mathbb{C}^N \qquad (7.1)$$

whereby $\mathbf{h}_{\mathrm{model}}$ is evaluated at the true $\mathbf{p}_{\mathrm{ag}}$ and $\mathbf{o}_{\mathrm{ag}}$. If the model is an adequate description of the physical reality then $\boldsymbol{\epsilon}$ will be small and seemingly random as it results from minor model inaccuracies as well as noise, interference and quantization. In this case $\mathbf{p}_{\mathrm{ag}}$ and $\mathbf{o}_{\mathrm{ag}}$ can be estimated by fitting $\mathbf{h}_{\mathrm{model}}$ to the observed $\mathbf{h}_{\mathrm{meas}}$. One possible way to do so is least-squares estimation: calculate the values $\mathbf{p}$ and $\mathbf{o}$ which minimize $\|\mathbf{h}_{\mathrm{meas}} - \mathbf{h}_{\mathrm{model}}(\mathbf{p}, \mathbf{o})\|^2$ (this and other computational options will be discussed in detail in Sec. 7.4). Such an approach will however yield poor results if $\mathbf{h}_{\mathrm{model}}$ is inaccurate, e.g., if it does not account for a relevant propagation mechanism or if the model parameters are poorly calibrated.

Throughout this chapter we use the deterministic channel model

$$h_{\mathrm{model},n} = \tilde{\alpha}_n \left(\mathbf{v}_{\mathrm{Dir},n} + \mathbf{v}_{\mathrm{MP},n}\right)^{\mathrm{T}} \mathbf{o}_{\mathrm{ag}}, \qquad\qquad n = 1 \ldots N \qquad (7.2)$$

whereby all occurring quantities are unitless. The model is equivalent to the free-space model (3.49) apart from the constant field vector $\mathbf{v}_{\mathrm{MP},n} \in \mathbb{C}^3$, which will be explained later. The unitless direct-path field vector $\mathbf{v}_{\mathrm{Dir},n} \in \mathbb{C}^3$ is given by[6]

$$\mathbf{v}_{\mathrm{Dir},n} = j e^{-jkr_n} \left(\left(\frac{1}{(kr_n)^3} + \frac{j}{(kr_n)^2} \right) \boldsymbol{\beta}_{\mathrm{NF},n} + \frac{1}{2kr_n} \boldsymbol{\beta}_{\mathrm{FF},n} \right) \qquad (7.3)$$

which uses the wavenumber k and link distance $r_n = \|\mathbf{p}_{\mathrm{ag}} - \mathbf{p}_n\|$. Furthermore the scaled near-field vector $\boldsymbol{\beta}_{\mathrm{NF},n} = \frac{1}{2}(3\,\mathbf{u}_n\mathbf{u}_n^{\mathrm{T}} - \mathbf{I}_3)\mathbf{o}_n$ by (2.19) and the scaled far-field vector $\boldsymbol{\beta}_{\mathrm{FF},n} = (\mathbf{I}_3 - \mathbf{u}_n\mathbf{u}_n^{\mathrm{T}})\mathbf{o}_n$ by (2.20), which use the direction vector $\mathbf{u}_n = \frac{1}{r_n}(\mathbf{p}_{\mathrm{ag}} - \mathbf{p}_n)$ from the center of the n-th anchor coil to the center of the agent coil. These quantities and the link geometry are illustrated in Fig. 7.2.

---

[6]At this point it might be useful to recall the relation to the actual physical quantities. We consider the case where the anchors are transmitting. According to this model, the complex phasor of the associated magnetic field (unit tesla) at the reference point and generated by the n-th anchor is given by $\mathbf{b}_n = \frac{\mu_0}{2\pi} i_{\mathrm{T}} A_n \mathring{N}_n k^3 (\mathbf{v}_{\mathrm{Dir},n} + \mathbf{v}_{\mathrm{MP},n})$, cf. Proposition 2.2. Thereby $i_{\mathrm{T}} = \sqrt{P_{\mathrm{T}}/R_n}$ is the phasor of the transmit coil current which follows from the active transmit power P_{T} and the coil resistance R_n. The actual magnetic field vector (unit tesla) follows as $\vec{B} = \sqrt{2}\,\mathrm{Re}\{\mathbf{b}e^{j\omega t}\}$ by (2.6). The magnetic dipole moment phasor of the n-th anchor coil is $i_{\mathrm{T}} A_n \mathring{N}_n \mathbf{o}_n$ when transmitting. The induced voltage phasor at the agent coil is $v_n = \mathbf{b}_n^{\mathrm{T}} \mathbf{o}_{\mathrm{ag}} A_{\mathrm{ag}} N_{\mathrm{ag}}$ and the active received power is $|v_n|^2/(4R_{\mathrm{ag}})$.

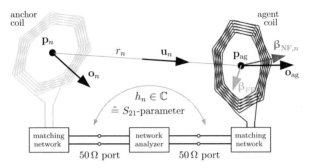

Figure 7.2: Link between the agent and an anchor with spiderweb coil geometries and description of the geometrical quantities. The employed tethered methodology for acquiring a measurement $h_{\mathrm{meas},n} \in \mathbb{C}$ of the channel coefficient is illustrated.

The technical parameters are subsumed in the complex coefficients

$$\tilde{\alpha}_n = \frac{\mu_0 A_{\mathrm{ag}} \mathring{N}_{\mathrm{ag}} A_n \mathring{N}_n}{\sqrt{4 R_{\mathrm{ag}} R_{\mathrm{an},n}}} k^3 f_c \cdot \xi_n , \qquad n = 1 \ldots N . \qquad (7.4)$$

The expression comprises the vacuum permeability μ_0, coil surface areas A_{ag} and A_n (the mean over all turns), coil turn numbers $\mathring{N}_{\mathrm{ag}}$ and $\mathring{N}_n$, and coil resistances R_{ag} and R_n. The term is an adaptation of (3.51) which now comprises a mismatch coefficient $\xi_n \in \mathbb{C}$ with $|\xi_n| \leq 1$. It models imperfect matching of the link involving the n-th anchor. In particular ξ_n models the compound mismatch of anchor and agent in terms of amplitude attenuation and phase shift. Due to ξ the values $\tilde{\alpha}_n$ must be calibrated; in the uncalibrated case we assume $\xi = 1$ (perfect lossless matching). Note that $\tilde{\alpha}_n$ does not depend on the link distance, in contrary to the prefactor α used in earlier chapters (here, the phase shift e^{-jkr_n} is part of $\mathbf{v}_{\mathrm{Dir},n}$ in (7.3), which simplifies the notation).

To understand the significance of various terms and calibration it is worthwhile to look at the encountered physical reality in Fig. 7.3. It shows network analyzer measurements $h_{\mathrm{meas},n}$ over r_n for a pair of coaxial coils ($\mathbf{o}_n = \mathbf{u}_n = \mathbf{o}_{\mathrm{ag}}$), one meter above the floor in an office corridor. Note that $h_{\mathrm{meas},n}$ is equal to the measured S_{21}-parameter as both coils are matched to the $50\,\Omega$ reference impedance. We first look at small distances $r_n < 2\,\mathrm{m}$ and observe that the uncalibrated model has an offset in magnitude and phase (the $90°$ phase shift stems from the law of induction). This is easily compensated by calibration of the coefficient $\tilde{\alpha}_n$, yielding a model that accurately fits the measurements for small r_n.

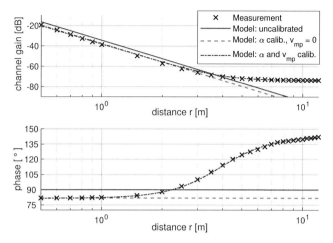

Figure 7.3: Channel coefficient $h \in \mathbb{C}$ of a coaxial link between two spider coils versus the link distance r. We compare network analyzer measurements at 500 kHz in an office corridor to the channel model (7.2) with various calibrations (which are based on fitting the measurements). For cases other than coaxial arrangement, analogous evaluations are given in [144].

For larger distances, about $r_n \geq 3\,\mathrm{m}$, the measured h levels off and does not follow a simple path-loss law anymore; it seems to approach a limit value instead. This effect is certainly not caused by the direct-path mid- or far-field (the terms $\frac{1}{(kr)^2}$ and $\frac{1}{2kr}$ in (7.3)) because the considered distance are near-field dominated ($kr \approx 0.03 \ll 1$). We attribute the effect to electromagnetic interaction with the environment giving rise to a superimposed magnetic field. To some extend this interaction comprises induced currents in nearby conductors as well as multipath wave propagation (radiated longwaves which are reflected and scattered at buildings, mountains, and the ground).[7] We assume that multipath wave propagation is the dominant cause which allows us to consider all $\mathbf{v}_{\mathrm{MP},n} \in \mathbb{C}^3$ as constant across our entire setup. This assumption is supported by the observed limit value $h_{\mathrm{model},n} \approx \tilde{\alpha}_n \mathbf{v}_{\mathrm{MP},n}^{\mathrm{T}} \mathbf{o}_{\mathrm{ag}}$ for larger r_n and by

---

[7]The following observations support that multipath wave propagation plays a significant role in this effect (although we can not make a firm statement based on the available measurements). From designs at different frequencies f_c we found that the limit value magnitude increases rapidly with f_c, which is expected for a radiation effect in this regime. Near-field distortions as cause would show strong path loss according to the image model and thus contradict the observed limit value. Still a near-field effect could play a role, e.g., the building's reinforcement bar mesh acting as passive relays. For localization, however, this aspect is secondary: the model is beneficial as it empirically improves the fit between measurement and model.

the well-known fact that spatial variations of a scatter field are on the order of a wavelength [36] (here $\lambda \approx 600\,\text{m}$ far exceeds our setup size). We do not attempt a geometry-based calculation of $\mathbf{v}_{\text{MP},n}$; instead it shall be *determined by calibration* (in the uncalibrated case we assume $\mathbf{v}_{\text{MP},n} = \mathbf{0}$). As seen in Fig. 7.3, a calibration of both $\tilde{\alpha}_n$ and $\mathbf{v}_{\text{MP},n}$ leads to a great fit between measurement and model at all distances of interest.

7.2 Position-Related Information in Measured Channel Coefficients

We shall briefly discuss the geometry-related information content in wireless received signals and its significance for low-frequency magneto-inductive localization based on an observed channel vector $\mathbf{h}_{\text{meas}}$.

If the channel spectrum $h_{\text{meas},n}(f)$ is observed over $f \in [f_c - \frac{B}{2}, f_c + \frac{B}{2}]$ with a sufficiently large bandwidth B, then the *time of arrival* can be estimated after an inverse Fourier transform. The time of arrival yields a distance estimate with the potential for particularly high accuracy with appropriate signaling: the error is on the order of $\lambda = c/f_c$ with coherent processing[8] (employing the carrier phase) or on the order of c/B otherwise [218]. This allows for cm-accuracy localization of aircraft by radar systems, for example operating at $f_c = 50\,\text{GHz}$. In a dense environment (e.g. indoors) such an approach faces severe problems from multipath propagation [50, 214, 219] but nevertheless can be fruitful if B is sufficiently large (so that multipath components can be resolved) [135]. In this chapter we use a different approach: we choose a very low frequency $f_c = 500\,\text{kHz}$ (and $B = 5\,\text{kHz}$) to minimize interaction with the environment. This signaling is very slow in relation to propagation delays, hence we can not use the time of arrival (cf. the disastrous time-of-arrival accuracy projections of $c/f_c = 600\,\text{m}$ or even $c/B = 60\,\text{km}$).

The observed *signal phase* can be particularly valuable when the typical distance r_n is on the order of λ, i.e. when the receiver is located in the transition region between near and far field. Then the interplay of the phase-shifted summands of (7.3) allows for an estimate of r_n from the phase of $h_{\text{meas},n}$. The use of such an approach for indoor localization is described in [220]. We aim for typical r_n of a few meters, however

---

[8]Establishing phase synchronization between distributed infrastructure can be a major challenge, for example for radar with extremely high frequency [218]. The henceforth considered magneto-inductive setup merely requires synchronization of a 500 kHz carrier among anchors which are a few meters apart, which is easily feasible.

choosing a similar λ would result in vast distortion and multipath propagation effects (cf. the discussion of $\mathbf{v}_{\mathrm{MP},n}$) that would interfere heavily with the direct path. Hence we choose a larger λ and abandon this specific approach.

Angle of arrival estimation from propagation delays would require a distributed array whose size is comparable to the wavelength. For the reasons stated above, our target λ is too large for such an approach to be meaningful.

An observed *amplitude attenuation* $|h_{\mathrm{meas},n}|$ contains information about the distance r_n because of the distance-dependent signal path loss. Such an approach to localization, usually termed received signal strength, is notoriously inaccurate at radio frequencies where the attenuation is subject to severe spatial and temporal fluctuation due to multipath fading and shadowing [51, 131, 214]. These effects are however specifically avoided by the low-frequency magnetic induction approach, which presents the prospect of drawing precise location information from amplitude attenuation. Yet, also in our case there is no direct relationship between $|h_{\mathrm{meas},n}|$ and r_n due to the effect of the unknown coil orientations relative to the link direction. In the following we provide an analytic description of this uncertainty, based on the theory of Cpt. 4.

Proposition 7.1 (ranging likelihood functions). *Assume that an observed channel coefficient $h_{\mathrm{meas},n}$ is accurately described by the employed direct-path channel model, i.e. $h_{\mathrm{meas},n} = h_{\mathrm{model},n}$ with $\mathbf{v}_{\mathrm{MP},n} = \mathbf{0}$. Furthermore assume that the direction vectors $\mathbf{u}_n$ and $\mathbf{o}_{\mathrm{ag}}$ are random with i.i.d. uniform distributions on the unit sphere.*

- *In the magnetoquasistatic regime[9] the likelihood function of distance r_n given $|h_{\mathrm{meas},n}|$ is*

$$L_n(r_n) = \frac{r_n^3}{|\tilde{\alpha}_n|\,/\,k^3} \cdot \begin{cases} 1 & r_n^3 \leq \frac{|\tilde{\alpha}_n|\,/\,k^3}{2\cdot|h_{\mathrm{meas},n}|} \\ 1 - \dfrac{\operatorname{arcosh}\left(r_n^3\,\frac{2\cdot|h_{\mathrm{meas},n}|}{|\tilde{\alpha}_n|\,/\,k^3}\right)}{\operatorname{arcosh}(2)} & \frac{|\tilde{\alpha}_n|\,/\,k^3}{2\cdot|h_{\mathrm{meas},n}|} < r_n^3 < \frac{|\tilde{\alpha}_n|\,/\,k^3}{|h_{\mathrm{meas},n}|} \\ 0 & \frac{|\tilde{\alpha}_n|\,/\,k^3}{|h_{\mathrm{meas},n}|} \leq r_n^3 \end{cases} \qquad (7.5)$$

which is shown in Fig. 7.4a. The maximum-likelihood distance estimate is

$$\hat{r}_n^{\mathrm{ML}} = \arg\max L_n = \left(\frac{1}{2}\cdot\frac{|\tilde{\alpha}_n|\,/\,k^3}{|h_{\mathrm{meas},n}|}\right)^{1/3}. \qquad (7.6)$$

---

[9]The r_n^{-2} mid-field term could be included but would vastly complicate the formula.

- *In the pure far field the likelihood function of distance r_n given $|h_{\mathrm{meas},n}|$ is*

$$L_n(r_n) = \frac{2kr_n}{|\tilde{\alpha}_n|}\left(\frac{\pi}{2} - \arcsin\left(\frac{2kr_n}{|\tilde{\alpha}_n|}|h_{\mathrm{meas},n}|\right)\right) \tag{7.7}$$

for $r_n \leq \frac{|\tilde{\alpha}_n|}{2k|h_{\mathrm{meas},n}|}$ and $L_n(r_n) = 0$ otherwise. The function is shown in Fig. 7.4b.

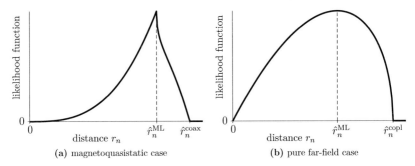

(a) magnetoquasistatic case   (b) pure far-field case

Figure 7.4: Likelihood function of the link distance r_n given an error-less link coefficient measurement $h_{\mathrm{meas},n}$, under the assumption that the coils on both ends have random orientation with uniform distribution on the 3D unit sphere. Comparable distance likelihood functions for multipath radio channels can be found in [221, Fig. 3].

The two cases of the likelihood function are illustrated in Fig. 7.4. The near-field-case ML estimate is equivalent to a distance calculation for two coils fixed in coplanar arrangement ($J_{\mathrm{NF}} = \frac{1}{2}$). This estimate is $\left(\frac{1}{2}\right)^{1/3}$ times the maximum possible distance given $h_{\mathrm{meas},n}$ (coaxial arrangement). We observe that the likelihood is more concentrated around the ML estimate in the magnetoquasistatic case, which is mainly caused by the much stronger path loss of this regime.

Proof. An error-less magnetoquasistatic channel coefficient is given by $h_{\mathrm{meas},n} = h_{\mathrm{model},n} = \frac{\tilde{\alpha}_n}{(kr_n)^3}J_{\mathrm{NF}}$. We take the absolute value and write the equation as $|h_{\mathrm{meas},n}| = \frac{|\tilde{\alpha}_n|/k^3}{r_n^3}|J_{\mathrm{NF}}|$. The PDF $f_{|J_{\mathrm{NF}}|}$ is according to (4.15). The random variable $|h_{\mathrm{meas},n}|$, which is just a deterministic scaling of $|J_{\mathrm{NF}}|$, has PDF $\frac{r_n^3}{|\tilde{\alpha}_n|/k^3}f_{|J_{\mathrm{NF}}|}\left(\frac{r_n^3}{|\tilde{\alpha}_n|/k^3}|h_{\mathrm{meas},n}|\right)$ which proves (7.5). In the pure far-field case without model error, $h_{\mathrm{meas},n} = \frac{\tilde{\alpha}_n}{2kr_n}J_{\mathrm{FF}}$. The likelihood function is derived analogously to above via the PDF (4.17). □

The remainder of the section states the Cramér-Rao lower bound (CRLB) on the root-mean-square (RMS) position error of the considered magneto-inductive localization problem. This quantity, also called the position error bound (PEB), is an established tool in radio localization [126]. It highlights the impact of the setup parameters, poses a benchmark for the position error achieved by practical localization algorithms and allows for projections of estimation accuracy in a simple fashion. In this context we assume that the model error ϵ is random with known statistics. Then any estimate $\hat{\mathbf{p}}_{\mathrm{ag}}$ returned by some localization algorithm is also random. If the estimation rule is unbiased then the CRLB applies to the error variances of the estimated components of $\mathbf{p}_{\mathrm{ag}}$. The PEB is the resulting lower bound on $\sqrt{\mathbb{E}[\|\hat{\mathbf{p}}_{\mathrm{ag}} - \mathbf{p}_{\mathrm{ag}}\|^2]}$. It is well-known that maximum-likelihood estimation attains the CRLB asymptotically for $N \to \infty$. So we can expect the PEB to be a tight lower bound for any well-designed localization algorithm when the number of anchors N is sufficiently large and especially when the SNR-like ratio $\mathbb{E}[\|\mathbf{h}_{\mathrm{model}}\|^2] / \mathbb{E}[\|\epsilon\|^2]$ is large (cf. [222]). Therefore, the PEB allows for meaningful accuracy projections. [126, 223]

As preparation for the CRLB results we note that for estimating $\mathbf{p}_{\mathrm{ag}}$ from $\mathbf{h}_{\mathrm{meas}}$ we must consider the unknown $\mathbf{o}_{\mathrm{ag}}$ as nuisance parameter because it affects the statistics of $\mathbf{h}_{\mathrm{meas}}$. Hence we engage in joint estimation of $\mathbf{p}_{\mathrm{ag}}$ and $\mathbf{o}_{\mathrm{ag}}$ (the same approach applies if an estimate of $\mathbf{o}_{\mathrm{ag}}$ is of interest to the application). Thereby the constraint $\|\mathbf{o}_{\mathrm{ag}}\| = 1$ causes some mathematical trouble but this can easily be bypassed with the use of the standard spherical parametrization[10]

$$
\mathbf{o}_{\mathrm{ag}} = \begin{bmatrix} \cos\phi_{\mathrm{ag}} \sin\theta_{\mathrm{ag}} \\ \sin\phi_{\mathrm{ag}} \sin\theta_{\mathrm{ag}} \\ \cos\theta_{\mathrm{ag}} \end{bmatrix} \tag{7.8}
$$

in terms of azimuth angle ϕ_{ag} and polar angle θ_{ag}. Thus, the considered estimation parameter is the vector

$$
\boldsymbol{\psi} := \begin{bmatrix} \mathbf{p}_{\mathrm{ag}} \\ \phi_{\mathrm{ag}} \\ \theta_{\mathrm{ag}} \end{bmatrix} \in \mathbb{R}^5 \tag{7.9}
$$

which constitutes a full description of the agent deployment. The discussion of estimation algorithms later in the chapter will show that the phase of $h_{\mathrm{meas},n}$ (especially the sign of $\mathrm{Im}(h_{\mathrm{meas},n})$) is essential for estimating $\mathbf{o}_{\mathrm{ag}}$. Hence we consider schemes that process the complex numbers $h_{\mathrm{meas},n}$ in their very form instead of drawing a simpler

---

[10]An alternative mathematical approach to incorporating a constraint such as $\|\mathbf{o}_{\mathrm{ag}}\| = 1$ into CRLB statements is provided by the theory in [224].

metric like $|h_{\text{meas},n}|$ that might discard location information.

Proposition 7.2 (Position Error Bound). *If $\hat{\boldsymbol{\psi}}$ is an unbiased estimate of $\boldsymbol{\psi}$ and the regularity condition of the Cramér-Rao lower bound is fulfilled (for the detailed condition we refer to [223]), then the Cramér-Rao lower bound on the estimation error variance is*

$$\text{var}[\hat{\psi}_i] = \mathbb{E}[(\hat{\psi}_i - \psi_i)^2] \geq \left(\boldsymbol{\mathcal{I}}_{\boldsymbol{\psi}}^{-1}\right)_{i,i}, \qquad i \in \{1,2,3,4,5\} \qquad (7.10)$$

where $\boldsymbol{\mathcal{I}}_{\boldsymbol{\psi}} \in \mathbb{R}^{5 \times 5}$ is the Fisher information matrix of $\boldsymbol{\psi}$.
The related position error bound (PEB) states that

$$\sqrt{\mathbb{E}[\|\hat{\mathbf{p}}_{\text{ag}} - \mathbf{p}_{\text{ag}}\|^2]} \geq \text{PEB}(\mathbf{p}_{\text{ag}}, \mathbf{o}_{\text{ag}}), \qquad (7.11)$$

$$\text{PEB}(\mathbf{p}_{\text{ag}}, \mathbf{o}_{\text{ag}}) = \sqrt{\text{tr}\left\{\left(\boldsymbol{\mathcal{I}}_{\boldsymbol{\psi}}^{-1}\right)_{1:3,1:3}\right\}} = \sqrt{\text{tr}\left\{\boldsymbol{\mathcal{I}}_{\mathbf{p}_{\text{ag}}}^{-1}\right\}}. \qquad (7.12)$$

The last expression uses the equivalent Fisher information matrix

$$\boldsymbol{\mathcal{I}}_{\mathbf{p}_{\text{ag}}} = \left(\boldsymbol{\mathcal{I}}_{\boldsymbol{\psi}}\right)_{1:3,1:3} - \left(\boldsymbol{\mathcal{I}}_{\boldsymbol{\psi}}\right)_{1:3,4:5}\left(\boldsymbol{\mathcal{I}}_{\boldsymbol{\psi}}\right)_{4:5,4:5}^{-1}\left(\boldsymbol{\mathcal{I}}_{\boldsymbol{\psi}}\right)_{4:5,1:3} \qquad (7.13)$$

which shows the effect of orientation uncertainty on position estimation. If the agent orientation is precisely known a-priori then $\boldsymbol{\mathcal{I}}_{\mathbf{p}_{\text{ag}}} = (\boldsymbol{\mathcal{I}}_{\boldsymbol{\psi}})_{1:3,1:3}$ applies.

Proof. The CRLB (7.10) for a vector parameter is a basic statement of estimation theory [223, Eq. 3.20]. The position error bound has been derived by [126]; it follows by forming the sum of (7.10) for $i \in \{1,2,3\}$ to obtain $\mathbb{E}[\sum_{i=1}^{3}(\hat{\psi}_i - \psi_i)^2] \geq \sum_{i=1}^{3}(\boldsymbol{\mathcal{I}}_{\boldsymbol{\psi}}^{-1})_{i,i}$ or rather $\mathbb{E}[\|\hat{\mathbf{p}}_{\text{ag}} - \mathbf{p}_{\text{ag}}\|^2] \geq \sum_{i=1}^{3}(\boldsymbol{\mathcal{I}}_{\boldsymbol{\psi}}^{-1})_{i,i}$. The equivalent Fisher information matrix follows from block-wise matrix inversion via the Schur complement; it was introduced by [126] in the context of position estimation with nuisance parameters. $\square$

The Fisher information matrix $\boldsymbol{\mathcal{I}}_{\boldsymbol{\psi}}$ is determined by the function $\mathbf{h}_{\text{model}}(\boldsymbol{\psi})$ in (7.2) and the distribution of $\boldsymbol{\epsilon}$. For a general definition please refer to [223]; the following proposition states the result for the case that the model error $\boldsymbol{\epsilon} \in \mathbb{C}^N$ has a zero-mean Gaussian distribution. The adequacy of a Gaussianity assumption is supported by Sec. 3.2 and the measurement results presented later in the chapter.

Proposition 7.3 (Fisher Information Matrix). *Let the model error $\boldsymbol{\epsilon} \in \mathbb{C}^N$ have circularly-symmetric complex Gaussian distribution with zero mean $\boldsymbol{\epsilon} \sim \mathcal{N}(\mathbf{0}, \mathbf{K}_\epsilon)$. Then the log-likelihood function of $\boldsymbol{\psi}$ given an observed $\mathbf{h}_{\mathrm{meas}}$ is*

$$L(\boldsymbol{\psi}) = \log f(\mathbf{h}_{\mathrm{meas}} \,|\, \boldsymbol{\psi}) =$$
$$- N \log(\pi) - \log \det(\mathbf{K}_\epsilon) - (\mathbf{h}_{\mathrm{meas}} - \mathbf{h}_{\mathrm{model}})^{\mathrm{H}} \mathbf{K}_\epsilon^{-1} (\mathbf{h}_{\mathrm{meas}} - \mathbf{h}_{\mathrm{model}}) \quad (7.14)$$

where $\mathbf{h}_{\mathrm{model}}$ (and possibly also $\mathbf{K}_\epsilon$) is a deterministic function of $\boldsymbol{\psi}$. The Fisher information matrix $\boldsymbol{\mathcal{I}}_\psi \in \mathbb{R}^{5 \times 5}$ associated with $\boldsymbol{\psi}$ is given by [223, Eq. 15.52]

$$(\boldsymbol{\mathcal{I}}_\psi)_{m,n} = 2 \operatorname{Re}\left(\frac{\partial \mathbf{h}_{\mathrm{model}}^{\mathrm{H}}}{\partial \psi_m} \mathbf{K}_\epsilon^{-1} \frac{\partial \mathbf{h}_{\mathrm{model}}}{\partial \psi_n} \right) + \operatorname{tr}\left(\mathbf{K}_\epsilon^{-1} \frac{\partial \mathbf{K}_\epsilon}{\partial \psi_m} \mathbf{K}_\epsilon^{-1} \frac{\partial \mathbf{K}_\epsilon}{\partial \psi_n} \right) \quad (7.15)$$

for elements $m, n \in \{1, \dots, 5\}$. The right-hand summand vanishes if $\mathbf{K}_\epsilon$ does not depend on $\boldsymbol{\psi}$, yielding the compact formula[11]

$$\boldsymbol{\mathcal{I}}_\psi = 2 \operatorname{Re}\left(\frac{\partial \mathbf{h}_{\mathrm{model}}^{\mathrm{H}}}{\partial \boldsymbol{\psi}} \mathbf{K}_\epsilon^{-1} \frac{\partial \mathbf{h}_{\mathrm{model}}}{\partial \boldsymbol{\psi}^{\mathrm{T}}} \right). \quad (7.16)$$

These statements require the $5 \times N$ Jacobian matrix of $\mathbf{h}_{\mathrm{model}}$ from (7.2) with respect to $\boldsymbol{\psi}$, which holds derivatives of $\mathbf{h}_{\mathrm{model},n}$ with respect to the components of $\mathbf{p}_{\mathrm{ag}}$ as well as ϕ_{ag} and θ_{ag}. These derivatives are given by the following involved proposition.

Proposition 7.4 (Geometric Gradient). *The spatial gradient of the channel coefficient is given by*

$$\frac{\partial h_{\mathrm{model},n}}{\partial \mathbf{p}_{\mathrm{ag}}} = \tilde{\alpha}_n \frac{\partial \mathbf{v}_{\mathrm{Dir},n}^{\mathrm{T}}}{\partial \mathbf{p}_{\mathrm{ag}}} \mathbf{o}_{\mathrm{ag}}, \quad (7.17)$$

$$\frac{\partial \mathbf{v}_{\mathrm{Dir},n}^{\mathrm{T}}}{\partial \mathbf{p}_{\mathrm{ag}}} = -jk\, \mathbf{u}_n \mathbf{v}_{\mathrm{Dir},n}^{\mathrm{T}} + je^{-jkr_n} \boldsymbol{\Gamma} \quad (7.18)$$

---

[11]We note that $N \geq 5$ is a necessary condition for $\boldsymbol{\mathcal{I}}_\psi$ in (7.16) being invertible. This is consistent with the fact that we would require at least five observations $h_{\mathrm{meas},n}$ to determine the five-dimensional $\boldsymbol{\psi}$ even in the error-less case.

where $\boldsymbol{\Gamma} \in \mathbb{C}^{3\times3}$ *is another Jacobian matrix with respect to position, given by*

$$\boldsymbol{\Gamma} = \frac{\partial}{\partial \mathbf{p}_{\mathrm{ag}}} \left(\left(\frac{1}{(kr_n)^3} + \frac{j}{(kr_n)^2} \right) \boldsymbol{\beta}_{\mathrm{NF},n}^{\mathrm{T}} + \frac{1}{2kr_n} \boldsymbol{\beta}_{\mathrm{FF},n}^{\mathrm{T}} \right) \qquad (7.19)$$

$$= \left(\frac{1}{(kr_n)^3} + \frac{j}{(kr_n)^2} \right) \frac{\partial \boldsymbol{\beta}_{\mathrm{NF},n}^{\mathrm{T}}}{\partial \mathbf{p}_{\mathrm{ag}}} + \frac{1}{2kr_n} \frac{\partial \boldsymbol{\beta}_{\mathrm{FF},n}^{\mathrm{T}}}{\partial \mathbf{p}_{\mathrm{ag}}}$$

$$- \left(\frac{3k}{(kr_n)^4} + \frac{2jk}{(kr_n)^3} \right) \mathbf{u}_n \boldsymbol{\beta}_{\mathrm{NF},n}^{\mathrm{T}} - \frac{k}{2(kr_n)^2} \mathbf{u}_n \boldsymbol{\beta}_{\mathrm{FF},n}^{\mathrm{T}} .$$

The above uses the spatial Jacobians of the scaled near- and far-field vectors

$$\frac{\partial \boldsymbol{\beta}_{\mathrm{NF},n}^{\mathrm{T}}}{\partial \mathbf{p}_{\mathrm{ag}}} = \frac{3}{2} \frac{1}{r_n} \left(\mathbf{o}_n \mathbf{u}_n^{\mathrm{T}} + (\mathbf{o}_n^{\mathrm{T}} \mathbf{u}_n)(\mathbf{I}_3 - 2\mathbf{u}_n \mathbf{u}_n^{\mathrm{T}}) \right), \qquad (7.20)$$

$$\frac{\partial \boldsymbol{\beta}_{\mathrm{FF},n}^{\mathrm{T}}}{\partial \mathbf{p}_{\mathrm{ag}}} = -\frac{1}{r_n} \left(\mathbf{o}_n \mathbf{u}_n^{\mathrm{T}} + (\mathbf{o}_n^{\mathrm{T}} \mathbf{u}_n)(\mathbf{I}_3 - 2\mathbf{u}_n \mathbf{u}_n^{\mathrm{T}}) \right). \qquad (7.21)$$

The gradients with respect to orientation are $\frac{\partial h_{\mathrm{model},n}}{\partial \phi_{\mathrm{ag}}} = \tilde{\alpha}_n \big(\mathbf{v}_{\mathrm{Dir},n} + \mathbf{v}_{\mathrm{MP},n} \big)^{\mathrm{T}} \frac{\partial \mathbf{o}_{\mathrm{ag}}}{\partial \phi_{\mathrm{ag}}}$ *with* $\frac{\partial \mathbf{o}_{\mathrm{ag}}}{\partial \phi_{\mathrm{ag}}} = [-\sin(\phi_{\mathrm{ag}})\sin(\theta_{\mathrm{ag}}), \cos(\phi_{\mathrm{ag}})\sin(\theta_{\mathrm{ag}}), 0]^{\mathrm{T}}$ *and* $\frac{\partial h_{\mathrm{model},n}}{\partial \theta_{\mathrm{ag}}} = \tilde{\alpha}_n (\mathbf{v}_{\mathrm{Dir},n} + \mathbf{v}_{\mathrm{MP},n})^{\mathrm{T}} \frac{\partial \mathbf{o}_{\mathrm{ag}}}{\partial \theta_{\mathrm{ag}}}$ *with* $\frac{\partial \mathbf{o}_{\mathrm{ag}}}{\partial \theta_{\mathrm{ag}}} = [\cos(\phi_{\mathrm{ag}})\cos(\theta_{\mathrm{ag}}), \sin(\phi_{\mathrm{ag}})\cos(\theta_{\mathrm{ag}}), -\sin(\theta_{\mathrm{ag}})]^{\mathrm{T}}.$

Proof Sketch. We note that the vectors $\mathbf{v}_{\mathrm{MP},n}$ are considered constant while $\mathbf{v}_{\mathrm{Dir},n}$ depend on $\mathbf{p}_{\mathrm{ag}}$ but not on ϕ_{ag} or θ_{ag}, leading to (7.17). Then (7.18) follows from the product rule, $\frac{\partial}{\partial \mathbf{p}_{\mathrm{ag}}} e^{-jkr_n} = -jk \frac{\partial r_n}{\partial \mathbf{p}_{\mathrm{ag}}} \cdot e^{-jkr_n}$, and $\frac{\partial r_n}{\partial \mathbf{p}_{\mathrm{ag}}} = \mathbf{u}_n$ which is easily shown by writing $r_n = (x^2 + y^2 + z^2)^{\frac{1}{2}}$ and computing $\frac{\partial r_n}{\partial x}, \frac{\partial r_n}{\partial y}, \frac{\partial r_n}{\partial z}$ (cf. [146]). Then (7.19) is expanded with basic rules. The near- and far-field Jacobians (7.20) and (7.21) are obtained by writing $\boldsymbol{\beta}_{\mathrm{NF},n}^{\mathrm{T}} = \frac{3}{2}(\mathbf{u}_n^{\mathrm{T}} \mathbf{o}_n)\mathbf{u}_n^{\mathrm{T}} - \frac{1}{2}\mathbf{o}_n^{\mathrm{T}}$ and $\boldsymbol{\beta}_{\mathrm{FF},n}^{\mathrm{T}} = -(\mathbf{u}_n^{\mathrm{T}} \mathbf{o}_n)\mathbf{u}_n^{\mathrm{T}} + \mathbf{o}_n^{\mathrm{T}}$, then noting $\frac{\partial \boldsymbol{\beta}_{\mathrm{NF},n}^{\mathrm{T}}}{\partial \mathbf{p}_{\mathrm{ag}}} = \frac{3}{2} \frac{\partial (\mathbf{u}_n^{\mathrm{T}} \mathbf{o}_n)\mathbf{u}_n^{\mathrm{T}}}{\partial \mathbf{p}_{\mathrm{ag}}}$ and $\frac{\partial \boldsymbol{\beta}_{\mathrm{FF},n}^{\mathrm{T}}}{\partial \mathbf{p}_{\mathrm{ag}}} = -\frac{\partial (\mathbf{u}_n^{\mathrm{T}} \mathbf{o}_n)\mathbf{u}_n^{\mathrm{T}}}{\partial \mathbf{p}_{\mathrm{ag}}}$. The common term $\frac{\partial (\mathbf{u}_n^{\mathrm{T}} \mathbf{o}_n)\mathbf{u}_n^{\mathrm{T}}}{\partial \mathbf{p}_{\mathrm{ag}}} = \frac{\partial \mathbf{u}_n^{\mathrm{T}}}{\partial \mathbf{p}_{\mathrm{ag}}} \mathbf{o}_n \mathbf{u}_n^{\mathrm{T}} + \mathbf{u}_n^{\mathrm{T}} \mathbf{o}_n \frac{\partial \mathbf{u}_n^{\mathrm{T}}}{\partial \mathbf{p}_{\mathrm{ag}}}$ is expanded with the use of $\frac{\partial \mathbf{u}_n^{\mathrm{T}}}{\partial \mathbf{p}_{\mathrm{ag}}} = \frac{1}{r_n}(\mathbf{I}_3 - \mathbf{u}_n \mathbf{u}_n^{\mathrm{T}})$. This last equality can be proven by writing $\mathbf{r}_n = [x\ y\ z]^{\mathrm{T}}$ and $\mathbf{u}_n = \frac{1}{r_n} \mathbf{r}_n$ to expand $\frac{\partial \mathbf{u}_n}{\partial x} = \frac{1}{r_n} \frac{\partial \mathbf{r}_n}{\partial x} - \frac{1}{r_n^2} \frac{\partial r_n}{\partial x} \mathbf{r}_n = \frac{1}{r_n}([1\ 0\ 0]^{\mathrm{T}} - (\mathbf{u}_n)_x \mathbf{u}_n)$ where $(\mathbf{u}_n)_x$ is the x-component of $\mathbf{u}_n$. Repeating the above for $\frac{\partial \mathbf{u}_n}{\partial y}$ and $\frac{\partial \mathbf{u}_n}{\partial z}$ and a few rearrangements conclude the derivation. The derivatives with respect to ϕ_{ag} and θ_{ag} follow from basic calculus. $\square$

Later in the chapter, in the evaluation of the practical system implementation in Sec. 7.5, we will encounter a distribution of $\boldsymbol{\epsilon}$ that does not exhibit circular symmetry, even for a thoroughly calibrated model. We will nevertheless require an appropriate formula for the Fisher information matrix, which is given in the following.

Proposition 7.5. *If the model error vector* $\epsilon \in \mathbb{C}^N$ *has a general complex Gaussian distribution with zero mean, modeled in terms of the real-valued stack vector*

$$\epsilon_{\text{stack}} = \begin{bmatrix} \text{Re } \epsilon \\ \text{Im } \epsilon \end{bmatrix} \sim \mathcal{N}(\mathbf{0}, \mathbf{K}_{\text{stack}}), \tag{7.22}$$

then the Fisher information matrix $\boldsymbol{\mathcal{I}}_\psi \in \mathbb{R}^{5 \times 5}$ *is given by [223, Eq. 3.31]*

$$(\boldsymbol{\mathcal{I}}_\psi)_{m,n} = \mathbf{s}_m^{\text{T}} \mathbf{K}_{\text{stack}}^{-1} \mathbf{s}_n + \frac{1}{2} \operatorname{tr}\left(\mathbf{K}_{\text{stack}}^{-1} \frac{\partial \mathbf{K}_{\text{stack}}}{\partial \psi_m} \mathbf{K}_{\text{stack}}^{-1} \frac{\partial \mathbf{K}_{\text{stack}}}{\partial \psi_n} \right) \tag{7.23}$$

for elements $m, n \in \{1, \dots, 5\}$. *Thereby* $\mathbf{s}_m \in \mathbb{R}^{2N}$ *is the* m-*th column of the stacked Jacobian* $\mathbf{S} \in \mathbb{R}^{2N \times 5}$ *stated below. If* $\mathbf{K}_{\text{stack}}$ *does not depend on* ψ *then*

$$\boldsymbol{\mathcal{I}}_\psi = \mathbf{S}^{\text{T}} \mathbf{K}_{\text{stack}}^{-1} \mathbf{S}, \qquad\qquad \mathbf{S} = \begin{bmatrix} \text{Re}(\partial \mathbf{h}_{\text{model}} / \partial \boldsymbol{\psi}^{\text{T}}) \\ \text{Im}(\partial \mathbf{h}_{\text{model}} / \partial \boldsymbol{\psi}^{\text{T}}) \end{bmatrix} \tag{7.24}$$

applies. The statements encompass circularly-symmetric distributions $\epsilon \sim \mathcal{CN}(\mathbf{0}, \mathbf{K}_\epsilon)$ *as a special case; the conversion is via*

$$\mathbf{K}_{\text{stack}} = \frac{1}{2} \begin{bmatrix} \text{Re}(\mathbf{K}_\epsilon) & -\text{Im}(\mathbf{K}_\epsilon) \\ \text{Im}(\mathbf{K}_\epsilon) & \text{Re}(\mathbf{K}_\epsilon) \end{bmatrix}. \tag{7.25}$$

7.3 CRLB-Based Study of Accuracy Regimes

In this section we evaluate the potential performance of low-frequency magneto-inductive localization for realistic technical parameters and a setup size of a few meters (e.g. indoor localization). The anchors are installed in the pattern indicated in Fig. 7.1 which is motivated by the compromise of establishing a large anchor spread in all three dimensions while maintaining small r_n to most positions in the volume (an extensive study of optimal anchor deployment is left for future work).

The following evaluation assumes a random agent deployment whereby $\mathbf{o}_{\text{ag}}$ has uniform distribution on the 3D unit sphere and $\mathbf{p}_{\text{ag}}$ has a uniform distribution on a cube with the same center position as the anchor setup and 80% of its side length (i.e. 2.4 m). This way we prevent that $\mathbf{p}_{\text{ag}}$ occurs very close to any $\mathbf{p}_n$. We are interested in the resulting statistics of the position error bound (7.12) in different setting.

All $\tilde{\alpha}_n$ are set such that the agent-to-anchor channel gain would be $-40\,\text{dB}$ over $r_n = 1\,\text{m}$ in coaxial arrangement (this choice is made for all considered wavelengths). For the model error we choose a circularly-symmetric Gaussian distribution $\epsilon \sim \mathcal{N}(\mathbf{0}, \mathbf{K}_\epsilon)$ for simplicity. This is supported by Sec. 3.2 and partially by observed measurements.

Choosing a sensible value for the error covariance matrix $\mathbf{K}_\epsilon$ is a major aspect of this evaluation and crucial for its meaningfulness.[12] Motivated by the error levels observed on the basis of our system implementation we set $\mathbf{K}_\epsilon = 10^{-10} \mathbf{I}_N$, corresponding to an error floor of $-100\,\mathrm{dB}$ on measurements of the channel coefficient.

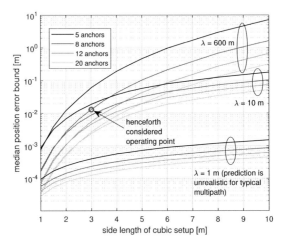

Figure 7.5: Median position error bound versus the side length of the cubic volume that constitutes the localization setup (see Fig. 7.1) for different numbers of anchors N and different wavelengths λ. The evaluation assumes the free-space model in (7.2) (with $\mathbf{v}_{\mathrm{MP},n} = 0$) and the distribution $\epsilon \sim \mathcal{CN}(\mathbf{0}, \mathbf{K}_\epsilon)$ with $\mathbf{K}_\epsilon = \mathbf{I}_N \cdot 10^{-10}$ (a choice motivated by practically observed error levels). The agent has uniformly distributed position within this volume and unknown random orientation with uniform distribution. For all λ the coil parameters are set such that a $1\,\mathrm{m}$ coaxial link has $-40\,\mathrm{dB}$ channel gain.

Fig. 7.5 shows the median PEB versus room side length for different numbers of anchors and different wavelengths. As expected, position errors are lowest in small

---

[12]One possible approach for setting $\mathbf{K}_\epsilon$ is the consideration of the SIMO signal model $\mathbf{y} = \mathbf{h}x + \mathbf{w}$ in the sense of (3.23), with $\mathbf{w} \sim \mathcal{CN}(\mathbf{0}, \mathbf{K})$ and a constant pilot signal $x = \sqrt{P_{\mathrm{T}}}$. Now observing $\frac{1}{\sqrt{P_{\mathrm{T}}}}\mathbf{y} = \mathbf{h} + \frac{1}{\sqrt{P_{\mathrm{T}}}}\mathbf{w}$ suggests that $\epsilon = \frac{1}{P_{\mathrm{T}}}\mathbf{K}$. This would yield model error levels on the order of magnitude of $\mathbb{E}[|\epsilon_n|^2] \approx \frac{1}{P_{\mathrm{T}}}B \cdot 10^{-20}$, e.g., $\approx 10^{-16}$ with $P_{\mathrm{T}} = 10\,\mathrm{mW}$ transmit power and $B = 100\,\mathrm{Hz}$ signaling bandwidth (corresponding to a measurement update rate of about 100 samples per second). This calculation is however highly optimistic (and contributed to the optimistic accuracy projections in our paper [128]) for the following reason. Whenever $\mathbf{h}_{\mathrm{meas}}$ has a appreciably large value, $\mathbf{h}_{\mathrm{meas}}$ and $\mathbf{h}_{\mathrm{model}}$ will differ by much more than measurement noise and interference due to imperfect calibration and unconsidered propagation effect in $\mathbf{h}_{\mathrm{model}}$. A channel model $\mathbf{h}_{\mathrm{model}}$ of surreal accuracy and an extremely high-resolution ADC would be required for measurement noise to become dominant.

rooms with many anchors. Remarkably, the projected accuracies for $\lambda = 600\,\mathrm{m}$ and $\lambda = 10\,\mathrm{m}$ are not vastly different for small setup sizes. This is because in either case the typical link distances hardly exceed the near field. For larger setups the smaller wavelength promises a better accuracy because of the reduced path loss in the mid- and far-field regimes. From a practical perspective however, the choice $\lambda = 10\,\mathrm{m}$ is disastrous because one must expect vast interaction with the environment, which interferes with the direct path and thus gives rise to huge model errors $\boldsymbol{\epsilon}$ (we observed this clearly in an experiment conducted at $13.56\,\mathrm{MHz} \Leftrightarrow \lambda = 22.1\,\mathrm{m}$).

The problem would be even more pronounced at $\lambda = 1\,\mathrm{m}$ where $\mathbf{h}_{\mathrm{meas}}$ would be governed by small-scale fading through rich scattering. The accuracy projections in combination with our practical experiments in the course of [143] show that $\lambda = 600\,\mathrm{m}$ is suitable for our magneto-inductive localization approach. We identify the $N = 8$ case with $3\,\mathrm{m}$ setup side length, which shows a median PEB of $13\,\mathrm{mm}$, as an attractive use case with reasonable infrastructure cost. We will use this operating point for the remainder of the chapter.

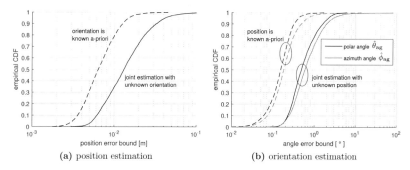

(a) position estimation (b) orientation estimation

Figure 7.6: The plots show the statistics of the CRLB on the RMS errors of the position estimate and the orientation angle estimates for random agent deployment in a $3\,\mathrm{m} \times 3\,\mathrm{m} \times 3\,\mathrm{m}$ room using 8 anchors and $\lambda = 600\,\mathrm{m}$ ($f_\mathrm{c} = 500\,\mathrm{kHz}$). Also shown are bounds which ignore the uncertainty in the respective other domain.

Fig. 7.6 shows the statistics of the PEB at the chosen operating point. We note that accuracies of $42\,\mathrm{mm}$ and $2.8°$ are feasible for 95% of all random agent deployments. If the agent orientation was known a-priori then the position accuracy would approximately double (the median PEB improves from $13\,\mathrm{mm}$ to $6.4\,\mathrm{mm}$). The plots of the RMS orientation angle errors show that azimuth and polar angle have similar error statistics. Here the accuracy would improve by about a factor of three if the position parameter was known a-priori. This indicates that orientation estimation is slightly more sensitive to position uncertainty than vice versa.

7.4   Localization Algorithm Design

We are interested in a localization algorithm that is fast, accurate, and robust (i.e. outliers in the sense of occasional very inaccurate estimates should not occur). In other words, we seek a computational procedure which reliably returns an accurate estimate $\hat{\mathbf{p}}_{ag}$ (and possibly also $\hat{\mathbf{o}}_{ag}$) given the observation $\mathbf{h}_{meas} \in \mathbb{C}^N$ and full knowledge of the anchor topology and technical parameters, based on a channel model $\mathbf{h}_{model}$. The latter is throughout considered as a function of an agent position hypothesis $\mathbf{p}$ and an orientation hypothesis $\mathbf{o}$ (as opposed to the true $\mathbf{p}_{ag}$ and $\mathbf{o}_{ag}$).

We shall review the related work on magneto-inductive localization algorithms. The joint estimation of position and orientation of a dipole-like magnet through distributed sensors, each measuring one field component, was studied by [122] and by [123, 124] for medical gastrointestinal applications. In [125], a medical microrobot estimates its position and orientation from voltages induced in its near-field antenna due to eight active anchors. For the 5D non-linear least squares problem associated with these works, the Levenberg-Marquardt (LM) algorithm was identified as a suitable solver [122,123]. The authors of [123] emphasized the importance of an accurate initial guess for LM because of local cost function minima. The magnetic field Jacobian was provided to the LM algorithm in [124] for performance enhancement. Most papers on near-field localization employ the dipole approximation, e.g., [58,59,61,122–124,217]. In distinction from planar coil setups, the use of tri-axial coil arrays at the anchors and/or the agent allows for simpler localization schemes [59,96,217,225]. In particular, [225] uses a simplified localization algorithm to initialize the LM solver applied to the original non-linear least squares problem.

In the following we review the standard approach(es) and the associated problems to then proceed with the design of more suitable localization algorithms. It shall be noted that we study location estimation from an instantaneous observation $\mathbf{h}_{meas}$ without temporal filtering (which could easily be applied to instantaneous estimates).

7.4.1   Least-Squares and Maximum-Likelihood Estimation

We consider $\mathbf{h}_{model} = [\,h_{model,1} \ldots h_{model,N}\,]^T$ as a function of $\mathbf{p}_{ag}$ and $\mathbf{o}_{ag}$ while the model parameters $\tilde{\alpha}_n$, $\mathbf{v}_{MP,n}$, $\mathbf{p}_n$, $\mathbf{o}_n$ are fixed $\forall n$. The least-squares estimate of $\mathbf{p}_{ag}$

and $\mathbf{o}_{\mathrm{ag}}$ is obtained by solving the optimization problem

$$(\hat{\mathbf{p}}_{\mathrm{ag}}, \hat{\mathbf{o}}_{\mathrm{ag}}) \in \underset{\mathbf{p}, \mathbf{o}}{\arg\min} \|\mathbf{h}_{\mathrm{meas}} - \mathbf{h}_{\mathrm{model}}\|^2 \quad \text{subject to} \quad \|\mathbf{o}\|^2 = 1. \tag{7.26}$$

The spherical parametrization (7.8) of $\mathbf{o}$ yields an unconstrained non-linear least squares problem in five dimensions,

$$\left(\hat{\mathbf{p}}_{\mathrm{ag}}, \hat{\phi}_{\mathrm{ag}}, \hat{\theta}_{\mathrm{ag}}\right) = \underset{\mathbf{p}, \phi, \theta}{\arg\min} \|\mathbf{h}_{\mathrm{meas}} - \mathbf{h}_{\mathrm{model}}\|_2^2 = \underset{\mathbf{p}, \phi, \theta}{\arg\min} \sum_{n=1}^{N} |h_{\mathrm{meas},n} - h_{\mathrm{model},n}|^2. \tag{7.27}$$

If the model error $\boldsymbol{\epsilon}$ has a circularly-symmetric Gaussian distribution $\boldsymbol{\epsilon} \sim \mathcal{CN}(\mathbf{0}, \mathbf{K}_{\boldsymbol{\epsilon}})$ then the maximum-likelihood estimate (MLE) is given by those values $\mathbf{p}, \phi, \theta$ which maximize the log-likelihood function (7.14). If furthermore $\mathbf{K}_{\boldsymbol{\epsilon}}$ does not depend on the agent location then the maximum-likelihood estimate becomes

$$\left(\hat{\mathbf{p}}_{\mathrm{ag}}, \hat{\phi}_{\mathrm{ag}}, \hat{\theta}_{\mathrm{ag}}\right) = \underset{\mathbf{p}, \phi, \theta}{\arg\min}(\mathbf{h}_{\mathrm{meas}} - \mathbf{h}_{\mathrm{model}})^{\mathrm{H}}\mathbf{K}_{\boldsymbol{\epsilon}}^{-1}(\mathbf{h}_{\mathrm{meas}} - \mathbf{h}_{\mathrm{model}}). \tag{7.28}$$

The extension to the case without circular symmetry is straightforward. If $\mathbf{K}_{\boldsymbol{\epsilon}}$ is a scaled identity matrix $\mathbf{K}_{\boldsymbol{\epsilon}} = \sigma^2 \mathbf{I}_N$ then the least-squares estimate (7.27) and the MLE (7.28) are equivalent.

Closed-form solutions are unavailable for these optimization problems because of the involved non-linear function $\mathbf{h}_{\mathrm{model}}(\mathbf{p}, \phi, \theta)$. Still, we can tackle them with numerical optimization methods. In particular we focus on iterative gradient-based methods which are suitable because the objective is a continuous and continuously differentiable function. To this effect, related optimization methods such as simulated annealing or particle swarm optimization (as used by [226] for magneto-inductive localization) are not considered. We apply a nonlinear least-squares solver (namely the Matlab function `lsqnonlin` with the `levenberg-marquardt` algorithm option [227])[13] to (7.28), using a certain $\boldsymbol{\psi}$ as initialization. We refer to this procedure as the ML5D algorithm. The objective function is non-convex (this can be seen at the example in Fig. 7.8) and thus the gradient search may converge to a local minimum [228]. In this case the

---

[13]The `trust-region-reflective` algorithm option results in very similar estimation accuracy as the `levenberg-marquardt` option but is a bit slower computationally, hence we prefer `levenberg-marquardt`. The solver is provided the $5 \times N$ Jacobian holding all the geometric error gradients, which follow directly from Proposition 7.4. In this context it shall be noted that during the gradient search there is no need to constrain the value range of ϕ and θ because any periodical ambiguity leads to the same $\mathbf{o}_{\mathrm{ag}}$. Such constraints could even cause a solver to get stuck at an interval boundary and thus impair its ability to find the global optimum. Likewise, we do not constrain the position search space in any way.

estimate will usually deviate severely from the actual MLE.[14] We will find that this happens frequently. Moreover, the ML5D approach has a large mean execution time which prevents real-time localization with a fast update rate. It is thus unsuitable for accurate real-time localization with a fast update rate. In order to improve on these shortcomings we propose and study alternative algorithms in the following.

7.4.2 Multilateration from Hard Distance Estimates

Given the measurements $h_{\text{meas},n}$ we compute the maximum-likelihood distance estimates $\hat{r}_n^{\text{ML}}$ with the simple formula (7.6) for all anchors $n = 1 \dots N$. These distance estimates are then used for multilateration, i.e. we estimate the agent position by applying a gradient-based solver (quasi-Newton method) to the least-squares problem

$$\hat{\mathbf{p}}_{\text{ag}} = \arg\min_{\mathbf{p}} \sum_{n=1}^{N} \left(r_n(\mathbf{p}) - \hat{r}_n^{\text{ML}} \right)^2 . \qquad (7.29)$$

Its appealing mathematical properties are discussed in [229, Sec. 4.4.1.2]. We will find that this method converges exceptionally fast but has very poor accuracy. We interpret this as a consequence of the rough distance estimates and of general shortcomings of the multilateration method for involved distance error statistics (e.g., see [51]).

7.4.3 Magnetic Eggs Algorithm

The above multilateration approach might be improvable by considering the soft distance information in the likelihood functions $L_{\text{NF},n}(r_n)$ from (7.5), e.g. by heuristically calculating $\mathbf{p}$ which minimizes $\prod_{n=1}^{N} L_{\text{NF},n}(r_n(\mathbf{p}))$, instead of using the hard estimates $\hat{r}_n^{\text{ML}}$. In the following we describe a similar yet mathematically simpler approach which pays more attention to the geometrical information per anchor.

First, we consider channel model (7.2), discard the far-field term, and assume that this model is without error ($\epsilon_n \equiv 0$). Then $h_{\text{meas},n} = h_{\text{model},n} = \tilde{\alpha}_n \left(\frac{1}{(kr)^3} + \frac{j}{(kr)^2} \right) J_{\text{NF},n}$ and

---

[14]By 'actual MLE' we refer to the formally true solution of the maximum-likelihood optimization problem (7.28), irregardless of any problems that may occur in computing it. In the following evaluations we emulate the actual MLE by running ML5D initialized at true agent location ψ. The meaningfulness of this emulation was confirmed by omitted experiments which showed that the bias and deviation between RMS error and position error bound are both negligibly small with this approach. Those properties are characteristic for the MLE when N is large [223] (our choice $N = 8$ is reasonably large). In this context note that if $\mathbf{h}_{\text{model}}$ was an affine map of ψ then $\hat{\psi} \sim \mathcal{N}(\psi, \mathcal{I}_\psi^{-1})$ would hold for the MLE; thus the position error bound would hold with equality. This is not the case, however the property holds in good approximation because $\mathbf{h}_{\text{model}}$ can be described by a first-order Taylor approximation (which is affine) in a vicinity of the true ψ (a small vicinity at high SNR).

we obtain $|h_{\text{meas},n}| = |\tilde{\alpha}_n|\sqrt{\frac{1}{(kr)^6} + \frac{1}{(kr)^4}}\,|J_{\text{NF},n}|$. For fixed $\mathbf{p}_{\text{ag}}$ but random $\mathbf{o}_{\text{ag}}$ (uniform distribution in 3D) we know from (4.13) that $|J_{\text{NF},n}|\big|\,\beta_{\text{NF},n} \sim \mathcal{U}(0,\beta_{\text{NF},n})$, associated with the PDF $\frac{1}{\beta_{\text{NF},n}}\mathbb{1}_{[0,\beta_{\text{NF},n}]}(|J_{\text{NF},n}|)$. A change of variables yields the PDF of $|h_{\text{meas},n}|$,

$$f\big(|h_{\text{meas},n}|;\mathbf{p}_{\text{ag}}\big) = \frac{\mathbb{1}_{[0,\beta_{\text{NF},n}g_n]}(|h_{\text{meas},n}|)}{\beta_{\text{NF},n}\,g_n}, \qquad g_n = |\tilde{\alpha}_n|\sqrt{\frac{1}{(kr_n)^6} + \frac{1}{(kr_n)^4}}. \qquad (7.30)$$

Note that the dependence on $\mathbf{p}_{\text{ag}}$ is via both r_n and $\beta_{\text{NF},n}$. We generalize the result to the case of an erroneous measurement according to the pragmatic model $|h_{\text{meas},n}|\big|\,|h_{\text{model},n}| \sim \mathcal{N}(|h_{\text{model},n}|,\sigma^2)$. By performing the convolution of the Gaussian PDF and the uniform PDF of $|h_{\text{model},n}|$ one can show that the resulting PDF is $\frac{1}{\beta_{\text{NF},n}\,g_n}\Big(Q\big(\frac{|h_{\text{meas},n}|-\beta_{\text{NF},n}g_n}{\sigma}\big) - Q\big(\frac{|h_{\text{meas},n}|}{\sigma}\big)\Big)$, reminiscent of a soft indicator function on the interval $[0,\beta_{\text{NF},n}g_n]$ and with smoothness parameter σ. We note that the non-sensical circumstance $|h_{\text{meas},n}| < 0$ can occur under this error model; the right-hand Q-function is an artifact thereof and is thus discarded. We obtain

$$f\big(|h_{\text{meas},n}|;\mathbf{p}_{\text{ag}}\big) = \frac{Q\big(\frac{|h_{\text{meas},n}|-\beta_{\text{NF},n}g_n}{\sigma}\big)}{\beta_{\text{NF},n}\,g_n}. \qquad (7.31)$$

Given measurements $h_{\text{meas},1},\ldots,h_{\text{meas},N}$ we consider the product of likelihood functions $L(\mathbf{p}) = \prod_{n=1}^{N} f(|h_{\text{meas},n}|;\mathbf{p})$ which is itself considered as a likelihood function. If the distributions $|J_{\text{NF},n}|\big|\,\beta_{\text{NF},n}$ (as well as the errors with variance σ^2) were statistically independent across $n = 1\ldots N$ then $L(\mathbf{p})$ would be a legitimate likelihood function, although we know from Cpt. 4 that this is not the case. Yet we use this heuristic construct for localization and refer to the minimization of $-\log L(\mathbf{p}_{\text{ag}})$ with a gradient-based solver (quasi-Newton method) as the Magnetic Eggs algorithm.

To best understand the concept and the naming we consider the magnetoquasistatic regime where $g_n = \frac{|\tilde{\alpha}_n|}{(kr_n)^3}$ and we furthermore set $\sigma = 0$, i.e. we use (7.30). We obtain

$$f\big(|h_{\text{meas},n}|;\mathbf{p}\big) = \begin{cases} \frac{r_n^3}{\beta_{\text{NF},n}|\tilde{\alpha}_n/k^3|} & 0 \leq r_n \leq \left|\frac{\beta_{\text{NF},n}\tilde{\alpha}_n/k^3}{h_{\text{meas},n}}\right|^{1/3} \\ 0 & \text{otherwise} \end{cases}. \qquad (7.32)$$

In the considered simple case it is supported inside an egg-shaped manifold in $\mathbb{R}^3$ (formally not an ellipsoid, but with very similar shape) which is illustrated in Fig. 7.7a. It is centered at the anchor position $\mathbf{p}_n$ and its major axis, which is $\sqrt[3]{2}$ times longer than the minor axes[15], is along $\mathbf{o}_n$. With $\sigma > 0$ we would see a smooth transition to

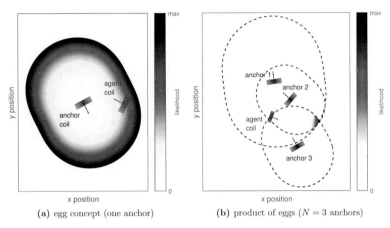

(a) egg concept (one anchor)　　　(b) product of eggs ($N = 3$ anchors)

Figure 7.7: Illustration of the likelihood function underlying the introduced Magnetic Eggs algorithm, evaluated for the magnetoquasistatic and error-free case described by (7.32). The same egg shape was indicated in [104] and a related distance bounding idea given in [26].

zero at the egg boundary instead of the immediate drop. The illustration in Fig. 7.7b shows an example of $\prod_{n=1}^{N} f(|h_{\mathrm{meas},n}| \, ; \, \mathbf{p})$ for $N = 3$, i.e. the intersection of three eggs. We observe that the true agent position deviates significantly from the position with maximum likelihood. This deviation is due to the wrong assumption of statistical independence between $J_{\mathrm{NF},1} \ldots J_{\mathrm{NF},N}$ that underlies the approach.

7.4.4　Dimensionality Reduction (ML3D Algorithm)

We attribute the convergence problems of ML5D to its high-dimensional parameter space, which is reduced to 3D in a rigorous fashion in the following. We write the employed channel model (7.2) as

$$\mathbf{h}_{\mathrm{model}} = \mathbf{V}_{\mathbf{p}}^{\mathrm{T}} \mathbf{o} \,, \qquad \mathbf{V}_{\mathbf{p}} = \left[\tilde{\alpha}_1 \left(\mathbf{v}_{\mathrm{Dir},1} + \mathbf{v}_{\mathrm{MP},1} \right), \, \ldots \, , \, \tilde{\alpha}_N \left(\mathbf{v}_{\mathrm{Dir},N} + \mathbf{v}_{\mathrm{MP},N} \right) \right] \qquad (7.33)$$

whereby $\mathbf{V}_{\mathbf{p}} \in \mathbb{R}^{3 \times N}$ is indicated as a function of position hypothesis $\mathbf{p}$. For the following approach we consider $\mathbf{p}$ fixed and for pragmatic reasons require $\mathbf{h}_{\mathrm{meas}} \approx \mathbf{V}_{\mathbf{p}}^{\mathrm{T}} \mathbf{o}$

---

[15]It is noteworthy that the maximum-likelihood distance estimate $\hat{r}_n^{\mathrm{ML}}$ in (7.6) is equal to the minor axis length of the egg shape. One can see how the likelihood function of r_n in Fig. 7.4a would emerges from the egg shape in Fig. 7.7a by marginalization of the anchor-to-agent direction $\mathbf{u}$ if it had uniform distribution in 3D.

in order to obtain an estimate $\hat{\mathbf{o}}_{\mathbf{p}}$ given $\mathbf{p}$. In particular we compute

$$\hat{\mathbf{o}}_{\mathbf{p}} = \arg\min_{\mathbf{o}} \ (\mathbf{h}_{\text{meas}} - \mathbf{V}_{\mathbf{p}}^{\text{T}}\mathbf{o})^{\text{H}}\mathbf{K}_{\epsilon}^{-1}(\mathbf{h}_{\text{meas}} - \mathbf{V}_{\mathbf{p}}^{\text{T}}\mathbf{o}) \ \text{ subject to } \ \|\mathbf{o}\|^2 = 1 \,. \qquad (7.34)$$

analogous to the maximum-likelihood rule (7.28). We demonstrate that this problem can be solved analytically. We use the error-whitened and real-valued reformulation

$$\hat{\mathbf{o}}_{\mathbf{p}} = \arg\min_{\mathbf{o}} \ \|\boldsymbol{\gamma} - \mathbf{A}_{\mathbf{p}}^{\text{T}}\mathbf{o}\|^2 \ \text{ subject to } \ \|\mathbf{o}\|^2 = 1 \qquad (7.35)$$

with $\boldsymbol{\gamma} = [\text{Re}(\mathbf{K}_{\epsilon}^{-1/2}\mathbf{h}_{\text{meas}})^{\text{T}}, \ \text{Im}(\mathbf{K}_{\epsilon}^{-1/2}\mathbf{h}_{\text{meas}})^{\text{T}}]^{\text{T}} \in \mathbb{R}^{2N}$ and, likewise, $\mathbf{A}_{\mathbf{p}}^{\text{T}} \in \mathbb{R}^{2N\times 3}$ is formed by stacking the real- and imaginary parts of $\mathbf{K}_{\epsilon}^{-1/2}\mathbf{V}_{\mathbf{p}}^{\text{T}}$. The optimization problem (7.35) is a linear least-squares problem with a quadratic equality constraint which is solved in the following according to the theory presented in [230]. By considering the stationary points of the Lagrange function associated with (7.35) we find

$$\hat{\mathbf{o}}_{\mathbf{p}}(\lambda) = \left(\mathbf{A}_{\mathbf{p}}\mathbf{A}_{\mathbf{p}}^{\text{T}} + \lambda\,\mathbf{I}_3\right)^{-1}\mathbf{A}_{\mathbf{p}}\,\boldsymbol{\gamma} \qquad (7.36)$$

as a function of the Lagrange multiplier λ (not to be confused with the wavelength). The solution to (7.34) is given by $\hat{\mathbf{o}}_{\mathbf{p}}(\lambda^*)$ where λ^* is the largest λ which satisfies the constraint $\|\hat{\mathbf{o}}_{\mathbf{p}}(\lambda)\|^2 = 1$ (we do not elaborate on special cases that occur with probability zero for our estimation problem from a random observation; for details please refer to [230]). In order to find λ^*, we use a reformulation [230]

$$\|\hat{\mathbf{o}}_{\mathbf{p}}(\lambda)\|^2 = \sum_{i=1}^{3} \frac{\mu_i c_i^2}{(\mu_i + \lambda)^2} = 1 \qquad (7.37)$$

based on the eigenvalue decomposition of 3×3 matrix $\mathbf{A}_{\mathbf{p}}\mathbf{A}_{\mathbf{p}}^{\text{T}} = \sum_{i=1}^{3} \mu_i \mathbf{a}_i \mathbf{a}_i^{\text{T}}$ and $c_i = \mathbf{a}_i^{\text{T}}\boldsymbol{\gamma}$. We can now compute λ^* efficiently by finding the largest real root of a sixth-order polynomial in λ which arises by multiplying (7.37) with its three denominators.

Using this rule for $\hat{\mathbf{o}}_{\mathbf{p}}$, we can compute a maximum-likelihood position estimate by solving an unconstrained optimization problem with 3D parameter space,

$$\hat{\mathbf{p}}_{\text{ag}} = \arg\min_{\mathbf{p}} \ \|\boldsymbol{\gamma} - \mathbf{A}_{\mathbf{p}}^{\text{T}}\hat{\mathbf{o}}_{\mathbf{p}}\|^2 \qquad (7.38)$$
$$= \arg\min_{\mathbf{p}} \ (\mathbf{h}_{\text{meas}} - \mathbf{V}_{\mathbf{p}}^{\text{T}}\hat{\mathbf{o}}_{\mathbf{p}})^{\text{H}}\mathbf{K}_{\epsilon}^{-1}(\mathbf{h}_{\text{meas}} - \mathbf{V}_{\mathbf{p}}^{\text{T}}\hat{\mathbf{o}}_{\mathbf{p}}) \,.$$

The interplay of (7.34) and (7.38) can be regarded as alternating minimization [231, Eq. 1.1] of the negative log-likelihood in (7.28), whereby the very fast and reliable rule for computing $\hat{\mathbf{o}}_{\mathbf{p}}$ however effectively reduces the problem from 5D to 3D ($\hat{\mathbf{o}}_{\mathbf{p}}$ can be regarded as a simple function of $\mathbf{p}$).

We define the ML3D algorithm as the application of the Levenberg-Marquardt solver to (7.38), using a certain initial $\mathbf{p}$. Later in the chapter we will see that ML3D has much faster convergence speed than ML5D, owing to the dimensionality reduction. The robustness improvement is minor though: just like ML5D, ML3D does not converge to the global optimum reliably. This flaw is tackled in the following.

7.4.5   Smoothing the Cost Function (WLS3D Algorithm)

The introduced ML5D and ML3D algorithms suffer from convergence to local cost function extrema. We attribute part of the problem to the high dynamic measurement range due to path loss: the values $h_{\mathrm{meas},n}$ with largest absolute value will usually dominate the squared-error term (7.38) at most position hypotheses. As a result, the majority of anchors is effectively ignored in early solver iterations, which hinders convergence to the global optimum. Apparently this problem could be avoided with a more balanced cost function.

We consider the distances between a hypothesis $\mathbf{p}$ and the anchor positions $\mathbf{p}_n$, collected in

$$\mathbf{R}_{\mathbf{p}} = \mathrm{diag}\left(r_1(\mathbf{p}), \ldots, r_N(\mathbf{p})\right), \qquad r_n(\mathbf{p}) = \|\mathbf{p} - \mathbf{p}_n\|. \qquad (7.39)$$

We use them in the computation of a weighted least-squares position estimate

$$\hat{\mathbf{p}}_{\mathrm{ag}} = \arg\min_{\mathbf{p}} \sum_{n=1}^{N} \left| r_n^3(\mathbf{p})\left(h_{\mathrm{meas},n} - h_{\mathrm{model},n}(\mathbf{p}, \hat{\mathbf{o}}_{\mathbf{p}})\right)\right|^2 \qquad (7.40)$$

$$= \arg\min_{\mathbf{p}} \|\mathbf{R}_{\mathbf{p}}^3(\mathbf{h}_{\mathrm{meas}} - \mathbf{V}_{\mathbf{p}}^{\mathrm{T}}\hat{\mathbf{o}}_{\mathbf{p}})\|^2$$

where $\mathbf{V}_{\mathbf{p}}$ is from (7.33). As corresponding orientation estimate we compute

$$\hat{\mathbf{o}}_{\mathbf{p}} = \arg\min_{\mathbf{o}} \|\mathbf{R}_{\mathbf{p}}^3(\mathbf{h}_{\mathrm{meas}} - \mathbf{V}_{\mathbf{p}}^{\mathrm{T}}\mathbf{o})\|^2 \ \text{ subject to } \ \|\mathbf{o}\|^2 = 1 \qquad (7.41)$$

which is solved analytically just like (7.35). We refer to the application of the Levenberg-Marquardt non-linear least squares solver to (7.40) as WLS3D algorithm.[16]

The idea of this distance-dependent weighting $r_n^3(\mathbf{p})$ is to map all observations

---

[16]In our implementation we multiply the cost function by the constant $N/\mathrm{tr}(\mathbf{K}_\epsilon)$ to avoid that the output values are many orders of magnitude smaller than those of ML5D and ML3D.

onto a common value range. To see this, consider the magnetoquasistatic model $h_{\mathrm{model},n} = \frac{\tilde{\alpha}_n}{(kr)^3} J_{\mathrm{NF},n}$ and the weighted term $r_n^3 h_{\mathrm{model},n} = \frac{\tilde{\alpha}_n}{k^3} J_{\mathrm{NF},n}$. This term should have similar order of magnitude across all n (at least at $\mathbf{p} = \mathbf{p}_{\mathrm{ag}}$); the same applies for the weighted observation $r_n^3 h_{\mathrm{meas},n}$. This prevents a degenerated value range of the cost function summands and achieves an effect comparable to a cost function relaxation in comparison to ML3D, as can be observed in Fig. 7.8.

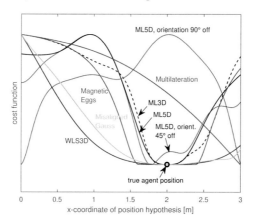

Figure 7.8: An example showing the spatial cost function evolution of the introduced gradient-based location estimation algorithms. The example considers agent position hypotheses $\mathbf{p}$ on a line inside the anchor setup of Fig. 7.1. The particular line is $\mathbf{p} = [x, \ 0.5\,\mathrm{m}, \ 1.5\,\mathrm{m}]^{\mathrm{T}}$ for $x \in [0\,\mathrm{m}, 3\,\mathrm{m}]$; it comprises the true $\mathbf{p}_{\mathrm{ag}}$ at $x = 2\,\mathrm{m}$. The evaluation shows three different choices for the orientation hypothesis required by the ML5D cost function: the true orientation orientations, one offset by by $45°$, and one orthogonal. The cost values were scaled and shifted to the same value range for visualization purposes.

Even in the case of global optimality, the WLS3D algorithm may exhibit slightly lower accuracy than ML3D and ML5D. This is because of the absence of noise whitening and because the weighting affects the model error components individually and, in consequence, the estimates (7.40) and (7.41) deviate from the maximum-likelihood estimate. This is however easily mitigated by cascading WLS3D and ML3D (initialized at the WLS3D estimate).

7.4.6 Misaligned Gauss Algorithm

We present a last algorithm design based on the same ideas as WLS3D, but now the dimensionality reduction is achieved via the random misalignment theory developed in

Cpt. 4. In this sense, we assume that $\mathbf{o}_{ag}$ is random with uniform distribution on the 3D unit sphere. We use $\mathbf{h}_{model} = \mathbf{V}_{\mathbf{p}}^T \mathbf{o}$ as defined by (7.33). Inspired by Sec. 7.4.4 we use the distant-dependent scaling $\boldsymbol{\delta} = \mathbf{R}_{\mathbf{p}}^3 \mathbf{h}_{model}$ with $\mathbf{R}_{\mathbf{p}} = \mathrm{diag}(r_1 \ldots r_N)$, leading to the model $\boldsymbol{\delta} = \mathbf{B}_{\mathbf{p}} \mathbf{o}_{ag}$ with $\mathbf{B}_{\mathbf{p}} = \mathbf{R}_{\mathbf{p}}^3 \mathbf{V}_{\mathbf{p}}^T \in \mathbb{C}^{N \times 3}$. From Proposition 4.8 we know that the random vector $\boldsymbol{\delta} = \mathbf{B}_{\mathbf{p}} \mathbf{o}_{ag}$ has zero mean and covariance matrix $\frac{1}{3} \mathbf{B}_{\mathbf{p}} \mathbf{B}_{\mathbf{p}}^H$.

We consider the scaled measurement $\boldsymbol{\zeta} = \mathbf{R}_{\mathbf{p}}^3 \mathbf{h}_{meas}$ and assume that $\boldsymbol{\zeta}$ and $\boldsymbol{\delta}$ deviate by a random error $\mathbf{e} = \boldsymbol{\zeta} - \boldsymbol{\delta}$ (statistically independent of $\mathbf{o}_{ag}$) with zero mean and covariance matrix $\boldsymbol{\Sigma}$. As a result the observation $\boldsymbol{\zeta} = \mathbf{B}_{\mathbf{p}} \mathbf{o}_{ag} + \mathbf{e}$ has zero mean and covariance matrix $\frac{1}{3} \mathbf{B}_{\mathbf{p}} \mathbf{B}_{\mathbf{p}}^H + \boldsymbol{\Sigma}$. To keep the mathematics tractable we assume a Gaussian distribution $\boldsymbol{\zeta} \sim \mathcal{CN}(\mathbf{0}, \frac{1}{3} \mathbf{B}_{\mathbf{p}} \mathbf{B}_{\mathbf{p}}^H + \boldsymbol{\Sigma})$. This certainly contradicts the specific PDF given by Proposition 4.2 with its ellipse-shaped support, although the color plots in Fig. 4.4 indicate that a Gaussian model with the same covariance matrix is at least a sensible approximation. For this Gaussian model we can easily state the PDF of $\boldsymbol{\zeta}$,

$$f(\boldsymbol{\zeta}; \mathbf{p}) = \frac{\exp\left(-\boldsymbol{\zeta}^H (\frac{1}{3} \mathbf{B}_{\mathbf{p}} \mathbf{B}_{\mathbf{p}}^H + \boldsymbol{\Sigma})^{-1} \boldsymbol{\zeta}\right)}{\pi^N \det\left(\frac{1}{3} \mathbf{B}_{\mathbf{p}} \mathbf{B}_{\mathbf{p}}^H + \boldsymbol{\Sigma}\right)} \tag{7.42}$$

which is parameterized by the position hypothesis $\mathbf{p}$. For a fixed measurement $\mathbf{h}_{meas}$ the expression becomes a likelihood function of $\mathbf{p}$. By writing the negative log-likelihood function and discarding irrelevant summands we obtain the cost function

$$C(\mathbf{p}) = \log \det\left(\frac{1}{3} \mathbf{B}_{\mathbf{p}} \mathbf{B}_{\mathbf{p}}^H + \boldsymbol{\Sigma}\right) + \boldsymbol{\zeta}^H \left(\frac{1}{3} \mathbf{B}_{\mathbf{p}} \mathbf{B}_{\mathbf{p}}^H + \boldsymbol{\Sigma}\right)^{-1} \boldsymbol{\zeta}. \tag{7.43}$$

We define the Misaligned Gauss algorithm as the application of a gradient-based solver (quasi-Newton method) for the three-dimensional minimization of this cost function.

We have yet to discuss the covariance matrix $\boldsymbol{\Sigma}$. Formally it should have the value $\mathbf{R}_{\mathbf{p}}^3 \mathbf{K}_{\boldsymbol{\epsilon}} \mathbf{R}_{\mathbf{p}}^3$ based on the covariance matrix $\mathbf{K}_{\boldsymbol{\epsilon}}$ of the model error $\boldsymbol{\epsilon} = \mathbf{h}_{meas} - \mathbf{h}_{model}$. However, we found that the algorithm becomes vastly more robust with a fixed choice, e.g. $\boldsymbol{\Sigma} = \sigma^2 \mathbf{I}_N$ with $\sigma = 10^{-6}$ as error level.[17] The considered key purpose of $\boldsymbol{\Sigma}$ is smoothing the cost function and ensuring that $\frac{1}{3} \mathbf{B}_{\mathbf{p}} \mathbf{B}_{\mathbf{p}}^H + \boldsymbol{\Sigma}$ has full rank rather than accurately reflecting the error statistics.

---

[17]With a scaled identity covariance matrix $\boldsymbol{\Sigma} = \sigma^2 \mathbf{I}_N$ the matrix inversion $(\frac{1}{3} \mathbf{B}_{\mathbf{p}} \mathbf{B}_{\mathbf{p}}^H + \boldsymbol{\Sigma})^{-1} = (\frac{1}{3} \mathbf{B}_{\mathbf{p}} \mathbf{B}_{\mathbf{p}}^H + \sigma^2 \mathbf{I}_N)^{-1}$ in (7.43) can equivalently be written as $\frac{1}{\sigma^2} \left(\mathbf{I}_N - \frac{1}{3} \mathbf{B}_{\mathbf{p}} (\frac{1}{3} \mathbf{B}_{\mathbf{p}}^H \mathbf{B}_{\mathbf{p}} + \sigma^2 \mathbf{I}_3)^{-1} \mathbf{B}_{\mathbf{p}}^H\right)$ via the Woodbury matrix identity. This reduces the computational effort to a 3×3 inversion.

7.4.7 Algorithm Performance Evaluation

We compare the introduced algorithms in terms of numerical accuracy, execution speed, and robustness (i.e. the odds of computing an estimate with useful accuracy). We consider the same parameters as the evaluation in Sec. 7.3, i.e. $N = 8$ anchors which surround a cubic volume of $3\,\mathrm{m}$ side length. We sample $\mathbf{p}_{\mathrm{ag}}$ and any initial solver $\mathbf{p}$ uniformly from a slightly smaller cubic volume ($2.4\,\mathrm{m}$ side length) with equal center position and $\mathbf{o}_{\mathrm{ag}}$ and any initial solver $\mathbf{o}$ uniformly from the 3D unit sphere.[18] The gradient solvers terminate when an iterative parameter update $\|\hat{\boldsymbol{\psi}}_{i+1} - \hat{\boldsymbol{\psi}}_i\|$ of less than 10^{-15} occurs or after $i > 1000$ iterations (which usually only takes effect for ML5D).

For a single run of the different algorithms from a random initialization, we observe the statistics of the position error in Fig. 7.9a and of the execution time in Fig. 7.11a. The robustness of ML5D is only 36%; it suffers from local convergence otherwise (as can be seen from the comparison to the actual MLE[19]). Furthermore its execution time frequently exceeds $1\,\mathrm{s}$. This slow converge prevents the use of a multi-start approach to solve the robustness issue for real-time applications.

The multilateration approach and the Magnetic Eggs algorithm do not provide a remedy since their accuracy is just terrible.[20] We attribute this to their strong reliance on formally wrong assumptions such as statistical independence of $\mathbf{h}_{\mathrm{meas},n}$ across $n = 1\ldots N$, a strong hint that simplified approaches are inadequate for the considered estimation problem.

The ML3D algorithm shows improved (yet still unsatisfactory) robustness of 50% and significantly faster and predictable execution time ($32\,\mathrm{ms}$ median) in comparison to ML5D. The WLS3D algorithm brings a vast improvement by raising the robustness to about 76%, with a fast and stable execution time of $27\,\mathrm{ms}$ in the median. The cascade of WLS3D and ML3D improves these numbers to about 77% and $42\,\mathrm{ms}$ (but sacrifices

---

[18]We tested several solver initialization heuristics such as choosing the SNR-weighted center of anchor positions as initial position. The resulting improvements over random initialization were appreciable but we considered them too insignificant for inclusion, for the sake of clarity.

[19]We use the MLE as benchmark instead of the position error bound because the latter applies to the RMS error which is not a suitable measure for estimation algorithms which are either very accurate or very inaccurate, depending on whether they found the global cost function minimum. In this context please recall the discussion in Sec. 7.4.1, especially Footnote 14, about the RMS error of the MLE being very close to the position error bound unless the error levels in $\mathbf{K}_\epsilon$ are very large.

[20]The Magnetic Eggs algorithm is furthermore rife with numerical problems, which can be observed in Fig. 7.8 and by comparing Fig. 7.9a and 7.11a with Fig. 7.9d and 7.11b. These problems are caused by the trade-off of choosing a large smoothness parameter σ to prevent ∞ cost occurring at the initial position (cf. the large region of zero likelihood in Fig. 7.7b) and choosing a small σ to not distort the original geometry-information-carrying egg shape too much. In the light of the poor performance of the approach even with perfect initialization we shall not discuss these issues any further.

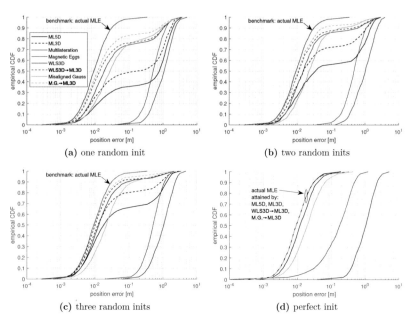

Figure 7.9: Statistics of the position error $\|\mathbf{p}_{ag} - \mathbf{p}_n\|$ of the different algorithms with $N = 8$ anchors and random initialization.

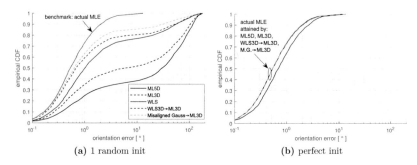

Figure 7.10: Statistics of the orientation error $\arccos(\hat{\mathbf{o}}_{ag}^{\mathrm{T}}\mathbf{o}_{ag})$ of the different algorithms with $N = 8$ anchors and random initialization.

some execution time stability). Even better performance numbers are demonstrated by the cascade of Misaligned Gauss and ML3D, which features 85% robustness at 40 ms

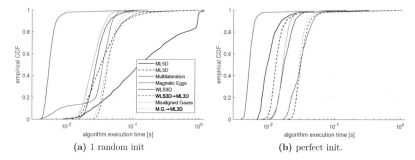

(a) 1 random init   **(b)** perfect init.

Figure 7.11: Statistics of the algorithm execution times with $N = 8$ anchors and random initialization. The experiments were conducted on a personal computer with an Intel Core i7-4790 processor (3.6 GHz) and 16 GB RAM.

median execution time (about 79% and 24 ms for the standalone Misaligned Gauss). A further advantage of the cascading approaches WLS3D $\rightarrow$ ML3D and Misaligned Gauss $\rightarrow$ ML3D is that they mitigate the slight accuracy deviations of the WLS3D and Misaligned Gauss standalone approaches from the theoretical limit (constituted by the MLE). The success of cascading indicates the cost function minimum associated with WLS3D or Misaligned Gauss is usually close to the position of maximum likelihood.

To some extend the remaining non-convexity issues[21] can be addressed by running an algorithm several times from different initializations and then picking the estimate with the smallest residual cost. Fig. 7.9c shows the error statistics for 3 random initializations. We observe a robustness of 94% for the WLS3D $\rightarrow$ ML3D cascade and even 96% for the Misaligned Gauss $\rightarrow$ ML3D cascade. These are very promising numbers for instantaneous estimates with no temporal filtering (yet). We conclude that these two algorithms are suitable for robust and real-time operation of a low-frequency magneto-inductive localization system for a single-coil agent.

---

[21]It is noteworthy that recent work has employed magneto-inductive passive relays (the topic of Cpt. 5) to improve magneto-inductive localization. In particular, they have been considered for resolving ambiguities in position estimation [26, 172] and to improve the localization accuracy [232].

7.5 Indoor Localization System Implementation

We present a system implementation for the purpose of accompanying the theoretical accuracy results with practical ones and investigating the practical performance limits.

7.5.1 System & Coil Design

We present a system implementation with $N = 8$ anchor coils, operating at $f_c = 500\,\text{kHz}$ ($\lambda = 600\,\text{m}$). The employed anchor topology is shown in Fig. 7.12; it is very similar to the topology considered throughout the theoretical results (minor differences are due to obstacles such as the depicted bookshelf and tripod adjustability). The system repeatedly acquires measurements $\mathbf{h}_{\text{meas}} \in \mathbb{C}^N$ of the agent-to-anchor channel coefficients. For the presented study this is implemented by connecting all nine matched coils (the agent and eight anchors) to a multiport network analyzer (Rohde & Schwarz ZNBT8) via coaxial cables and measuring the respective S-parameters. This way we establish phase synchronization, which would be a challenge (and potential source of error) if done wirelessly. The network analyzer is configured to use a 6 dBm probing signal and 5 kHz measurement bandwidth.

At each node, we use a spiderweb coil wounded on a plexiglas body with 10 turns each, i.e. $\mathring{N}_{\text{ag}} = \mathring{N}_{\text{an}} = 10$. The inner and outer coil diameters are 100 mm and 130 mm, respectively. We use fairly thick wire with 1 mm diameter (including insulation) to keep ohmic resistance low. The operating frequency of 500 kHz is well below the coil self-resonance frequency of about 8.7 MHz. Single-link measurements between such coils have been presented earlier in Fig. 7.3. The design frequency $f_c = 500\,\text{kHz}$ was chosen based on the trade-off between achieving large channel gains at smaller distances (f_c should be large) and suppressing interaction with the environment (f_c should be small).[22] Since the wire is orders of magnitude shorter than the 600 m wavelength, it is safe to assume a spatially constant current distribution which is necessary for our channel model (7.2) to hold. The coils have a measured resistance of $0.68\,\Omega$ and a self-inductance of $17.3\,\mu\text{H}$; the measured coil impedance is plotted versus frequency in Fig. 7.13.

---

[22]This was concluded by comparison to a 1 MHz design. When doubling f_c in this regime, the decaying graph at smaller r in Fig. 7.3 gains 3 dB (+6 dB from doubled induced voltages and −3 dB from the skin effect increasing the coil resistance) but for the leveled graph at larger r we suspect a 15 dB gain (it seems to scale like the far-field term in (7.3)). This effectively reduces the reach of the decaying graph which however holds particularly valuable location information. For an experiment at the 13.56 MHz ISM frequency, radiation was dominant and localization was impossible.

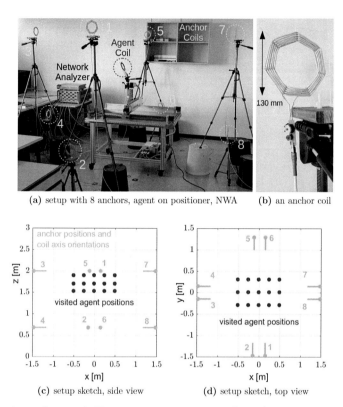

(a) setup with 8 anchors, agent on positioner, NWA   (b) an anchor coil

(c) setup sketch, side view   (d) setup sketch, top view

Figure 7.12: Setup with $N = 8$ stationary anchor coils (yellow) and a mobile agent coil (blue). The anchor-to-agent measurements are acquired with the multiport network analyzer shown in (a). The anchors are distributed on the border of a $3\,\text{m} \times 3\,\text{m}$ area as shown in (a), (c) and (d). Note the alternating elevation of the anchors: the anchors $n \in \{1, 3, 5, 7\}$ are $2\,\text{m}$ above the floor, those with $n \in \{2, 4, 6, 8\}$ are at $68\,\text{cm}$. This is to establish ample anchor spread in all three dimensions. In (c) and (d), the 45 predefined agent positions $\mathbf{p}_{\text{ag}}$ which are visited via the positioner device are shown in blue. Not shown are the six different agent orientations $\mathbf{o}_{\text{ag}}$ that will be assumed.

At each coil we use a two-port network to match the coil impedance to the connecting $50\,\Omega$ coaxial cable. In particular, each matching network is an L-structure of two capacitors, cf. Fig. 3.6. We use capacitors with a high Q-factor to prevent that their resistance impairs the Q-factor of the matched coil.

The agent coil is mounted on a positioner device (HIGH-Z S-1000 three-axis posi-

169

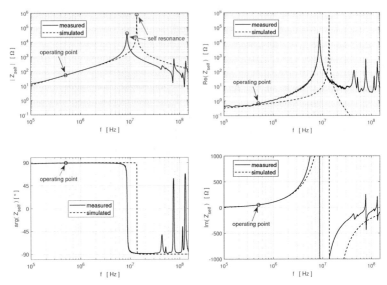

Figure 7.13: Measured impedance of a spiderweb coil versus frequency. The data is compared with a coil impedance simulation in the sense of Sec. 2.2, for a single-layer solenoid coil of comparable coil diameter ($D_c = 117$ mm), coil length ($l_c = 25$ mm) and wire diameter ($D_w = 0.5$ mm) and the same turn number ($\mathring{N} = 10$). The simulation predicts a larger self-resonance frequency because the self-capacitance C is underestimated; the actual C is possibly increased by the permittivity of the insulation and plexiglas body and the close turn proximity of the spiderweb geometry. The simulation is unreliable for larger frequencies of about $f > 8$ MHz where the coil is not electrically small (when $\lambda/10$ is shorter than the wire).

tioner) which is controlled via Matlab. This allows for accurate and automated adjustment of $\mathbf{p}_{ag}$. In particular, the agent coil is mounted to the positioner with a 2-DoF joint. This allows to adjust orientation $\mathbf{o}_{ag}$, whereby the resultant change of coil center position $\mathbf{p}_{ag}$ is duly considered. The measurement acquisition time is about 14 ms, which is a bit faster than the typical execution times of the proposed algorithms.

7.5.2 Accuracy of the Calibrated System

The following presents an evaluation of the localization accuracy of the system after thorough calibration. The positioner device was placed in the middle of the anchors (see Fig. 7.12) and was used to visit 45 different positions $\mathbf{p}_{ag}$ of the mounted agent coil, as illustrated in Fig. 7.12c and 7.12d. These positions were visited six times with six different orientations $\mathbf{o}_{\mathrm{ag}}$. In particular, we chose for $\mathbf{o}_{\mathrm{ag}}$ the canonical x, y and z directions as well as directions at $45°$ to the axes in the xy, yz and xz planes. In total, we establish $45 \cdot 6 = 270$ different agent deployments $(\mathbf{p}_{\mathrm{ag}}, \mathbf{o}_{\mathrm{ag}})$. At each of these 270 deployments the S-Parameters were measured with the network analyzer, resulting in a set of channel vector measurements $\mathbf{h}_{\mathrm{meas},i}$ for $i = 1, \ldots, 270$. In order to avoid overfitting we partition the measurements into an evaluation subset $i \in \{1, 3, 5, \ldots\}$ and a calibration subset $i \in \{2, 4, 6, \ldots\}$.

We consider an *essential calibration* which adjusts all $\tilde{\alpha}_n$ and $\mathbf{v}_{\mathrm{MP},n}$ and a *full calibration* which additionally tunes the anchor positions and orientations to compensate minor inaccuracies of the installation. Hence, essential and full calibration adjust 8 and 13 real-valued parameters per anchor, respectively. All $\tilde{\alpha}_n$ and $\mathbf{v}_{\mathrm{MP},n}$ parameters are calibrated with least-squares estimation per anchor. Afterwards, if a full calibration is conducted, each individual anchor position and orientation is calibrated individually by maximum a posteriori (MAP) estimation with informative priors. This procedures are described in full detail in [144].

The system employs the cascade of WLS3D and ML3D as localization algorithm, in a multi-start fashion from 3 random initializations. Thereby ML3D uses the empirical covariance matrix $\mathbf{K}_\epsilon$ for whitening. This algorithm is applied individually to each channel vector in the measurement evaluation subset. Fig. 7.14 shows the resulting localization error statistics. We observe a median position error of about $53\,\mathrm{mm}$ with essential calibration. Full calibration even achieves $29.9\,\mathrm{mm}$ median error and a 90% confidence to be below $65\,\mathrm{mm}$. The median error of orientation estimation is below $3°$ for both calibrations. Finally it shall be noted that the whitening operation does yield a slight accuracy improvement but is certainly not crucial.

7.6 Investigation of Practical Performance Limits

While the achieved accuracy may be sufficient for various applications, one might expect better results from a thoroughly calibrated system (see also item 2 of Sec. 1.2). The system seems to face a similar bottleneck as the related work and we want to investigate the cause. As discussed below (7.1), the key to accurate localization via parameter estimation is a small model error $\epsilon = \mathbf{h}_{\mathrm{meas}} - \mathbf{h}_{\mathrm{model}}$ (with $\mathbf{h}_{\mathrm{model}}$ evaluated at the true $\mathbf{p}_{\mathrm{ag}}, \mathbf{o}_{\mathrm{ag}}$,). Therefore, the observed residue model error is a key quantity for the study of accuracy limits as it reflects the unconsidered effect(s). Finding its dominant contribution corresponds to isolating the system's performance bottleneck

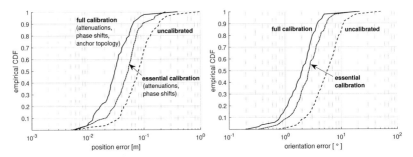

Figure 7.14: Performance of the 3D localization system in terms of position and orientation estimation error. The results are shown in terms of cumulative distribution function (CDF) of the localization error over the different visited agent deployments $(\mathbf{p}_{ag}, \mathbf{o}_{ag})$ (the 135 deployments of the evaluation set) and for different calibrations.

and might allow to even further improve the accuracy. Fig. 7.15a shows and discusses the realizations of the model error ϵ_i after full calibration for the different agent deployments i (for brevity we depict only the first component, i.e. $n = 1$). We analyze the relative model error $|\epsilon_{n,i}| \big/ |h_{\mathrm{meas}:n,i}| = |h_{\mathrm{meas}:n,i} - h_{\mathrm{model}:n,i}| \big/ |h_{\mathrm{meas}:n,i}|$ over the entire evaluation set and find a median relative error of 0.055 and 90th percentile of 0.302.

As first possible cause we investigate measurement noise. Its statistics can be observed in the fluctuations of channel gain measurements about their empirical mean for a stationary agent as illustrated in Fig. 7.15b. By comparing the deviation magnitudes in Fig. 7.15a and Fig. 7.15b, it can be seen that measurement noise is not the limiting factor for our system and that the system is not SINR limited. This also means that noise averaging of the measurement error would not improve the accuracy significantly whereas it would harm the real-time capabilities of the system.

Another possible performance bottleneck are the errors due to the assumptions that underlie the signal model of Sec. 7.1. Even in free space, the employed model is exact only between two dipoles or between a thin-wire single-turn circular loop antenna and an infinitesimally small coil. The actual coil apertures and spiderweb geometry are however neglected by the model. We evaluate the associated performance impact with the following procedure. Instead of acquiring $\mathbf{h}_{\mathrm{meas},i}$ by measurement, we synthesize it with a simulation that would be exact between thin-wire spiderweb coils in free space. In particular, we solve the double line integral of Proposition 2.1 numerically based on a 3D model of the coil geometry and the feed wire seen in Fig. 7.12b. Subsequently, we apply the same calibration and evaluation routines to the synthesized $\mathbf{h}_{\mathrm{meas},i}$ as

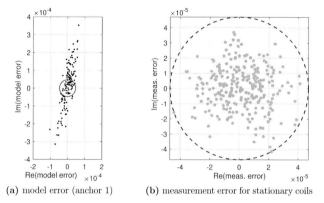

(a) model error (anchor 1) (b) measurement error for stationary coils

Figure 7.15: Scatter plot (a) shows the first component (anchor $n = 1$) of the model error $\epsilon_i = \mathbf{h}_{\mathrm{meas},i} - \mathbf{h}_{\mathrm{model},i}$ of the fully calibrated system. Thereby, i indexes the visited agent deployments (evaluation subset $i = 1, 3, 5, \ldots, 269$). The points are close to a line at about $80°$ to the real axis, an angle that can also be observed in Fig. 7.3 at small distances: in the near-field the phase is always close to $\pm 90°$ depending on polarity, with a small offset due to mismatch ξ. Now ξ multiplies $\mathbf{h}_{\mathrm{meas},i}$ and $\mathbf{h}_{\mathrm{model},i}$ and thus also ϵ_i, resulting in the observed angle of the cloud. Scatter plot (b) shows the errors of channel coefficient measurements acquired over time between stationary coils whereby the mean was subtracted. This shows the error magnitude due to noise, interference, and quantization. We observe that the standard deviation of this error carries over to the model error in (a) and how it spreads the associated cloud (red circle). Please note the different axis scales and that (a) and (b) were obtained with completely different methodology. Thus there is no one-to-one correspondence between data points of the two plots.

previously. This results in a relative model error with median 0.0114 and a 90th percentile of 0.0615, which are significantly lower than the practically observed values. Hence, this aspect does not pose the performance bottleneck.

As third possible cause investigate the model error due to unconsidered nearby conductors which react with the generated magnetic field. To this effect, we tested the impact that the ferro-concrete building structures of the setup room have on single link measurements. Indeed, moving the coils closer to a wall or the floor can affect $\mathbf{h}_{\mathrm{meas}}$ considerably. In an experiment of two coplanar coils (parallel to the floor, i.e. vertical orientation vectors) at $r = 2\,\mathrm{m}$ link distance, we compared the impact of different elevations by first choosing $0.5\,\mathrm{m}$ and then $1\,\mathrm{m}$ elevation above the floor for both coils. Although the links should be equal according to the free space model, we observed a relative deviation of 0.11 between the two h_{meas}. This exceeds the 0.055

median relative model error of our fully-calibrated system, which is a strong hint that nearby ferro-concrete building structures pose a significant performance bottleneck for our localization system.

The remainder of the section presents accuracy projections under more ideal circumstances, based on the position error bound (PEB). We evaluate PEB($\mathbf{p}_{ag}, \mathbf{o}_{ag}$) for the same agent deployments as in Sec. 7.5.2, for different possible values of $\mathbf{K}_\epsilon$ of technical relevance. In particular we determine the value of $\mathbf{K}_\epsilon$ for the following cases.

1. Empirically from model errors ϵ_i after full calibration.

2. Empirically from the measurement fluctuations observed while the agent is stationary.

3. Same as 2) but the transmitting agent coil was disconnected and replaced by a 50 Ω termination.

4. Same as 1) but the observations $\mathbf{h}_{meas,i}$ were obtained by free-space EM simulation for thin-wire spiderweb coils (using the coupling model from Proposition 2.1) instead of actual measurements.

5. Thermal noise and typical background noise picked up by the anchors at $f_c = 500\,\text{kHz}$, as described by Sec. 3.2. We use the same spatial correlation model as Cpt. 6, using a Bessel function of kd times the inner product of the two associated anchor orientations.

6. Independent thermal noise of power $N_0 B$ at each anchor. We assume the minimum noise spectral density at room temperature, i.e. $N_0 = -174\,\text{dBm}$ per Hz. The bandwidth is $B = 5\,\text{kHz}$ as specified in Sec. 7.5.1.

The results are shown in Fig. 7.16. First of all, the PEB for case 1 matches the practically achieved accuracy in Fig. 7.14 well (again it shall be noted that the PEB applies to the RMSE and not to single error realizations). The results for the cases 2 and 3 indicate the performance limit assuming that noise, interference and/or quantization determine the achievable accuracy. This case would allow for sub-cm accuracy. The PEB-results for case 4 also exhibit sub-cm accuracy in most cases but are slightly worse than cases 2 and 3. This case 4 is particularly important as it represents the accuracy limit of parametric location estimation based on an analytical signal model such as (7.2). This indicates that sub-cm accuracy is infeasible for this approach and a system of our scale, even in a distortion- and interference-free environment. The cases 5 and 6 use noise power estimates from communication theory and yield projections between 20

174

and 200 μm. As discussed earlier, these are vastly optimistic because they would require an extremely high-resolution ADC and a precise and well-calibrated signal model that accounts for any appreciable physical detail.

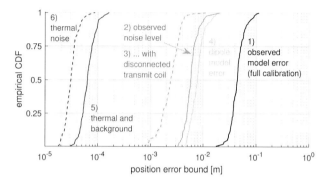

Figure 7.16: Distribution of the position error bound (PEB) over various agent deployments within the 3 m × 3 m room, evaluated for six different hypotheses on the statistics of the model error $\epsilon = \mathbf{h}_{\mathrm{meas}} - \mathbf{h}_{\mathrm{model}}$. The results serve as accuracy projections for various (idealistic) circumstances. We suspect that the significant differences between cases 2 and 3 are caused by the way the network analyzer adapts the probing signal to the link, e.g. adding dither noise to mitigate ADC non-linearity and quantization error.

Chapter 8

Distance Estimation from UWB Channels to Observer Nodes

In the previous chapter we studied an approach which operated at a very low frequency f_c in order to avoid significant interaction with the environment to begin with. Thereby, the large associated wavelength λ and small signal bandwidth B limited our options in location estimation. Now we shift our focus to a different (and more established) paradigm for localization: the use of a signal bandwidth B in the GHz range. Such ultra-wideband (UWB) signaling allows for extraction of the time of arrival from a received signal, as discussed in Sec. 7.2. Now the inevitably large f_c will cause significant interaction with the environment, namely multipath propagation (reflection, scattering and diffraction) of the radiated wave. In the UWB regime, however, there is a saving grace: if the duration of a transmitted pulse (which is about $1/B$) is much shorter than the delay spread of the propagation channel, then individual multipath components can be resolved in the received signal [36, 233]. This comprises the line-of-sight path (i.e. direct path) if it is unobstructed.

In this domain, this chapter proposes and studies a novel approach in the localization context. We study the estimation of distance r between two wireless nodes by means of their wideband channels to a *third* node, called observer. The motivating principle is that the channel impulse responses are similar for small r and drift apart when r increases. In particular we make the following contributions.

- In Sec. 8.1 we propose specific distance estimators based on the differences of path delays of the extractable multipath components. In particular, we derive such estimators for rich multipath environments and various important cases: with and without clock synchronization as well as errors on the extracted path delays (e.g. due to limited bandwidth). The estimators readily support (and benefit from) the presence of multiple observers.

- Sec. 8.2 presents an error analysis and, using ray tracing in an exemplary indoor environment, shows that the estimators perform well in realistic conditions.

- Sec. 8.3 describes possible localization applications of the proposed scheme and highlights its major advantages: it requires neither precise synchronization nor line-of-sight connection. This could make wireless user tracking feasible in dynamic indoor settings with reasonable infrastructure requirements.

We shall position our approach in relation to the state of the art in indoor localization. Most proposals for wireless localization systems rely on distance estimates to fixed infrastructure nodes (anchors) to determine the position of a mobile node [120], e.g. via trilateration. Cooperative network localization furthermore employs the distances between different mobile nodes [120, 121, 129, 130]. A simple way to obtain such inter-node distance estimates is from the received signal strength (RSS) but the resultant accuracy is usually very poor due to shadowing, small-scale fading, and antenna patterns [51, 131]. A much more sophisticated method measures the time-of-arrival (TOA) with wideband signaling and a round-trip protocol for synchronization [50, 120].

TOA-based localization schemes require involved hardware at both ends and suffer from synchronization errors and processing delays [50, 132–134]. Yet the main problem is ensuring a sufficient number of anchors in line of sight (LOS) to all relevant mobile positions [135]. TOA thus exhibits a large relative error at short distances and is not well-suited for dense and crowded settings such as lobbies, metro stations, access gates, and large events. These however entail important use cases (e.g., see [234]). The related time difference of arrival (TDOA) scheme does not offer a solution because it suffers the same non-LOS problem as TOA, requires precise synchronization between the anchors (which hinders their distribution and coverage), and cannot be used for inter-mobile distance estimation.

We propose and study an alternative paradigm for inter-node distance estimation (which, to the best of our knowledge, has not received attention so far) with the aim of alleviating the outlined problems of wireless localization systems. To begin with, we abandon the notion that an estimate of the distance r between two nodes A and B should be based on a direct measurement such as the TOA or RSS between them. Instead, we consider the presence of *another* node, henceforth called *observer* node. We furthermore assume the availability of the channel impulse response (CIR) $h_A(\tau)$ of the channel between node A and the observer as well as CIR $h_B(\tau)$ between node B and the observer. The CIRs can be obtained via channel estimation at the observer after transmitting wideband training sequences at A and B [233]. The basic setup is shown in Fig. 8.1a. The starting point of this paper is the observation that the CIRs $h_A(\tau)$ and $h_B(\tau)$ are similar for small r and that this *similarity vanishes* steadily with increasing r. A good metric for the similarity between the CIRs could give rise to an accurate estimate $\hat{d}$ as a function of this metric, with the prospect of particularly good performance at short distances (due to the focus on local channel variations) and no requirements for LOS connections.

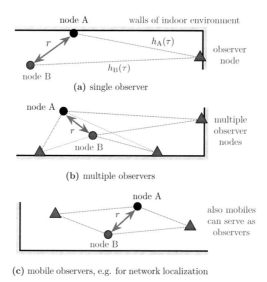

(a) single observer

(b) multiple observers

(c) mobile observers, e.g. for network localization

Figure 8.1: Proposed scheme for estimation of the distance r between two wireless nodes A and B in different possible setups. The estimation shall be based on the similarity of the CIRs $h_A(\tau)$ and $h_B(\tau)$ to an observer node (or the similarity of all their respective CIRs to multiple observer nodes). The gray walls indicate indoor environments with rich multipath propagation.

From an application perspective (details follow in Sec. 8.3), the setup in Fig. 8.1a evaluates proximity to a stationary node, e.g. some point of interest. If distance estimates to multiple stationary nodes at known positions are obtained, trilateration of the mobile position can be performed. Fig. 8.1b and 8.1c are concerned with inter-mobile distances, e.g. for network localization. They also show the possibility of using multiple observers, which can be fixed infrastructure (8.1b) or other mobiles (8.1c).

To tap the great potential of the proposed paradigm, this chapter focuses on a specific realization that is based on the multipath delay structure of the CIRs.

8.1 Distance Estimates from Delay Differences

We consider the setup in Fig. 8.2 with the nodes A and B with distance r and an observer node located in a multipath propagation environment. We express $r = \|\mathbf{r}\|$ in terms of the displacement vector $\mathbf{r} = \mathbf{p}_B - \mathbf{p}_A \in \mathbb{R}^3$ from node A at position $\mathbf{p}_A \in \mathbb{R}^3$ to node B at $\mathbf{p}_B \in \mathbb{R}^3$. The unit vectors $\mathbf{e}_k \in \mathbb{R}^3$ denote the multipath directions of departure at $\mathbf{p}_A$.

Given $h_A(\tau)$ and $h_B(\tau)$ from nodes A and B to the observer, we want to determine r by a comparison of the CIRs. If those CIRs are estimated with large bandwidth, several multipath components (MPCs) are usually resolvable and can be extracted [233]. We consider only the subset of MPCs that occur in both CIRs (propagation paths that emerge from both $\mathbf{p}_A$ and $\mathbf{p}_B$ to the observer, cf. [135]) and that were successfully extracted from both. We denote $\tau_{A,k}$ and $\tau_{B,k}$ for the MPC path delays, whereby indexation $k = 1 \ldots K$ is such that delays of equal k arise from the same propagation path[1] (e.g. via the same reflector or scatterer). K is the number of MPCs that were extracted from both CIRs.

The node displacement causes delay differences[2]

$$\Delta_k = \tau_{B,k} - \tau_{A,k} \ , \qquad\qquad k = 1, \ldots, K \qquad (8.1)$$

over equal propagation paths, as illustrated in Fig. 8.2b. The enabling fact for our approach is that all Δ_k are subject to the bounds[3] $-r \leq c\Delta_k \leq r$ due to propagation at the speed of light c. Because of this geometric significance we consider Δ_k as key observable quantity for distance estimation: each value yields a lower bound $r \geq c|\Delta_k|$ on the distance. With all observations considered we get $r \geq c \cdot \max_k |\Delta_k|$, a tight bound whenever the direction of $\mathbf{r}$ is similar to $\mathbf{e}_k$ or $-\mathbf{e}_k$ for any k. This is highly probable when K is large and the MPCs have diverse directions, which is characteristic for dense indoor or urban environments. In this case, we can compute an accurate

---

[1]We note from Fig. 8.2b that the association between the MPCs across the two CIRs (comprising the problem of finding the subset of common MPCs) is a non-trivial task; a nearest neighbor scheme will usually fail unless r is very small. Such association problems however have been studied thoroughly, e.g., for a single temporal snapshot in [235] and for temporal tracking in [135, and references therein]. In this paper we assume perfect association and leave an evaluation of the cited methods in this context to future work.

[2]The same delay-difference quantities have been employed for microphone synchronization in audio engineering [236].

[3]To obtain these bounds formally, denote $\mathbf{p}_k \in \mathbb{R}^3$ for the k-th MPC virtual sink position, e.g., the observer position mirrored at the wall(s) of a reflection [135]. Write $c\Delta_k = \|\mathbf{p}_B - \mathbf{p}_k\| - \|\mathbf{p}_A - \mathbf{p}_k\|$ and, using $r = \|\mathbf{p}_B - \mathbf{p}_A\|$, obtain $c\Delta_k \geq -r$ and $c\Delta_k \leq r$ from the triangle inequality.

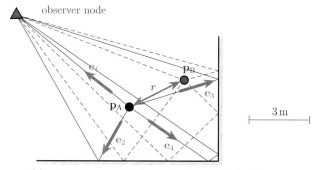

(a) multipath propagation from nodes A and B to the observer

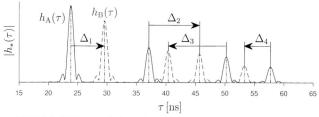

(b) CIR $h_\mathrm{A}(\tau)$ from node A to observer node;
CIR $h_\mathrm{B}(\tau)$ from node B to observer node

Figure 8.2: Concept of distance estimation between two nodes A and B by comparing (the path delays of) their wideband CIRs to an observer node. The upper plot depicts this approach in an indoor environment with two walls, also showing the significant propagation paths. The plot below shows the two corresponding CIRs (responses to raised-cosine pulses of $1\,\mathrm{GHz}$ bandwidth) with $K = 4$ MPCs and illustrates the delay differences concept.

distance estimate

$$\hat{r}^{(\mathrm{sync})} = c \cdot \max_k |\Delta_k|. \tag{8.2}$$

Measuring the values Δ_k however requires precise time synchronization between the two nodes (sub-ns precision) which can hardly be achieved with mobile consumer electronics. An alternative is to consider asynchronous delay differences

$$\tilde{\Delta}_k = \Delta_k + \epsilon \tag{8.3}$$

as observations, subject to an *unknown clock offset* ϵ (the same for all k). In this case estimation rule (8.2) cannot be applied. Yet we can find a meaningful distance estimate by looking at the value range $c\epsilon - r \leq c\widetilde{\Delta}_k \leq c\epsilon + r$, again an interval of width $2r$. For large K and diverse MPC directions we expect $c \cdot \min_k \widetilde{\Delta}_k$ and $c \cdot \max_k \widetilde{\Delta}_k$ close to the lower and upper bounds, respectively. We are hence able to compute a distance estimate from asynchronous observations

$$\hat{r}^{(\text{asyn})} = \frac{c}{2}\left(\max_k \widetilde{\Delta}_k - \min_k \widetilde{\Delta}_k\right). \tag{8.4}$$

So far our approach has been heuristic and we like to formalize it by means of estimation theory. For this we need to establish statistics for the observations Δ_k. We do so with the following *assumptions* on the MPC directions $\mathbf{e}_k$:

I: The MPC directions $\mathbf{e}_k$ are the same at $\mathbf{p}_A$ and $\mathbf{p}_B$.

II: $\mathbf{e}_k$ is random and all directions are equiprobable, i.e. $\mathbf{e}_k$ has uniform distribution on the 3D unit sphere.

III: The directions $\mathbf{e}_k$ and $\mathbf{e}_l$ of different paths $k \neq l$ are statistically independent.

By I we assume a locally constant MPC geometry.[4] This is equivalent to a plane-wave approximation if the observer was transmitting and is supported by the example in Fig. 8.2a to a large extend. Therewith we can relate the delay differences to projections $c\Delta_k = -\mathbf{e}_k^{\mathrm{T}}\mathbf{r}$ of the displacement vector.[5] By $c\Delta_k = -\mathbf{e}_k^{\mathrm{T}}\mathbf{r}$, the assumptions II and III, as well as Lemma 4.1 from page 75, the resultant observation statistic is the uniform distribution

$$c\Delta_k \overset{\text{i.i.d.}}{\sim} \mathcal{U}(-r, +r). \tag{8.5}$$

We are now ready for an estimation-theoretic study of the proposed distance estimation scheme. In the following we state our key findings for four relevant cases.

8.1.1 Delays Extracted Without Error; Synchronous Clocks

We assume the delay differences Δ_k are available exactly as defined in (8.1). This requires that (i) the delays $\tau_{A,k}$ and $\tau_{B,k}$ were extracted from the respective CIRs without error, e.g. by using a very large bandwidth, and (ii) the clocks of node A

---

[4]In detail, assumption I is valid when r is much smaller than the distances from $\mathbf{p}_A$ and $\mathbf{p}_B$ to the virtual sink of the MPC in question (cf. [135]).

[5]If the directions $\mathbf{e}_k$ were known, $\mathbf{r}$ could be determined from the linear system of K equations $c\Delta_k = -\mathbf{e}_k^{\mathrm{T}}\mathbf{r}$, but this would require specific knowledge about the environment as in multipath-assisted localization [127, 135] and is not possible with our statistical description of the $\mathbf{e}_k$.

and B are perfectly synchronous. For this case and the assumed MPC statistics, we find that $\hat{r}^{\text{(sync)}}$ in (8.2) is the maximum likelihood estimate (MLE) of r. It is an underestimate with probability 1 (because any $\mathbf{e}_k$ hits the exact direction of $\mathbf{r}$ or $-\mathbf{r}$ with probability 0) and the bias is $\mathbb{E}[\hat{r}^{\text{(sync)}}] - r = -\frac{r}{K+1}$. A simple bias correction of (8.2) leads to an unbiased estimate

$$\hat{r}_{\text{UMVUE}}^{\text{(sync)}} = \frac{K+1}{K}\, c \cdot \max_k |\Delta_k| \tag{8.6}$$

which is in fact the uniform minimum-variance unbiased estimate (UMVUE) for this problem.

Proof. To estimate r from observed $c\Delta_k \overset{\text{i.i.d.}}{\sim} \mathcal{U}(-r, r)$ we can equivalently consider $c|\Delta_k| \overset{\text{i.i.d.}}{\sim} \mathcal{U}(0, r)$. The MLE (8.2) is easily found by maximizing the conditional PDF.

To analyze the statistics of the estimates, consider $x_k = \frac{c}{r}|\Delta_k| \overset{\text{i.i.d.}}{\sim} \mathcal{U}(0, 1)$. We employ the order statistics [237] of x_k with notation $x_{(k)}$ such that $x_{(1)} \le x_{(2)} \le \ldots \le x_{(K)}$. The key consequence is $x_{(k)} \sim \text{Beta}(k, K-k+1)$ and thus

$$\mathbb{E}[x_{(k)}] = \frac{k}{K+1}\,, \qquad \text{var}[x_{(k)}] = \frac{k(K-k+1)}{(K+1)^2(K+2)}\,. \tag{8.7}$$

Now $\mathbb{E}[\hat{r}^{\text{(sync)}}] = r\,\mathbb{E}[x_{(K)}]$ and $\text{std}[\hat{r}^{\text{(sync)}}] = r\,\text{var}[x_{(K)}]^{1/2}$ with (8.7) yield the remaining results, including that (8.6) is unbiased. It is thus the UMVUE by the Lehmann–Scheffé theorem as $\max\{|\Delta_k|\}$ is a complete sufficient statistic. $\square$

8.1.2 Delays Extracted Without Error; Asynchronous Clocks

We consider the case where time synchronization is not established or required but asynchronous delay differences $\tilde{\Delta}_k$ are available as defined in (8.3). The estimate $\hat{r}^{\text{(asyn)}}$ in (8.4) is the MLE for the assumed MPC statistics. The bias is $\mathbb{E}[\hat{r}^{\text{(asyn)}}] - r = -\frac{2r}{K+1}$ and, therefrom, an unbiased estimate

$$\hat{r}_{\text{UMVUE}}^{\text{(asyn)}} = \frac{K+1}{K-1} \cdot \frac{c}{2}\left(\max_k \tilde{\Delta}_k - \min_k \tilde{\Delta}_k\right) \tag{8.8}$$

is obtained, which is the UMVUE for this problem.

It is worth noting the associated clock offset estimate

$$\hat{\epsilon}_{\text{UMVUE}}^{\text{(asyn)}} = \frac{1}{2}\left(\max_k \tilde{\Delta}_k + \min_k \tilde{\Delta}_k\right) \tag{8.9}$$

which could be useful by itself for distributed synchronization in dense multipath. It is both the MLE and the UMVUE.[6]

Proof. From the observations $c\tilde{\Delta}_k \overset{\text{i.i.d.}}{\sim} \mathcal{U}(c\epsilon - r, c\epsilon + r)$, the joint MLE (8.4), (8.9) of r and ϵ is found by careful maximization of the conditional PDF. To prove the bias and RMSE results, consider the i.i.d. $\tilde{x}_k = \frac{1}{2}\left(\frac{c}{r}\Delta_k + 1\right) \sim \mathcal{U}(0,1)$ and their order statistics $\tilde{x}_{(k)}$ with mean and variance in (8.7). With $\tilde{\Delta}_k = \Delta_k + \epsilon$, we find $\max \tilde{\Delta}_k - \min \tilde{\Delta}_k = \max \Delta_k - \min \Delta_k = \frac{2r}{c}(\tilde{x}_{(K)} - \tilde{x}_{(1)})$ and further $\mathrm{E}[\hat{r}^{(\text{asyn})}] = r\left(\mathrm{E}[\tilde{x}_{(K)}] - \mathrm{E}[\tilde{x}_{(1)}]\right) = r\frac{K-1}{K+1}$. Thus $\hat{r}_{\text{UMVUE}}^{(\text{asyn})}$ is unbiased. It is the UMVUE because it uses the minimum and maximum sample which form a complete sufficient statistic of the uniform distribution. The RMSE follows from $\mathrm{var}\left[\hat{r}^{(\text{asyn})}\right] = \mathrm{var}\left[r\left(\tilde{x}_{(K)} - \tilde{x}_{(1)}\right)\right] = r^2\left(\mathrm{var}[\tilde{x}_{(K)}] + \mathrm{var}[\tilde{x}_{(1)}] - 2\,\mathrm{cov}[\tilde{x}_{(K)}, \tilde{x}_{(1)}]\right)$. We argue $\mathrm{cov}[\tilde{x}_{(K)}, \tilde{x}_{(1)}] \approx 0$ for sufficiently large K and through (8.7) obtain $\mathrm{var}[\hat{r}^{(\text{asyn})}] \approx r^2\frac{2K}{(K+1)^2(K+2)}$ and finally (8.20) by expanding $\mathrm{std}[\hat{r}_{\text{UMVUE}}^{(\text{asyn})}] = \frac{K+1}{K-1}\mathrm{var}[\hat{r}^{(\text{asyn})}]^{1/2}$. For $\hat{\epsilon}_{\text{UMVUE}}^{(\text{asyn})}$ the RMSE and zero bias follow analogously. □

8.1.3 General Case With Synchronous Clocks

When the path delays are measured with error (e.g., due to limited bandwidth), the distance estimates introduced so far might get distorted heavily: they are very susceptible to outliers since they regard only the maximum and minimum delay difference. It is thus sensible to include such errors in the statistical model and derive according distance estimates.

We first consider the case of perfectly synchronous nodes but with observed delay differences $T_k = \Delta_k + n_k$ subject to random errors n_k (as a result of delay extraction errors). We assume that the distribution of n_k is known and furthermore that n_k and n_l are statistically independent for $k \neq l$. The resulting distance MLE is given by the optimization problem

$$\hat{r}^{(\text{sync,gen})} \in \arg\max_r \frac{1}{r^K}\prod_{k=1}^{K} I_k(T_k, r), \qquad (8.10)$$

$$I_k(T_k, r) = F_{n_k}(T_k + r/c) - F_{n_k}(T_k - r/c) \qquad (8.11)$$

where F_{n_k} is the cumulative distribution function (CDF) of the observation error n_k.

---

[6]It can be shown that (8.2), (8.4), (8.9) are the MLE also for the respective 2D cases with analogous assumptions on $\mathbf{e}_k$. Instead of (8.5), this case features $f(\Delta_k|r) = \frac{c}{\pi}(r^2 - c^2\Delta_k^2)^{-1/2}$ as observation PDF. The details are omitted.

Hence I_k can be regarded as a soft indicator function of $cT_k \in [-r, r]$. If the errors have Gaussian distribution[7] $n_k \sim \mathcal{N}(0, \sigma_k^2)$ we can use the Q-function to write

$$I_k(T_k, r) = Q\left(\frac{T_k - r/c}{\sigma_k}\right) - Q\left(\frac{T_k + r/c}{\sigma_k}\right). \tag{8.12}$$

Estimate (8.10) is biased in general. This is seen by the example of errorless extraction: $n_k \equiv 0$ results in $I_k(T_k, r) = \mathbb{1}_{[-r,r]}(cT_k)$ (the actual indicator function) and consequently (8.10) yields (8.2) as a special case (the proof is straightforward) which we know is a biased estimate.

Proof. The likelihood function (LHF) of r from one observation $T_k = \Delta_k + n_k$ is the conditional PDF given by the convolution

$$f(T_k \,|\, r) = \int_{\mathbb{R}} f_{n_k}(n) \, f_{\Delta_k | r}(T_k - n \,|\, r) \, dn. \tag{8.13}$$

With $\Delta_k | r \sim \mathcal{U}(-r/c, r/c)$ from (8.5) we furthermore obtain

$$f(T_k | r) = \frac{c}{2r} \int_{T_k - r/c}^{T_k + r/c} f_{n_k}(n) \, dn = \frac{c}{2r} I_k(T_k, r) \tag{8.14}$$

where we use definition (8.11). The LHF of r given $T_1, \ldots, T_K$ is the product of the individual $f(T_k | r)$, i.e.

$$L(r) = f(T_1, \ldots, T_K \,|\, r) = \left(\frac{c}{2r}\right)^K \prod_{k=1}^{K} I_k(T_k, r) \tag{8.15}$$

because the observations are assumed statistically independent. A distance r that maximizes $L(r)$ is an MLE, giving (8.16). $\qquad\square$

8.1.4 General Case With Asynchronous Clocks

Finally, we consider the case where erroneous asynchronous delay differences $\tilde{T}_k = T_k + \epsilon = \Delta_k + n_k + \epsilon$ are observed, i.e. subject to a clock offset ϵ and an extraction

---

[7]A Gaussian error on the delay differences could be the result of the delays $\tau_{A,k}$ and $\tau_{B,k}$ being extracted subject to uncorrelated Gaussian errors. This model is suggested by [238] for high SNR.

error n_k. The joint MLE of distance and clock offset is given by

$$\left(\hat{r}^{(\text{asyn,gen})}, \hat{\epsilon}^{(\text{asyn,gen})}\right) \in \arg\max_{r,\epsilon} \frac{1}{r^K} \prod_{k=1}^{K} I_k(\tilde{T}_k - \epsilon, r). \tag{8.16}$$

This distance estimate is biased in general (an unbiased estimate remains as an open problem). This is seen at the special case of error-less extraction, analogous to Sec. 8.1.3: for $n_k \equiv 0$ it can be shown that (8.16) yields the biased (8.4).

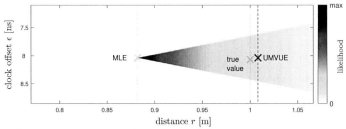

(a) Likelihood function (8.18) and associated estimates for delay differences without extraction errors, i.e. for $n_k \equiv 0$ and $I_k(T_k, r) = \mathbb{1}_{[-r,r]}(cT_k)$

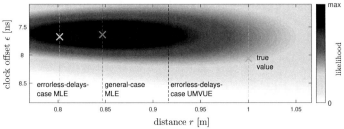

(b) Likelihood function (8.18) and various estimates for delay differences with Gaussian errors n_k, i.e. using (8.12)

Figure 8.3: Color plots of examples for the asynchronous-case likelihood function $\tilde{L}(r, \epsilon)$. Plot (a) is based on observations $\tilde{\Delta}_k$ without extraction errors and (b) is based on $\tilde{T}_k \sim \mathcal{N}(\tilde{\Delta}_k, \sigma^2)$. We assumed a true distance $r = 1\,\text{m}$, an error level of $c\sigma/r = 0.2$, and a total of $K = 15$ MPCs.

Proof. The only difference to the above case are the asynchronous $\tilde{T}_k = T_k + \epsilon$. As ϵ is modeled non-random and $T_k = \tilde{T}_k - \epsilon$,

$$f_{\tilde{T}_k | r, \epsilon}(\tilde{T}_k | r, \epsilon) = f_{T_k | r}(\tilde{T}_k - \epsilon | r) = \frac{c}{2r} I_k(\tilde{T}_k - \epsilon, r). \tag{8.17}$$

186

The joint MLE (8.16) given $T_1, \ldots, T_K$ follows from the likelihood function

$$\tilde{L}(r, \epsilon) = \left(\frac{c}{2r}\right)^K \prod_{k=1}^{K} I_k(\tilde{T}_k - \epsilon, r) \tag{8.18}$$

of which two examples are illustrated in Fig. 8.3. □

8.1.5 Technical Aspects and Comments

A subtle but important aspect is that the presented estimators can *incorporate multiple observer nodes* without further ado by simply considering the MPCs from nodes A and B to all observer nodes (use index k on this set). The increased number of observations can improve performance considerably.

We never assumed or required knowledge of the observer positions, synchronization between observer(s) and the other two nodes, or synchronization among multiple observers as such circumstances would not even improve the scheme. These are the *key complexity advantages* of our proposal.

Another fortunate aspect is that assumption I is almost superfluous because any MPC relevant to estimation (an MPC where $|\Delta_k|$ is large) fulfills it quite naturally: large $|\Delta_k|$ corresponds to $\mathbf{e}_k$ being similar to the direction of $\mathbf{r}$ or $-\mathbf{r}$ and, thus, $\mathbf{e}_k$ hardly changes when moving by $\mathbf{r}$.

The properties of optimization problems (8.10) and (8.16) depend on the error CDFs F_{n_k}. With Gaussian error statistics (8.12) the problems are non-convex (because the Q-function is non-convex) yet very amenable: in all conducted experiments, the likelihood function was unimodal and the problems could be solved with very few iterations of a gradient-based solver.

8.2 Performance Evaluation

This section discusses various sources of error and their effect on the accuracy of the proposed distance estimates.

8.2.1 Impact of Unknown MPC Directions

An important source of error are the unknown $\mathbf{e}_k$ which determine the observed delay differences. Because of the mathematical simplicity of the observation statistics (8.5), the resulting estimation error statistics can be described in closed form (for derivations see the appendix).

The root-mean-squared error (RMSE) of the synchronous-case UMVUE (8.6) is given by its standard deviation

$$\text{std}\!\left[\hat{r}_{\text{UMVUE}}^{\text{(sync)}}\right] = \frac{r}{\sqrt{K(K+2)}} \tag{8.19}$$

while the RMSE of asynchronous estimates (8.8) and (8.9) is characterized by the large-K approximations

$$\text{std}\!\left[\hat{r}_{\text{UMVUE}}^{\text{(asyn)}}\right] \approx \frac{r}{K-1}\sqrt{\frac{2K}{K+2}}, \tag{8.20}$$

$$\text{std}\!\left[\hat{\epsilon}_{\text{UMVUE}}^{\text{(asyn)}}\right] \approx \frac{r/c}{K+1}\sqrt{\frac{2K}{K+2}} \tag{8.21}$$

which are accurate for about $K \geq 5$. All errors are proportional to r/K asymptotically. In other words, the error of distance estimation based on unknown MPC directions increases linearly with distance.

For reliable estimation of the distance between synchronous nodes, a single $\mathbf{e}_k$ similar to the direction of *either* $\mathbf{r}$ or $-\mathbf{r}$ suffices. The asynchronous case, in contrary, requires $\mathbf{e}_k$ similar to the directions of *both* $\mathbf{r}$ and $-\mathbf{r}$ to occur and is thus more reliant on diverse $\mathbf{e}_k$. This fundamental difference is due to the unknown clock offset ϵ and is apparent when comparing (8.2) to (8.4). The performance difference can be quantified as $\text{std}[\hat{r}_{\text{UMVUE}}^{\text{(asyn)}}] \approx \sqrt{2}\,\text{std}[\hat{r}_{\text{UMVUE}}^{\text{(sync)}}]$ under our assumptions.

8.2.2 Impact of Delay Extraction Errors

The errors n_k on the delay differences, which stem from delay extraction errors, cause additional performance degradation. We will now evaluate how the relative error $(\hat{r}-r)/r$ of various distance estimates is affected by independent Gaussian errors $n_k \sim \mathcal{N}(0,\sigma^2)$ while the statistics of Δ_k are according to the assumptions in Sec. 8.1. We do not assume a specific setup geometry but instead specify the ratio of $c\sigma$ (the distance-translated error standard deviation) to r as it determines the statistics of the relative error. To implement the general-case MLEs we use (8.12) in (8.10) and (8.16) and solve the respective optimization problems with a gradient-based solver (quasi-Newton method). Fig. 8.4 shows the impact of σ and K on the relative error.

When $c\sigma/r$ is considerably large, we observe that the estimators designed for error-

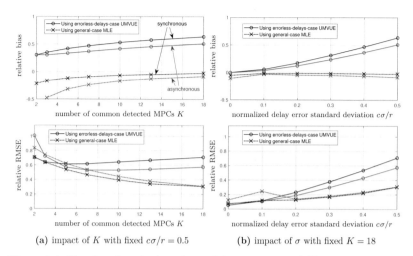

(a) impact of K with fixed $c\sigma/r = 0.5$ (b) impact of σ with fixed $K = 18$

Figure 8.4: The plots show the dependence of the relative bias $\mathrm{E}[\hat{r}]/r - 1$ and the relative RMSE $\mathrm{std}[\hat{r}]/r$ on the number of common detected delays K and the standard deviation σ of the Gaussian error on each Δ_k. The black and red graphs represent synchronous and asynchronous estimates, respectively.

less delay extraction are heavily distorted. This effect is even amplified with increasing K (giving rise to more outliers). In this case, the asynchronous estimate outperforms the synchronous estimate as it uses not one but two delay-differences (the extrema) which amounts to some error averaging.

In the high $c\sigma/r$ regime, the general-case estimators perform much better because they are tailored to the observation statistics at hand. We observe that bias and RMSE converge to zero with increasing K and that the bias is very small even at high error levels. If $c\sigma$ is very small, e.g. with a capable ultra-wideband system, then $c\sigma/r$ is significant only for small r. This short-range regime is of particular practical interest though, as argued in the introduction of the chapter.

We conclude that the simple proposed estimates (8.6) and (8.8) perform well if $c\sigma/r$ is less than about 0.1. For larger error levels the general-case estimates (8.10) and (8.16) should instead be used, e.g. at close proximity or with small bandwidth.

8.2.3   Impact of Realistic MPC Directions

We will now evaluate how the proposed estimates, which were designed for the propagation assumptions of Sec. 8.1, perform in realistic indoor propagation conditions and with a signaling bandwidth of 1 GHz for estimation of the CIRs. In particular, we consider a room with the floor plan shown in Fig. 8.5a and a static observer node, a static node A, and a mobile node B that can be located anywhere in the room. We employ ray tracing to simulate reflection paths of up to three bounces whereby each

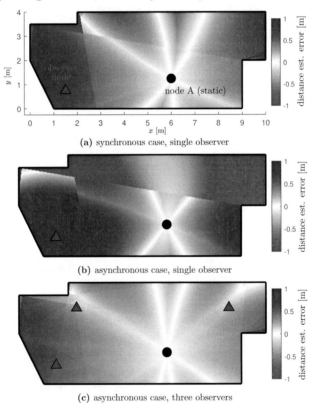

(a) synchronous case, single observer

(b) asynchronous case, single observer

(c) asynchronous case, three observers

Figure 8.5: Distance estimation error between a static node A (•) and a mobile node B (anywhere in the room) in an indoor environment of the shown floor plan. This experiment assumes error-less delay extraction (and uses the according UMVUEs), yet estimation errors occur because of the reliance on unknown MPC directions. We use an MPC detection threshold SINR ≥ 0 dB assuming diffuse multipath and additive noise at 1 GHz bandwidth.

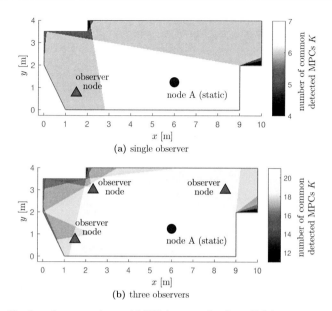

Figure 8.6: Number of common detected MPCs between the channel(s) from static node A (•) to the observer(s) and the channel(s) from mobile node B (anywhere in the room) to the observer(s). The figure is associated with the SINR-based MPC detection criterion described in Sec. 8.2.3 with the same floor plan and node setup as in Fig. 8.5.

bounce is assumed to cause 3 dB attenuation [127,135]. We consider reflections via the side walls as well as the floor and ceiling. The assumed room height is 3 m and all devices are 1.2 m above the floor.

To obtain practically meaningful results we need to define a criterion for the detection of a MPC. We use the detection threshold $SINR_k \geq 0$ dB based on the signal-to-interference-plus-noise ratio whereby the interference is due to diffuse multipath propagation. In particular, we employ the definition $SINR_k = |a_k|^2/(N_0 + T_p S_\nu(\tau_k))$ from [127, Eq. 14] where a_k is the k-th path amplitude (which is subject to free-space path loss), N_0 the single-sided noise spectral density, T_p the effective pulse duration (inversely proportional to bandwidth), and $S_\nu(\tau)$ the power delay profile of the diffuse multipath portion in the CIRs. Following the proposal of [127, Tab. 1], we choose a double-exponential $S_\nu(\tau)$ with 5 ns rise time, 20 ns decay time constant, and $1.16 \cdot 10^{-6}$ normalized power.

Fig. 8.5 shows the distance estimation error (the color at any point x, y in the room marks the error when the mobile node B is at that position) for the UMVUEs in the absence of extraction errors. We observe a significant performance advantage with synchronization in Fig. 8.5a over the asynchronous case in Fig. 8.5b. The reason is that with a single observer, K is small and the MPC directions tend to be similar to the LOS direction rather than uniformly distributed. This heavily impairs the asynchronous estimate. With the three observer deployment of Fig. 8.5c however, K increases vastly (from 6 or 7 to about 20 for most positions) and the $\mathbf{e}_k$ are spread more evenly, which results in great performance even in asynchronous mode.

For the three-observer setup, Fig. 8.7 shows the RMSE as a function of r around the static node (computed from error realizations on a circle of radius r). We observe almost constant slopes, consistent with the scaling behavior described in Sec. 8.2.2. At $r = 3\,\mathrm{m}$, we measure a relative RMSE of 4.39% (synchronous) and 8.51% (asynchronous) which compare to analytical projections of 4.77% and 7.10%, respectively, from (8.19) and (8.20). We infer that this setup faces no performance degradation due to non-uniform $\mathbf{e}_k$ with established synchronization and just a slight degradation in the asynchronous case.

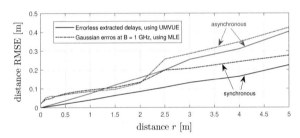

Figure 8.7: For the three-observer setup of Fig. 8.5c, the plot shows the RMSE of distance estimation as a function of distance r around the static node. The plot compares distance estimation performance under error-less and erroneous delay extraction. We assume $1\,\mathrm{GHz}$ bandwidth and use the MPC detection criterion $\mathrm{SINR}_k \geq 0\,\mathrm{dB}$. The values of the near-constant slopes conform with the predictions of Sec. 8.2.1 (the RMSE $\propto r$ for error-less extraction).

Fig. 8.7 also shows the performance for erroneous MPC extraction and using the general-case MLEs. The chosen error model is $n_k \sim \mathcal{N}(0, \sigma_k^2)$ where $\sigma_k^2 = \sigma_{\mathrm{A},k}^2 + \sigma_{\mathrm{B},k}^2$ is the sum of the variances of assumed independent Gaussian errors on $\tau_{\mathrm{A},k}$ and $\tau_{\mathrm{B},k}$, respectively, which are set to the minimum variance according to the Cramér-Rao lower bound (CRLB) for delay extraction (neglecting path overlap) in diffuse multipath and noise as presented in [127]. For the details we refer to [127, Sec. III.-B]. The resulting

$c\sigma_k$ are between 5 cm and 10 cm and thus the extraction errors should have a significant impact for about $r \leq 1$ m (i.e. $c\sigma_k/r \leq 0.1$), which agrees with the numerical results in large part. Extraction errors obviously impair the performance but do not change the order of magnitude of the estimation errors, which stay below 20 cm in a circle of at least $r \leq 2$ m around the static node. We conclude that the proposed distance estimators are viable in realistic conditions.

8.3 Technological Comparison and Opportunities

A performance comparison to related distance estimation schemes is in order. With the parameters of our evaluation in Fig. 8.7, a TOA distance estimate to fixed infrastructure (e.g. from node A to an observer position) would have an RMSE lower bound [127] of $c\sigma_{A,1} \approx 2.7$ cm, which implies high accuracy but requires LOS and perfect time synchronization. When a synchronization error ϵ occurs, a TOA estimate suffers a distance error $c\epsilon$ (e.g. $c\epsilon = 30$ cm error from just $\epsilon = 1$ ns) which is particularly severe for short distances. Yet, to the best of our knowledge, distributed synchronization with sub-ns precision is currently not feasible with reasonable complexity. Our scheme compares well to reported TOA ranging errors of up to 2 m in [239] (500 MHz bandwidth) and [133] (1 GHz) or up to 10 m in [129] (125 MHz) and [133] (200 MHz), although a thorough comparison is out of scope.

As indicated in the chapter introduction, our proposal has various promising applications in indoor localization. Due to the conceptual individuality, a direct performance comparison to existing schemes is not possible at this point. Instead we highlight the major technological opportunities and benefits in the following.

The absence of synchronization and LOS requirements qualifies the proposal for localization in dense and crowded settings. It allows for accurate ranging between low-complexity nodes, which only need to transmit pilot sequences and do not require high-resolution wideband receivers (only the observers do). The scheme is thus a prime candidate for estimating distances between mobiles for the purpose of network localization and, at that, does not require interaction of the mobiles. Thereby, the fact that mobiles can be observers (as knowledge of observer positions is not required) promises particularly great performance scaling with network density: $N - 2$ out of N mobiles can be observers for each distance estimation.

As described earlier, the proposal can be used for localization via trilateration when distances to multiple stationary nodes (henceforth called beacons) are obtained. The low complexity requirements allow for battery-powered transmit-only beacons without a wired connection. They can thus be deployed easily and in vast numbers. This is a major advantage over state-of-the-art systems, e.g., TDOA systems which require precisely synchronized anchor infrastructure with LOS coverage. A vast amount of distance estimates between many beacons and mobiles together with all the inter-mobile distances promises accurate and robust network localization. The mobiles can also be of low complexity, as all processing and hardware complexity could be pushed to fixed observer infrastructure.

The single-beacon ranging application of Fig. 8.1a is similar to wideband location fingerprinting [240] but does not rely on offline training: The beacon-to-observer CIR can be estimated online, enabling robust operation in dynamic environments.

Our proposal utilizes multipath propagation without using any specific knowledge about the environment. This is in contrary to multipath-assisted localization [127, 135] which uses an a-priori known floor plan or online learning in order to utilize reflected paths for localization (which softens requirements on LOS conditions and number of anchors).

Chapter 9

Summary

This thesis focused on *low-frequency magnetic induction* for wireless communication, power transfer and localization in the context of wireless sensor networks. The use of magnetic induction is motivated by attractive properties in terms of material penetration and field predictability, the ability to realize strong mid-range links between size-limited devices, as well as powerful opportunities for passive communication and co-operation (load modulation and passive relaying). Considering the high industrial relevance of the topic (e.g., wireless charging, RFID, NFC), the existing communication-theoretic corpus of knowledge is small and leaves important questions unanswered. This thesis shall fill a considerable portion of these academic white spaces.

A major focus was put on *general modeling* of magneto-inductive systems and the consequent application of theoretical tools for the study of the performance limits and behavior of such systems. A first aspect of this general approach concerned the previously open problem of an analytically tractable model for coil coupling that does consider radiation and the near-far-field transition. We derived two such models in Cpt. 2: a slight adaption of the mutual inductance double-line-integral formula of Neumann and a particularly simple dipole-type formula. At the core of Cpt. 3 is a general frequency-dependent model of the channel matrix $\mathbf{H}$ and noise covariance matrix $\mathbf{K}$ of a magneto-inductive link. This description encompasses MIMO, MISO, SIMO and SISO links from either a narrowband or a broadband perspective. This model enabled a study of the *fundamental performance limits* of magneto-inductive links in terms of power transfer efficiency and achievable data rate. The approach is based on concepts that should be familiar to the radio communication theorist, namely channel and noise modeling, achievable rates and channel capacity, array techniques, degrees of freedom in frequency and space, and different paradigms for matching circuits and transmit signaling. The exposition is supplemented by novel channel capacity results for resonant channels (active transmission) and load modulation systems (with a single passive tag or multiple cooperating passive tags).

In Cpt. 4 we presented the first study of the *random channel* between two coils (or two dipoles) with random arrangement, in the form of fully *random coil orientations* on both ends. We derived the channel statistics in terms of the closed-form

PDF of the channel coefficient in the pure near-field case, the near-far-field transition, and the pure far-field case. The performance implications were analyzed with familiar communication-theoretic tools, namely outage probability, diversity order, and outage capacity. We demonstrated the particularly *intense fading* and resulting terrible reliability in the pure near- and far-field cases, whose outage behavior is resembled by a channel coefficient that is just the real part of a Rayleigh-distributed random variable. Accordingly, the derived diversity orders are 1/2 for active transmission (and even 1/4 for load modulation, where the channel applies twice like in backscatter communication). The near-far-field transition at least exhibits the outage behavior of a Rayleigh channel (diversity order 1) because two linearly-independent field components with a phase shift arise. A resulting wireless design guideline is that this powerful *polarization diversity effect* can be utilized to the fullest if the typical link distance is close to 0.3747 times the employed wavelength. The chapter furthermore studied tri-axial coil arrays together with diversity combining schemes and the resulting channel statistics and performance. We concluded with auxiliary implications of the developed theory of random orientations, namely a capacity result between coil arrays in the massive MIMO limit and a spatial correlation result which gave rise to a novel localization-knowledge-based beamforming scheme in Cpt. 6 and a novel localization algorithm in Cpt. 7.

The topic of magneto-inductive *passive relaying* was reconsidered in Cpt. 5 from the perspective of arbitrary *random arrangements*. In an introductory discourse we decomposed the effect of passive relays into *two opposing effects*: (i) the losses from the relay coil resistances and (ii) the gain from increased mutual impedance between transmitter and receiver. We demonstrate that this mutual impedance is a noncoherent sum of phasors which causes *frequency-selective fading*. We decompose the relaying gain into two major effects: the gain from the transmitter-receiver mutual impedance change and the loss from increased encountered coil resistance (due to coupling with lossy relays). We proceeded with a simulation of one passive relay randomly placed near the transmitter or receiver of a randomly arranged link. We found that the relay allows for significant improvement of the channel gain at the design frequency. These gains stem mostly from misalignment mitigation, i.e. the relay recovers the link from a random-orientation-induced deep fade, and require *optimization of the relay load capacitance* (or equivalently: the relay resonance frequency). An unoptimized passive relay hardly affects the channel statistics with random placement. Finally, we studied random swarms of passive relays around a node and characterized the resultant frequency-selective fading channel in terms of coherence bandwidth and affected bandwidth. Exploiting these fluctuations via transmit-side channel knowledge (i.e. using frequency tuning or waterfilling) allows for reliable performance gains. To draw even larger gains from a set of randomly arranged relays we propose a low-complexity optimization scheme based on load switching. The method yielded a 10.1 dB gain over an unoptimized relay swarm , used at the design frequency, and a 4.8 dB gain over frequency tuning in terms of median channel gain, in a setting with high local relay density. Load switching at passive relays is thus a promising means to enable reliable communication and power transfer in dense magneto-inductive sensor networks.

The exposition in Cpt. 6 presented a thorough evaluation of *medical microsensors* which receive power and transmit data via low-frequency magnetic induction. Due consideration was given to practice-oriented modeling, tissue attenuation, scaling behavior, arbitrary arrangement, array design and spacing, matching circuits, choice of operating frequency and the highly asymmetric frequency-dependent links. For a wireless-powered sensor 5 cm deep into muscle tissue, the determined minimum coil size for copper wire is about 0.35 mm. Such size allows for a mean uplink data rate in excess of 1 Mbit/s. This result is however critically dependent on the sensor depth and associated with full channel knowledge and low reliability requirements; a vastly larger coil is required otherwise. In the same context, load modulation at passive sensors was shown to be a very promising approach to realize mediocre uplink data rates from very small sensors. Thereby, the minimum passive sensor size is crucially affected by the fidelity of the external RFID-reader-type receiver, whose practical aspects should receive attention by future work. A subsequent study of transmit cooperation between distributed sensor nodes, either via active transmission or load modulation, showed that the potential rate and reliability improvements should be well worth the technical effort. We furthermore showed that the passive relaying effect in dense swarms of resonant nodes can lead to performance improvement, although realizing significant gains requires relay load optimization and adaptive matching as well as adaptation of the signaling to the spectral and spatial channel fluctuations. Future research should extend the evaluation of medical microrobots to advanced materials with very high conductivity (e.g., iron cores, graphene, carbon nanotubes) and conduct a critical comparison between magnetic induction, acoustic propagation, and optical propagation.

Cpt. 7 presented various contributions to magneto-inductive localization, in particularly on 3D localization on the indoor scale with flat anchor coils and an unknown arbitrary agent orientation. The discussion of the associated 5D estimation problem is the first to consider radiation and (to some extend) multipath propagation. We opened with a qualitative comparison between magneto-inductive localization and other wireless localization approaches. After deriving the Cramér-Rao lower bound (CRLB) on the position and orientation error, we studied its dependence on room size and anchor count and found cm-accuracy being feasible in a square room of 3 m side length. We investigated the opportunities in terms of localization algorithm design and, in the process, we highlighted the problems of standard approaches: multilateration and distance bounding approaches are vastly inadequate and a 5D least-squares approach struggles with the complicated non-convex cost function. These problems were remedied by the derivation of a tailored weighted least-squares algorithm and an equally capable algorithm called Magnetic Gauss, which is based on the random orientation theory from Cpt. 4. The algorithms' parameter spaces are only 3D as they get rid of the complexity associated with estimating the agent orientation as nuisance parameter. Both algorithms perform near the CRLB with high robustness and consistently low computational cost. The developed theory was then applied to a system implementation with 8 flat anchor coils which localize an arbitrarily oriented agent coil. After thorough calibration we achieve an accuracy better than 10 cm in 92% of cases in an office environment. A quantitative investigation of the potential accuracy-limiting factors identified distortions due to conductive building structures to be dominant. We project that 1 cm accuracy is possible in a distortion-free environment or by accurately modeling any appreciable impact of nearby conductors (which would however vastly complicate the system deployment). Much better than 1 cm accuracy seems to be infeasible for magneto-inductive indoor localization via parameter estimation based on an analytical signal model.

The exposition on ulta-wideband radio localization in Cpt. 8 is an outlier in this otherwise magneto-inductive thesis but has mathematical synergies with Cpt. 4. It proposes a novel paradigm to estimate the distance between two wireless nodes by a comparison of the impulse responses of their channels to auxiliary observer nodes. Based on the multipath delay structure of the CIRs, we derived distance estimators and their properties for different relevant cases. A numerical evaluation showed that an accuracy of 20 cm can be achieved over large parts of a typical-size office room when using three observers (which could be other mobiles), 1 GHz signaling bandwidth, and no synchronization requirements whatsoever. The scheme could improve indoor localization in various important use cases: (i) spacious buildings because the distributed infrastructure could be mostly simple transmit-only beacons, (ii) crowded settings because it does not rely on line of sight, or (iii) network localization because it promises to be well-suited for the estimation of small distances between mobiles. Future work should evaluate the performance of the scheme with practical extraction and association techniques for multipath components.

Appendix A

Fields Generated by a Circular Loop: Vector Formula

This appendix derives a linear-algebraic formula for the magnetic field $\mathbf{b}$ generated by a driven single-turn circular loop that is electrically small (i.e. the circumference is much shorter than the employed wavelength) in free space. In particular, we consider a loop with axis orientation $\mathbf{o}_{\mathrm{T}}$ (unit vector) and a reference point at distance r and direction $\mathbf{u}$ (unit vector) from the loop center position. We show that the magnetic field in tesla at the reference point is

$$\mathbf{b} = \frac{\mu_0 m k^3}{2\pi} e^{-jkr} \left(\left(\frac{1}{(kr)^3} + \frac{j}{(kr)^2} \right) \boldsymbol{\beta}_{\mathrm{NF}} + \frac{1}{2kr} \boldsymbol{\beta}_{\mathrm{FF}} \right) , \tag{A.1}$$

$$\boldsymbol{\beta}_{\mathrm{NF}} = \frac{1}{2} \left(3\mathbf{u}\mathbf{u}^{\mathrm{T}} - \mathbf{I}_3 \right) \mathbf{o}_{\mathrm{T}} , \tag{A.2}$$

$$\boldsymbol{\beta}_{\mathrm{FF}} = \left(\mathbf{I}_3 - \mathbf{u}\mathbf{u}^{\mathrm{T}} \right) \mathbf{o}_{\mathrm{T}} \tag{A.3}$$

and the electric field (unit V/m) at the reference point is

$$\mathbf{e} = \frac{Z_{\mathrm{w}} m k^3}{4\pi} e^{-jkr} \left(\frac{j}{(kr)^2} - \frac{1}{kr} \right) \mathbf{u} \times \mathbf{o}_{\mathrm{T}} . \tag{A.4}$$

Thereby $m = A_{\mathrm{T}} i_{\mathrm{T}}$ is the magnetic dipole moment of the driven loop due to current i_{T} (complex phasor, effective value) and enclosed area $A_{\mathrm{T}} = \frac{D_{\mathrm{c}}^2 \pi}{4}$. Furthermore $Z_{\mathrm{w}} = \sqrt{\mu_0/\epsilon_0} = \mu_0 c_0 \approx 377\,\Omega$ is the wave impedance of free space. The vectors $\mathbf{b}$ and $\mathbf{e}$ are complex phasor representations of the physical fields $\vec{B} = \sqrt{2}\,\mathrm{Re}\{\mathbf{b}\,e^{j\omega t}\}$ and $\vec{E} = \sqrt{2}\,\mathrm{Re}\{\mathbf{e}\,e^{j\omega t}\}$. In the front matter the result is extended to related important cases such a non-circular multi-turn loops, which is achieved with superposition and large-r arguments.

The derivation is based on trigonometric field formulas from Balanis [62] and proceeds as follows. We first derive the statement in a coordinate system that is fixed such that the axis orientation $\mathbf{o}_{\mathrm{T}} = [0\ 0\ 1]^{\mathrm{T}}$ is along the z-axis. We represent the field phasors in the orthonormal base

$$\mathbf{b} = b_r \mathbf{c}_r + b_\phi \mathbf{c}_\phi + b_\theta \mathbf{c}_\theta \,, \tag{A.5}$$

$$\mathbf{e} = e_r \mathbf{c}_r + e_\phi \mathbf{c}_\phi + e_\theta \mathbf{c}_\theta \tag{A.6}$$

consisting of the radial, azimuthal, and polar unit vectors [241]

$$\mathbf{c}_r = \begin{bmatrix} \sin\theta \cos\phi \\ \sin\theta \sin\phi \\ \cos\theta \end{bmatrix}, \qquad \mathbf{c}_\phi = \begin{bmatrix} -\sin\phi \\ \cos\phi \\ 0 \end{bmatrix}, \qquad \mathbf{c}_\theta = \begin{bmatrix} \cos\theta \cos\phi \\ \cos\theta \sin\phi \\ -\sin\theta \end{bmatrix} \tag{A.7}$$

whereby ϕ is the azimuth angle of direction vector $\mathbf{u}$ and $\theta = \arccos(\mathbf{u}^\mathsf{T}\mathbf{o}_\mathsf{T})$ its polar angle.

We employ exact descriptions of these field components stated in [62, Eq.5-18 and Eq.5-19]. In fact b_ϕ, e_r and e_θ are zero and the description simplifies to

$$\mathbf{b} = b_r \mathbf{c}_r + b_\theta \mathbf{c}_\theta \,, \tag{A.8}$$

$$\mathbf{e} = e_\phi \mathbf{c}_\phi \,, \tag{A.9}$$

$$b_r = \frac{\mu_0 m k^3}{2\pi} \left(\frac{1}{(kr)^3} + \frac{j}{(kr)^2} \right) \cos(\theta)\, e^{-jkr} \,, \tag{A.10}$$

$$b_\theta = \frac{\mu_0 m k^3}{4\pi} \left(\frac{1}{(kr)^3} + \frac{j}{(kr)^2} - \frac{1}{kr} \right) \sin(\theta)\, e^{-jkr} \,, \tag{A.11}$$

$$e_\phi = -\frac{Z_0 m k^3}{4\pi} \left(\frac{j}{(kr)^2} - \frac{1}{kr} \right) \sin(\theta)\, e^{-jkr} \,. \tag{A.12}$$

These expressions can be written as

$$b_r = T_\mathrm{NF}\, \cos(\theta)\,, \qquad b_\theta = \left(\frac{1}{2} T_\mathrm{NF} - T_\mathrm{FF} \right) \sin(\theta)\,, \qquad e_\phi = T_\mathbf{e}\, \sin(\theta) \tag{A.13}$$

with the short-hand notations

$$T_\mathrm{NF} := A \left(\frac{1}{(kr)^3} + \frac{j}{(kr)^2} \right), \qquad T_\mathrm{FF} := A\, \frac{1}{2kr}\,, \tag{A.14}$$

$$A := \frac{\mu_0 m k^3}{2\pi} e^{-jkr}\,, \qquad T_\mathbf{e} := -\frac{Z_\mathrm{w} m k^3}{4\pi} \left(\frac{j}{(kr)^2} - \frac{1}{kr} \right) e^{-jkr}\,. \tag{A.15}$$

After these preparations, we can expand the electric field

$$\mathbf{e} = e_\phi \mathbf{c}_\phi = T_\mathbf{e} \sin(\theta) \begin{bmatrix} -\sin\phi \\ \cos\phi \\ 0 \end{bmatrix} = -T_\mathbf{e} \begin{bmatrix} \sin\theta\cos\phi \\ \sin\theta\sin\phi \\ \cos\theta \end{bmatrix} \times \begin{bmatrix} 0 \\ 0 \\ 1 \end{bmatrix} = -T_\mathbf{e} \, \mathbf{u} \times \mathbf{o}_\mathrm{T} \,.$$

$$(A.16)$$

For the magnetic field we write

$$\mathbf{b} = T_\mathrm{NF} \cos(\theta) \begin{bmatrix} \sin\theta\cos\phi \\ \sin\theta\sin\phi \\ \cos\theta \end{bmatrix} + \left(\frac{1}{2}T_\mathrm{NF} - T_\mathrm{FF}\right)\sin(\theta) \begin{bmatrix} \cos\theta\cos\phi \\ \cos\theta\sin\phi \\ -\sin\theta \end{bmatrix}$$

$$= T_\mathrm{NF}\,\boldsymbol{\beta}_\mathrm{NF} + T_\mathrm{FF}\,\boldsymbol{\beta}_\mathrm{FF}$$

$$(A.17)$$

whereby the vector quantities hold involved trigonometric expressions. These can be rearranged to

$$\boldsymbol{\beta}_\mathrm{NF} = \begin{bmatrix} \frac{3}{2}\sin\theta\cos\theta\cos\phi \\ \frac{3}{2}\sin\theta\cos\theta\sin\phi \\ \cos^2\theta - \frac{1}{2}\sin^2\theta \end{bmatrix} = \frac{3}{2}\begin{bmatrix} \sin\theta\cos\phi \\ \sin\theta\sin\phi \\ \cos\theta \end{bmatrix}\cos(\theta) - \frac{1}{2}\begin{bmatrix} 0 \\ 0 \\ 1 \end{bmatrix}$$

$$= \frac{3}{2}\mathbf{u}\cos(\theta) - \frac{1}{2}\mathbf{o}_\mathrm{T} = \frac{3}{2}\mathbf{u}(\mathbf{u}^\mathrm{T}\mathbf{o}_\mathrm{T}) - \frac{1}{2}\mathbf{o}_\mathrm{T} = \frac{1}{2}\left(3\mathbf{u}\mathbf{u}^\mathrm{T} - \mathbf{I}_3\right)\mathbf{o}_\mathrm{T} \,, \qquad (A.18)$$

$$\boldsymbol{\beta}_\mathrm{FF} = \begin{bmatrix} -\sin\theta\cos\theta\cos\phi \\ -\sin\theta\cos\theta\sin\phi \\ \sin^2\theta \end{bmatrix} = \begin{bmatrix} 0 \\ 0 \\ 1 \end{bmatrix} - \begin{bmatrix} \sin\theta\cos\phi \\ \sin\theta\sin\phi \\ \cos\theta \end{bmatrix}\cos\theta$$

$$= \mathbf{o}_\mathrm{T} - \mathbf{u}(\mathbf{u}^\mathrm{T}\mathbf{o}_\mathrm{T}) = \left(\mathbf{I}_3 - \mathbf{u}\mathbf{u}^\mathrm{T}\right)\mathbf{o}_\mathrm{T} \,.$$

$$(A.19)$$

Substituting the final expressions into $\mathbf{b} = T_\mathrm{NF}\,\boldsymbol{\beta}_\mathrm{NF} + T_\mathrm{FF}\,\boldsymbol{\beta}_\mathrm{FF}$ concludes the derivation for $\mathbf{o}_\mathrm{T} = [0\ 0\ 1]^\mathrm{T}$.

The formulas however hold in any Cartesian coordinate system, which can be seen by the following argument. First, fix the coordinate system such that $\mathbf{o}_\mathrm{T} = [0\ 0\ 1]^\mathrm{T}$ and consider $\mathbf{u}, \mathbf{b}, \mathbf{e}$. A change of coordinate system by a rotation matrix $\mathbf{Q}$ gives rise to the corresponding quantities $\tilde{\mathbf{o}}_\mathrm{T} = \mathbf{Q}\mathbf{o}_\mathrm{T}$, $\tilde{\mathbf{u}} = \mathbf{Q}\mathbf{u}$, $\tilde{\mathbf{e}} = \mathbf{Q}\mathbf{e}$, $\tilde{\mathbf{b}} = \mathbf{Q}\mathbf{b}$, $\tilde{\boldsymbol{\beta}}_\mathrm{NF} = \mathbf{Q}\boldsymbol{\beta}_\mathrm{NF}$,

$\tilde{\boldsymbol{\beta}}_{\mathrm{FF}} = \mathbf{Q}\boldsymbol{\beta}_{\mathrm{FF}}$. Now

$$\tilde{\mathbf{e}} = \mathbf{Q}\mathbf{e} = \mathbf{Q}(-T_e\mathbf{u} \times \mathbf{o}_{\mathrm{T}})$$
$$= -T_e\,\mathbf{Q}(\mathbf{u} \times \mathbf{o}_{\mathrm{T}}) = -T_e\,(\mathbf{Q}\mathbf{u}) \times (\mathbf{Q}\mathbf{o}_{\mathrm{T}}) = -T_e\,\tilde{\mathbf{u}} \times \tilde{\mathbf{o}}_{\mathrm{T}} \qquad (\mathrm{A}.20)$$

by a property of a rotation matrix applied to a cross product. This is just the same formula in the new coordinate system. For the magnetic field we analogously obtain

$$\tilde{\mathbf{b}} = \mathbf{Q}(T_{\mathrm{NF}}\,\boldsymbol{\beta}_{\mathrm{NF}} + T_{\mathrm{FF}}\,\boldsymbol{\beta}_{\mathrm{FF}}) = T_{\mathrm{NF}}\,\tilde{\boldsymbol{\beta}}_{\mathrm{NF}} + T_{\mathrm{FF}}\,\tilde{\boldsymbol{\beta}}_{\mathrm{FF}}\,. \qquad (\mathrm{A}.21)$$

This also leads to an equivalent formula because the expressions

$$\tilde{\boldsymbol{\beta}}_{\mathrm{FF}} = \mathbf{Q}\boldsymbol{\beta}_{\mathrm{FF}} = \mathbf{Q}(\mathbf{I}_3 - \mathbf{u}\mathbf{u}^{\mathrm{T}})\mathbf{o}_{\mathrm{T}} = (\mathbf{Q} - \mathbf{Q}\mathbf{u}\mathbf{u}^{\mathrm{T}})\mathbf{o}_{\mathrm{T}}$$
$$= (\mathbf{Q}\mathbf{Q}^{\mathrm{T}} - \mathbf{Q}\mathbf{u}\mathbf{u}^{\mathrm{T}}\mathbf{Q}^{\mathrm{T}})\mathbf{Q}\mathbf{o}_{\mathrm{T}} = (\mathbf{I}_3 - \tilde{\mathbf{u}}\tilde{\mathbf{u}}^{\mathrm{T}})\tilde{\mathbf{o}}_{\mathrm{T}}\,, \qquad (\mathrm{A}.22)$$
$$\tilde{\boldsymbol{\beta}}_{\mathrm{NF}} = \mathbf{Q}\boldsymbol{\beta}_{\mathrm{NF}} = \frac{1}{2}\mathbf{Q}(3\mathbf{u}\mathbf{u}^{\mathrm{T}} - \mathbf{I}_3)\mathbf{o}_{\mathrm{T}} = \frac{1}{2}(3\mathbf{Q}\mathbf{u}\mathbf{u}^{\mathrm{T}} - \mathbf{Q})\mathbf{o}_{\mathrm{T}}$$
$$= \frac{1}{2}(3\mathbf{Q}\mathbf{u}\mathbf{u}^{\mathrm{T}}\mathbf{Q}^{\mathrm{T}} - \mathbf{Q}\mathbf{Q}^{\mathrm{T}})\mathbf{Q}\mathbf{o}_{\mathrm{T}} = \frac{1}{2}(3\tilde{\mathbf{u}}\tilde{\mathbf{u}}^{\mathrm{T}} - \mathbf{I}_3)\tilde{\mathbf{o}}_{\mathrm{T}} \qquad (\mathrm{A}.23)$$

are equivalent to the definitions in the other coordinate system, cf. (2.19) and (2.20).

Appendix B

Coupling Formulae Correspondences and Propagation Modes

This appendix discusses mathematical correspondences between the different formulas for the mutual impedance Z_{RT} between two coils that were introduced in Cpt. 2. The analysis also points out the mathematical cause of the different propagation modes (reactive near field and radiated far field) in detail.

We start with the magnetoquasistatic regime, where $Z_{\mathrm{RT}} = j\omega M$ holds. The mutual inductance M is given by the Neumann formula (2.17),

$$M = \frac{\mu_0}{4\pi} \int_{\mathcal{C}_{\mathrm{R}}} \int_{\mathcal{C}_{\mathrm{T}}} \frac{d\boldsymbol{\ell}_{\mathrm{T}} \cdot d\boldsymbol{\ell}_{\mathrm{R}}}{r} \, . \tag{B.1}$$

We also introduced an approximation for flat-turn coils (2.26), given by

$$\tilde{M} = \frac{\mu_0 A_{\mathrm{T}} \mathring{N}_{\mathrm{T}} A_{\mathrm{R}} \mathring{N}_{\mathrm{R}}}{2\pi} \frac{J_{\mathrm{NF}}}{r^3} \tag{B.2}$$

where r is the center-to-center distance. A comparison prompts the question why (B.2) decays with r^{-3} even though the integrand of (B.1) scales as r^{-1}. The reason is the alternating sign of the integrand $\frac{d\boldsymbol{\ell}_{\mathrm{T}} \cdot d\boldsymbol{\ell}_{\mathrm{R}}}{r}$, which is due to the inner product $d\boldsymbol{\ell}_{\mathrm{T}} \cdot d\boldsymbol{\ell}_{\mathrm{R}}$. Over wires $\mathcal{C}_{\mathrm{T}}, \mathcal{C}_{\mathrm{T}}$ that resemble closed loops, the integrand will have a fairly symmetric distribution around zero when the wire separation is large (in which case $\frac{1}{r}$ is almost constant). Thus, the value of the full integral will be very small. This circumstance is similar to forming the expected value of a random variable with significant variance but a very small mean value.

To investigate this aspect formally, we write the distance r between wire points as

$$r = r_{\min} + \delta \, , \qquad r_{\min} = \min \left\{ \|\mathbf{p}_{\mathrm{R}} - \mathbf{p}_{\mathrm{T}}\| \,\middle|\, \mathbf{p}_{\mathrm{T}} \in \mathcal{C}_{\mathrm{T}}, \mathbf{p}_{\mathrm{R}} \in \mathcal{C}_{\mathrm{R}} \right\} \, . \tag{B.3}$$

Note that $\delta \geq 0$ holds $\forall \mathbf{p}_{\mathrm{R}}, \mathbf{p}_{\mathrm{T}}$. We expand the Taylor series $\frac{1}{r} \approx \frac{1}{r_{\min}} - \frac{\delta}{r_{\min}^2} + \frac{\delta^2}{r_{\min}^3}$ to approximate the integral (B.1), yielding

$$M \approx \frac{\mu_0}{4\pi} \int_{\mathcal{C}_R} \int_{\mathcal{C}_T} \left(\frac{1}{r_{\min}} - \frac{\delta}{r_{\min}^2} + \frac{\delta^2}{r_{\min}^3} \right) d\boldsymbol{\ell}_T \cdot d\boldsymbol{\ell}_R . \tag{B.4}$$

The first summand is zero under the integral if $\mathcal{C}_T$ is a closed loop or $\mathcal{C}_R$ is a closed loop (or both). In this case,

$$M \approx \frac{\mu_0}{4\pi} \left(\frac{1}{r_{\min}^3} \int_{\mathcal{C}_R} \int_{\mathcal{C}_T} \delta^2 \, d\boldsymbol{\ell}_T \cdot d\boldsymbol{\ell}_R - \frac{1}{r_{\min}^2} \int_{\mathcal{C}_R} \int_{\mathcal{C}_T} \delta \, d\boldsymbol{\ell}_T \cdot d\boldsymbol{\ell}_R \right) . \tag{B.5}$$

The weighting kernels $\delta^2 \geq 0$ and $\delta \geq 0$ of the integrals attenuate contributions from the closest pairs of points ($\delta \approx 0$) while pairs that are far apart (large δ) are weighted most. This mitigates the discussed cancellation effect under the integral. Thereby, δ^2 is the more aggressive weighting kernel. The linear kernel δ apparently does not have a strong mitigation effect: $\int_{\mathcal{C}_R} \int_{\mathcal{C}_T} \delta \, d\boldsymbol{\ell}_T \cdot d\boldsymbol{\ell}_R$ decays with at least $r_{\min}^{-1}$ for closed loops and large $r_{\min}$. Otherwise M would decay slower than $r_{\min}^{-3}$, which would contradict the approximation $\tilde{M}$ in (B.2). We conclude the investigation on the path loss of the mutual inductance between closed loops with the statement

$$M \approx \frac{\mu_0}{4\pi} \frac{1}{r_{\min}^3} I_{NF} , \qquad I_{NF} = \int_{\mathcal{C}_R} \int_{\mathcal{C}_T} \delta(\delta - r_{\min}) \, d\boldsymbol{\ell}_T \cdot d\boldsymbol{\ell}_R . \tag{B.6}$$

and the understanding that I_{NF} converges to a finite value for $r_{\min} \to \infty$. This finite value is different from zero for most coil arrangements.

We proceed with an investigation of the correspondences between formulas for the mutual impedance Z_{RT} in the general case (i.e. without a magnetoquasistatic assumption). We introduced the integral formulation (2.16) for arbitrary wires, given by

$$Z_{RT} = \frac{j\omega\mu_0}{4\pi} \int_{\mathcal{C}_R} \int_{\mathcal{C}_T} e^{-jkr} \frac{d\boldsymbol{\ell}_T \cdot d\boldsymbol{\ell}_R}{r} , \tag{B.7}$$

as well as the approximation (2.23) for electrically small flat-turn coils (here r is again the center-to-center distance), given by

$$\tilde{Z}_{RT} = \frac{j\omega\mu_0 A_T \mathring{N}_T A_R \mathring{N}_R k^3}{2\pi} e^{-jkr} \left(\left(\frac{1}{(kr)^3} + \frac{j}{(kr)^2} \right) J_{NF} + \frac{1}{2kr} J_{FF} \right) . \tag{B.8}$$

An inspection of the mathematical structure prompts another important question: how does the subtle retardation term e^{-jkr} in (B.7) give rise to propagation modes that decay with r^{-2} and r^{-1}? After all, such modes do not arise from the magnetoquasistatic formulation in (B.1). Again, we conduct an investigation with the decomposition $r = r_{\min} + \delta$ and a Taylor approximation

$$e^{-jkr} = e^{-jkr_{\min}} e^{-jk\delta} \approx e^{-jkr_{\min}} \left(1 - jk\delta - \frac{(k\delta)^2}{2} \right), \qquad \delta \geq 0. \tag{B.9}$$

which is accurate for electrically small antennas because they fulfill $k\delta \ll 1$ for any occurring δ. By using (B.9) in (B.7) and doing a few rearrangements we find that

$$Z_{\mathrm{RT}} = \frac{j\omega\mu_0}{4\pi} e^{-jkr_{\min}} \left(\underbrace{\iint\limits_{\mathcal{C}_R \mathcal{C}_T} \frac{d\boldsymbol{\ell}_{\mathrm{T}} \cdot d\boldsymbol{\ell}_{\mathrm{R}}}{r}}_{\substack{\text{near field,}\\ \text{cubic decay}}} - jk \underbrace{\iint\limits_{\mathcal{C}_R \mathcal{C}_T} \delta\, \frac{d\boldsymbol{\ell}_{\mathrm{T}} \cdot d\boldsymbol{\ell}_{\mathrm{R}}}{r}}_{\substack{\text{transition,}\\ \text{quadratic decay}}} - \frac{k^2}{2} \underbrace{\iint\limits_{\mathcal{C}_R \mathcal{C}_T} \delta^2\, \frac{d\boldsymbol{\ell}_{\mathrm{T}} \cdot d\boldsymbol{\ell}_{\mathrm{R}}}{r}}_{\substack{\text{far field,}\\ \text{linear decay}}} \right)$$

$$\tag{B.10}$$

which now exhibits a striking mathematical correspondence to (B.8). A comparison suggests that the integrals decay with r^{-3}, r^{-2} and r^{-1}, respectively, which highlights the effect of the positive weighting kernels δ and δ^2. The leftmost integral is directly from the Neumann formula (B.1) whose rapid decay was discussed above.

To reveal the mathematical correspondence between the two formulas (B.7) and (B.8) in more detail we use a more elaborate approach: for the integrand of (B.7) we consider the second-order Taylor approximation

$$\frac{e^{-jkr}}{r} \approx \frac{e^{-jkr_{\min}}}{r_{\min}} \left(1 - \left(\frac{1}{r_{\min}} + jk \right) \delta + \left(\frac{1}{r_{\min}^2} + \frac{jk}{r_{\min}} - \frac{1}{2}k^2 \right) \delta^2 \right) \tag{B.11}$$

$$= \frac{e^{-jkr_{\min}}}{r_{\min}} \left(1 + \frac{1}{r_{\min}} \left(\frac{\delta^2}{r_{\min}} - \delta \right) + jk \left(\frac{\delta^2}{r_{\min}} - \delta \right) - \frac{k^2}{2}\delta^2 \right)$$

$$= \frac{e^{-jkr_{\min}}}{r_{\min}} + k^3 e^{-jkr_{\min}} \left(\left(\frac{1}{(kr_{\min})^3} + \frac{j}{(kr_{\min})^2} \right) \delta(\delta - r_{\min}) + \frac{-\delta^2}{2kr_{\min}} \right).$$

We use this expression in formula (B.7) for Z_{RT}. The leftmost summand of (B.12) is again constant and thus zero under a closed line integral. We thus obtain

$$Z_{\mathrm{RT}} \approx \frac{j\omega\mu_0 k^3}{4\pi} e^{-jkr_{\mathrm{min}}} \left(\left(\frac{1}{(kr_{\mathrm{min}})^3} + \frac{j}{(kr_{\mathrm{min}})^2} \right) \int_{\mathcal{C}_{\mathrm{R}}} \int_{\mathcal{C}_{\mathrm{T}}} \delta(\delta - r_{\mathrm{min}})\, d\boldsymbol{\ell}_{\mathrm{T}} \cdot d\boldsymbol{\ell}_{\mathrm{R}} \right.$$

$$\left. + \frac{1}{2kr_{\mathrm{min}}} \int_{\mathcal{C}_{\mathrm{R}}} \int_{\mathcal{C}_{\mathrm{T}}} -\delta^2\, d\boldsymbol{\ell}_{\mathrm{T}} \cdot d\boldsymbol{\ell}_{\mathrm{R}} \right) \quad \text{(B.12)}$$

$$= \frac{j\omega\mu_0 k^3}{4\pi} e^{-jkr_{\mathrm{min}}} \left(\left(\frac{1}{(kr_{\mathrm{min}})^3} + \frac{j}{(kr_{\mathrm{min}})^2} \right) I_{\mathrm{NF}} + \frac{1}{2kr_{\mathrm{min}}} I_{\mathrm{FF}} \right) \quad \text{(B.13)}$$

with I_{NF} from (B.6), i.e. the same factor that multiplies mutual inductance in the magnetoquasistatic case, and with $I_{\mathrm{FF}} = \int_{\mathcal{C}_{\mathrm{R}}} \int_{\mathcal{C}_{\mathrm{T}}} -\delta^2\, d\boldsymbol{\ell}_{\mathrm{T}} \cdot d\boldsymbol{\ell}_{\mathrm{R}}$. The expression is now completely analogous to (B.8), which is what we wanted to achieve. A comparison between (B.8) and (B.13) suggests that the alignment factors fulfill

$$J_{\mathrm{NF}} = \frac{1}{2A_{\mathrm{T}}\mathring{N}_{\mathrm{T}}A_{\mathrm{R}}\mathring{N}_{\mathrm{R}}} \lim_{r_{\mathrm{min}} \to \infty} I_{\mathrm{NF}}, \qquad I_{\mathrm{NF}} = \int_{\mathcal{C}_{\mathrm{R}}} \int_{\mathcal{C}_{\mathrm{T}}} \delta(\delta - r_{\mathrm{min}})\, d\boldsymbol{\ell}_{\mathrm{T}} \cdot d\boldsymbol{\ell}_{\mathrm{R}}, \quad \text{(B.14)}$$

$$J_{\mathrm{FF}} = \frac{1}{2A_{\mathrm{T}}\mathring{N}_{\mathrm{T}}A_{\mathrm{R}}\mathring{N}_{\mathrm{R}}} \lim_{r_{\mathrm{min}} \to \infty} I_{\mathrm{FF}}, \qquad I_{\mathrm{FF}} = \int_{\mathcal{C}_{\mathrm{R}}} \int_{\mathcal{C}_{\mathrm{T}}} -\delta^2\, d\boldsymbol{\ell}_{\mathrm{T}} \cdot d\boldsymbol{\ell}_{\mathrm{R}} \quad \text{(B.15)}$$

which we were able to confirm in numerical experiments.

In the process, we revealed the mathematical background of how the retardation term e^{-jkr} gives rise to a radiative r^{-1} propagation mode and a transition mode decaying with r^{-2}. The above also explains why the near-field mode and the transition mode are subject to the same alignment factor J_{NF} (they are both multiplied by $I_{\mathrm{NF}} \approx 2A_{\mathrm{T}}\mathring{N}_{\mathrm{T}}A_{\mathrm{R}}\mathring{N}_{\mathrm{R}}J_{\mathrm{NF}}$) but the far-field mode to a different alignment factor J_{FF}.

Appendix C

Maximum PTE over a Two-Port Network: Z-Parameter Formula

Consider the simultaneous power matching problem depicted in Fig. C.1: given a two-port network with 2×2 impedance matrix $\mathbf{Z}$, we want to find the source impedance Z_G and load impedance Z_L which establish power matching (conjugate matching) $Z_G = Z_{\text{in}}^*$ and $Z_L = Z_{\text{out}}^*$ simultaneously. This problem is non-trivial because Z_G affects Z_{out} and Z_L affects Z_{in}, hence Z_G and Z_L must be optimized jointly. In this thesis the two-port network of interest is a magneto-inductive link with impedance matrix $\mathbf{Z}_C$ among the coil ports. A related goal is the calculation of the power transfer efficiency η from source to load in the case of successful simultaneous power matching (which maximizes the power into the load). These problems can be solved with established tools such as the transducer power gain formula for S-parameters [242, Eq. 28] [52, Eq. 12.13] or with generalized S-parameters [242, Eq. 29] (then $\eta = |S_{21}|^2$). Those approaches however obfuscate the dependence of η on intuitive technical quantities such as coil mutual impedance and self-impedances: the transducer power gain formula is quite complicated and the conversion to generalized S-parameters uses a matrix determinant. For this reason we prefer a formula for η in terms of Z-parameters, which directly reflect the intuitive quantities.

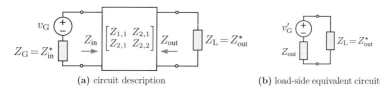

(a) circuit description   (b) load-side equivalent circuit

Figure C.1: Simultaneous power matching problem for a reciprocal two-port network.

In particular, in this appendix we derive a novel Z-parameter formula for the power transfer efficiency

$$\eta = \frac{\zeta^2 + \chi^2}{\left(\sqrt{1 - \zeta^2} + \sqrt{1 + \chi^2}\right)^2} \overset{\zeta^2 + \chi^2 \ll 1}{\approx} \frac{\zeta^2 + \chi^2}{4} = \frac{|Z_{2,1}|^2}{4\,\mathrm{Re}(Z_{1,1})\,\mathrm{Re}(Z_{2,2})} \qquad \text{(C.1)}$$

in the case of simultaneous power matching for a reciprocal two-port network (i.e. $\mathbf{Z}$ is symmetric). Thereby $\zeta \in [-1, +1]$ and $\chi \in \mathbb{R}$ are metrics of $\mathbf{Z}$, defined as

$$\zeta = \frac{\mathrm{Re}(Z_{2,1})}{\sqrt{\mathrm{Re}(Z_{1,1})\,\mathrm{Re}(Z_{2,2})}}\,, \qquad\qquad \chi = \frac{\mathrm{Im}(Z_{2,1})}{\sqrt{\mathrm{Re}(Z_{1,1})\,\mathrm{Re}(Z_{2,2})}}\,. \qquad \text{(C.2)}$$

In the process we show that simultaneous power matching of a reciprocal two-port network is established with the impedances

$$Z_{\mathrm{G}} = Z_{\mathrm{in}}^* = \left(\sqrt{1 - \zeta^2}\sqrt{1 + \chi^2} + j\zeta\chi\right)\mathrm{Re}(Z_{1,1}) - j\mathrm{Im}(Z_{1,1})\,, \qquad \text{(C.3)}$$

$$Z_{\mathrm{L}} = Z_{\mathrm{out}}^* = \left(\sqrt{1 - \zeta^2}\sqrt{1 + \chi^2} + j\zeta\chi\right)\mathrm{Re}(Z_{2,2}) - j\mathrm{Im}(Z_{2,2})\,. \qquad \text{(C.4)}$$

These results are derived in the following.

From Proposition 2.6 and the requirements $Z_{\mathrm{G}} = Z_{\mathrm{in}}^*$ and $Z_{\mathrm{L}} = Z_{\mathrm{out}}^*$ we obtain

$$Z_{\mathrm{in}} = Z_{1,1} - \frac{Z_{2,1}^2}{Z_{2,2} + Z_{\mathrm{L}}} \qquad \Longrightarrow \qquad Z_{\mathrm{in}} \overset{!}{=} Z_{1,1} - \frac{Z_{2,1}^2}{Z_{2,2} + Z_{\mathrm{out}}^*}\,, \qquad \text{(C.5)}$$

$$Z_{\mathrm{out}} = Z_{2,2} - \frac{Z_{2,1}^2}{Z_{1,1} + Z_{\mathrm{G}}} \qquad \Longrightarrow \qquad Z_{\mathrm{out}}^* \overset{!}{=} Z_{2,2}^* - \frac{(Z_{2,1}^2)^*}{Z_{1,1}^* + Z_{\mathrm{in}}}\,. \qquad \text{(C.6)}$$

The two right-hand side equations determine the two unknown complex variables Z_{in} and Z_{out}^*. By rearrangements and substitution we obtain two individual quadratic equations for Z_{in} and Z_{out},

$$(Z_{\mathrm{in}} - Z_{1,1})\left(2\,\mathrm{Re}(Z_{2,2})(Z_{1,1}^* + Z_{\mathrm{in}}) - (Z_{2,1})^*\right) + Z_{2,1}^2(Z_{1,1}^* + Z_{\mathrm{in}}) = 0\,, \qquad \text{(C.7)}$$

$$(Z_{\mathrm{out}} - Z_{2,2})\left(2\,\mathrm{Re}(Z_{1,1})(Z_{2,2}^* + Z_{\mathrm{out}}) - (Z_{2,1})^*\right) + Z_{2,1}^2(Z_{2,2}^* + Z_{\mathrm{out}}) = 0\,. \qquad \text{(C.8)}$$

They are solved with the standard formula for quadratic equations. By also canceling and merging terms (the process is omitted), we find the solutions (C.3) and (C.4) of Z_{in} and Z_{out}. The other solutions are physically meaningless due to negative real parts.

The Helmholtz-Thévenin equivalent source voltage v_{T}' of the load-side equivalent circuit as in Fig. C.1b is given by $v'_\text{T} = v_\text{T} \cdot Z_{2,1} / (Z_{1,1} + Z_\text{in}^*)$. Therewith we can formulate the ratio of the power wave emitted by the source to the power wave into the load,

$$\frac{v'_\text{G} / \sqrt{4 \operatorname{Re} Z_\text{out}}}{v_\text{G} / \sqrt{4 \operatorname{Re} Z_\text{in}}} = \frac{\zeta + j\chi}{1 + \sqrt{1 - \zeta^2} \sqrt{1 + \chi^2} + j\zeta\chi} =: h \tag{C.9}$$

which is interpreted as link coefficient $h \in \mathbb{C}$, comprising magnitude attenuation and a phase shift. Its squared absolute value is the power transfer efficiency

$$\eta = |h|^2 = \frac{\zeta^2 + \chi^2}{\left(1 + \sqrt{1 - \zeta^2} \sqrt{1 + \chi^2}\right)^2 + \zeta^2 \chi^2} \tag{C.10}$$

$$= \frac{\zeta^2 + \chi^2}{1 + (1 - \zeta^2)(1 + \chi^2) + 2\sqrt{1 - \zeta^2}\sqrt{1 + \chi^2} + \zeta^2 \chi^2} \tag{C.11}$$

$$= \frac{\zeta^2 + \chi^2}{(1 - \zeta^2) + (1 + \chi^2) + 2\sqrt{1 - \zeta^2}\sqrt{1 + \chi^2}} \tag{C.12}$$

which proves (C.1) after applying the binomial formula $x^2 + y^2 + 2xy = (x + y)^2$.

A noteworthy aspect is that the power wave ratio (C.9) suggests that h is the generalized S_{21}-parameter of the two-port network. In fact, after lengthy rearrangements of the S-parameter results in [242], we find that the associated scattering matrix is

$$\mathbf{S} = (\mathbf{A} - \gamma^* \mathbf{I}_2)(\mathbf{A} + \gamma \mathbf{I}_2)^{-1} \tag{C.13}$$

whereby the occurring variables relate to our formalism according to

$$\mathbf{A} = \begin{bmatrix} 1 & \zeta + j\chi \\ \zeta + j\chi & 1 \end{bmatrix}, \qquad \gamma = \sqrt{1 - \zeta^2}\sqrt{1 + \chi^2} + j\zeta\chi . \tag{C.14}$$

An interesting related statement is

$$\mathbf{Z} = \begin{bmatrix} \sqrt{\operatorname{Re}(Z_{1,1})} & 0 \\ 0 & \sqrt{\operatorname{Re}(Z_{2,2})} \end{bmatrix} \mathbf{A} \begin{bmatrix} \sqrt{\operatorname{Re}(Z_{1,1})} & 0 \\ 0 & \sqrt{\operatorname{Re}(Z_{2,2})} \end{bmatrix} . \tag{C.15}$$

Appendix D

Resonant SISO Channels: Power Allocation and Capacity

We study the fundamental limits of information transfer from a resonant transmitting coil to a resonant receive coil with reception in additive white Gaussian noise (AWGN). We work with the frequency-dependent channel coefficient $h(f)$ and the single-sided power spectral density (PSD) $S_{ww}(f) = N_0$ of the white Gaussian noise process $w(t)$. In particular, the contents of this appendix are as follows.

- We exploit the mathematical structure of $|h(f)|^2$ for a resonant channel to formulate a detailed closed-form expression for the waterfilling solution of the transmit PSD,

$$
S_{xx}(f) = \begin{cases} \frac{N_0}{|h_{\text{res}}|^2} \left(\left(\frac{B/2}{B_{\text{R}}} \right)^2 - \left(\frac{f - f_{\text{res}}}{B_{\text{R}}} \right)^2 \right) & \text{if } |f - f_{\text{res}}| \leq B/2 \\ 0 & \text{otherwise} \end{cases}
\tag{D.1}
$$

  where B_{R} is the 3-dB bandwidth of the receive coil and B is the effectively used bandwidth, given by

$$
B = B_{\text{R}} \left(\frac{6 |h_{\text{res}}|^2 P_{\text{T}}}{N_0 B_{\text{R}}} \right)^{1/3} .
\tag{D.2}
$$

  The fact $B \propto \sqrt[3]{P_{\text{T}}}$ shows that a considerable increase of transmit power is required to warrant an increase of the used bandwidth.

- For the resulting channel capacity we derive the formula

$$
C = \frac{2}{\log(2)} \left(B - 2B_{\text{R}} \arctan \left(\frac{B}{2B_{\text{R}}} \right) \right) .
\tag{D.3}
$$

- For the above formula we derive the upper bound

$$
C \leq \frac{B_{\text{R}}}{6 \cdot \log(2)} \left(\frac{B}{B_{\text{R}}} \right)^3 = \frac{|h_{\text{res}}|^2 P_{\text{T}}}{\log(2) N_0} ,
\tag{D.4}
$$

which agrees with the well-known capacity upper bound that is obtained with log-linearization in the power-limited regime over a small bandwidth (where the channel is flat).

The results rely on an approximation that is accurate for coils with a large Q-factor.

Continuous-Frequency Waterfilling in General

First consider a general channel $h(f)$ with single-sided noise PSD $S_{ww}(f)$. The transmit signal $x(t)$ has PSD $S_{xx}(f)$ and power $\int_0^\infty S_{xx}(f)\, df = P_{\mathrm{T}}$. This channel can be modeled as a set of parallel AWGN channels, each with an infinitesimally bandwidth df and channel capacity $df \cdot \log_2(1 + \mathrm{SNR}(f))$. The information rate [83, 85]

$$D = \int_0^\infty \log_2\left(1 + \frac{|h(f)|^2\, S_{xx}(f)}{S_{ww}(f)}\right) df \qquad (\mathrm{D.5})$$

in bit/s is achievable over this channel. Clearly D depends on the spectral power allocation $S_{xx}(f)$. The channel capacity C (the largest achievable rate D) is obtained when $S_{xx}(f)$ is allocated with waterfilling [90, Eq. 5.43] [243], i.e. by setting

$$S_{xx}(f) = \max\left\{ 0,\ \mu - \frac{S_{ww}(f)}{|h(f)|^2} \right\}. \qquad (\mathrm{D.6})$$

The so-called water level μ is found by solving the equation

$$\int_0^\infty \max\left\{ 0,\ \mu - \frac{S_{ww}(f)}{|h(f)|^2} \right\} df = P_{\mathrm{T}}. \qquad (\mathrm{D.7})$$

for μ. Since the integral is monotonically increasing with μ, the problem is easily solved numerically for general $S_{ww}(f)$ and $h(f)$, e.g, for the complicated $h(f)$ of a wideband outdoor radio channel. In our case, we are able to evaluate C and the associated $S_{xx}(f)$ analytically because of the specific resonant shape of $h(f)$ and constant $S_{ww}(f) = N_0$.

Considered Resonant Link Model

We study a communication system described by the equivalent circuit in Fig. D.1. The series impedance of the transmit circuit $Z_{\mathrm{T}} = 2R_{\mathrm{T}} + j\omega L_{\mathrm{T}} + \frac{1}{j\omega C_{\mathrm{T}}}$ and of the receive circuit $Z_{\mathrm{R}} = 2R_{\mathrm{R}} + j\omega L_{\mathrm{R}} + \frac{1}{j\omega C_{\mathrm{R}}}$ whereby both circuits are tuned to the same resonance frequency $2\pi f_{\mathrm{res}} = \omega_{\mathrm{res}} = 1/\sqrt{L_{\mathrm{T}} C_{\mathrm{T}}} = 1/\sqrt{L_{\mathrm{R}} C_{\mathrm{R}}}$. The coil quality factors are $Q_{\mathrm{T}} = \frac{\omega_{\mathrm{res}} L_{\mathrm{T}}}{R_{\mathrm{T}}}$ and $Q_{\mathrm{R}} = \frac{\omega_{\mathrm{res}} L_{\mathrm{R}}}{R_{\mathrm{R}}}$. We consider the power wave $x = \sqrt{R_{\mathrm{T}}}\, i_{\mathrm{T}} = \frac{\sqrt{R_{\mathrm{T}}}\, v_{\mathrm{T}}}{Z_{\mathrm{T}}}$ as transmit

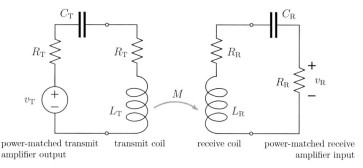

power-matched transmit   transmit coil   receive coil   power-matched receive
amplifier output   amplifier input

Figure D.1: Equivalent circuit description of the considered magneto-inductive SISO link. For mathematical simplicity the model comprises series capacitances for resonant matching instead of two-port networks in L- or T-structure.

signal. Thereby $|x|^2$ is the active power into the transmit coil. Likewise $y = \frac{v_R}{\sqrt{R_R}}$ is the considered received signal. For the weak-coupling case (unilateral assumption, i.e. the presence of the transmitter does not affect the receive-side electrical properties and vice versa) it is easy to show that signal propagation is according to $y = hx$ with

$$h(f) = \frac{f}{f_{\text{res}}} \frac{2R_R}{Z_R} h_{\text{res}}, \qquad h_{\text{res}} = \frac{j\omega_{\text{res}}M}{\sqrt{4R_T R_R}} = \frac{jk\sqrt{Q_T Q_R}}{2} \qquad (D.8)$$

where $k = \frac{M}{\sqrt{L_T L_R}}$ is the coupling coefficient and $Z_R = 2R_R + j2\pi f L_R + \frac{1}{j2\pi f C_R}$ is the serial impedance of the receive-side circuit. The term $\frac{2R_R}{Z_R}$ represents receive-side resonance mismatch that occurs for $f \neq f_{\text{res}}$. At $f = f_{\text{res}}$ the value $\frac{2R_R}{Z_R} = 1$ is attained.

As a preparation for the following derivation we consider the ratio

$$\frac{|h_{\text{res}}|^2}{|h(f)|^2} = \frac{f_{\text{res}}^2}{f^2} \left(\frac{|Z_R|}{2R_R} \right)^2 . \qquad (D.9)$$

With $Z_R = 2R_R + j2\pi f L_R + \frac{1}{j2\pi f C_R} = 2R_R \left(1 + \frac{jQ_R}{2} \frac{f^2 - f_{\text{res}}^2}{f f_{\text{res}}} \right)$ we obtain

$$\frac{|h_{\text{res}}|^2}{|h(f)|^2} = \frac{f_{\text{res}}^2}{f^2} \left| 1 + \frac{jQ_R}{2} \frac{(f - f_{\text{res}})(f + f_{\text{res}})}{f f_{\text{res}}} \right|^2 \qquad (D.10)$$

which shows a quite complicated dependence on f. To simplify the term drastically we assume that the value of f is close to f_{res}, which will typically be the case for operation

213

with a high-Q coil. We obtain

$$\frac{|h_{\mathrm{res}}|^2}{|h(f)|^2} \approx \frac{f_{\mathrm{res}}^2}{f_{\mathrm{res}}^2}\left|1 + \frac{jQ_{\mathrm{R}}}{2}\frac{(f-f_{\mathrm{res}})(2f_{\mathrm{res}})}{f_{\mathrm{res}}^2}\right|^2 = \left|1 + jQ_{\mathrm{R}}\frac{f-f_{\mathrm{res}}}{f_{\mathrm{res}}}\right|^2 = 1 + \left(\frac{f-f_{\mathrm{res}}}{B_{\mathrm{R}}}\right)^2.$$
(D.11)

Derivation of the Core Results

The above consideration yields $\frac{S_{ww}(f)}{|h(f)|^2} = \frac{N_0}{|h_{\mathrm{res}}|^2}\left(1 + \left(\frac{f-f_{\mathrm{res}}}{B_{\mathrm{R}}}\right)^2\right)$ and the water level equation (D.7) becomes

$$\int_0^\infty \max\left\{0, \mu - \frac{N_0}{|h_{\mathrm{res}}|^2}\left(1 + \left(\frac{f-f_{\mathrm{res}}}{B_{\mathrm{R}}}\right)^2\right)\right\} df = P_{\mathrm{T}}.$$
(D.12)

We observe that the integrand is symmetric about $f = f_{\mathrm{res}}$ and monotonically decreasing with $|f - f_{\mathrm{res}}|$. Therefore, the integrand must vanish right at values $f = f_{\mathrm{res}} \pm B/2$ where B is the effectively used bandwidth. Based thereon we determine the water level $\mu = \frac{N_0}{|h_{\mathrm{res}}|^2}\left(1 + \left(\frac{B/2}{B_{\mathrm{R}}}\right)^2\right)$ which is used in the above integral (under due consideration of the support of the integrand) to obtain

$$\int_{f_{\mathrm{res}}-B/2}^{f_{\mathrm{res}}+B/2}\left(\mu - \frac{N_0}{|h_{\mathrm{res}}|^2}\left(1 + \left(\frac{f-f_{\mathrm{res}}}{B_{\mathrm{R}}}\right)^2\right)\right) df = \mu B - \frac{N_0}{|h_{\mathrm{res}}|^2}\left(B + \frac{1}{B_{\mathrm{R}}^2}\int_{-B/2}^{+B/2}s^2 ds\right)$$

$$= \frac{N_0}{|h_{\mathrm{res}}|^2}\left(B + \frac{B^3}{4B_{\mathrm{R}}^2} - B - \frac{1}{B_{\mathrm{R}}^2}\int_{-B/2}^{+B/2}s^2 ds\right) = \frac{N_0}{|h_{\mathrm{res}}|^2}\left(\frac{B^3}{4B_{\mathrm{R}}^2} - \frac{B^3}{12B_{\mathrm{R}}^2}\right)$$

$$= \frac{N_0}{|h_{\mathrm{res}}|^2}\frac{B^3}{B_{\mathrm{R}}^2}\left(\frac{1}{4} - \frac{1}{12}\right) = \frac{N_0}{|h_{\mathrm{res}}|^2}\frac{B^3}{6B_{\mathrm{R}}^2}.$$

This last expression must be equal to the available power P_{T}. As a result we obtain the effectively used bandwidth $B = B_{\mathrm{R}}\sqrt[3]{\frac{6|h_{\mathrm{res}}|^2 P_{\mathrm{T}}}{N_0 B_{\mathrm{R}}}}$ which proves (D.2). Thus we allocate the transmit PSD according to $S_{xx}(f) = \frac{N_0}{|h_{\mathrm{res}}|^2}\left(\left(\frac{B/2}{B_{\mathrm{R}}}\right)^2 - \left(\frac{f-f_{\mathrm{res}}}{B_{\mathrm{R}}}\right)^2\right)$ if $|f - f_{\mathrm{res}}| \le B/2$ and $S_{xx}(f) = 0$ outside that band. Thereby we have proven (D.1).

We use this allocation in the capacity integral (D.5), resulting in

$$C = \int_{f_{res}-B/2}^{f_{res}+B/2} \log_2\left(1 + \frac{|h(f)|^2 S_{xx}(f)}{N_0}\right) df$$

$$= \int_{f_{res}-B/2}^{f_{res}+B/2} \log_2\left(1 + \frac{|h(f)|^2}{N_0}\left(\mu - \frac{N_0}{|h(f)|^2}\right)\right) df = \int_{f_{res}-B/2}^{f_{res}+B/2} \log_2\left(\mu \frac{|h(f)|^2}{N_0}\right) df$$

$$= \int_{f_{res}-B/2}^{f_{res}+B/2} \log_2\left(\frac{N_0}{|h_{res}|^2}\left(1 + \left(\frac{B/2}{B_R}\right)^2\right)\frac{|h(f)|^2}{N_0}\right) df$$

$$= B \log_2\left(1 + \left(\frac{B/2}{B_R}\right)^2\right) - \int_{f_{res}-B/2}^{f_{res}+B/2} \log_2\left(1 + \left(\frac{f - f_{res}}{B_R}\right)^2\right) df$$

$$= B \log_2\left(1 + \frac{(B/2)^2}{B_R^2}\right) - 2 \int_0^{B/2} \log_2\left(1 + \frac{s^2}{B_R^2}\right) ds \qquad (D.13)$$

and by solving the integral

$$\int_0^{B/2} \log_2\left(1 + \frac{s^2}{B_R^2}\right) ds = \frac{B}{2} \log_2\left(1 + \frac{(B/2)^2}{B_R^2}\right) - \frac{B}{\log(2)} + \frac{2B_R}{\log(2)} \arctan\left(\frac{B/2}{B_R}\right)$$

we obtain the channel capacity $C = \frac{2}{\log(2)}\left(B - 2B_R \arctan\left(\frac{B/2}{B_R}\right)\right)$ which proves (D.3).

The Taylor series $\arctan(x) \approx x - \frac{1}{3}x^3$ is a lower bound $\arctan(x) \geq x - \frac{1}{3}x^3$ for positive x and thus, after canceling terms, yields $C \leq \frac{B_R}{6 \cdot \log(2)}\left(\frac{B}{B_R}\right)^3$ as in (D.4).

A Comment on Transmit-Side Mismatch

In the above exposition the transmit-side mismatch does not have an impact because, throughout this thesis and in related literature such as [53], the transmit power $P_T = |x|^2$ is defined as the active power into the transmit coil. Thereby we do not consider power that is possibly reflected back into or dissipated in the transmit amplifier. These effects are certainly of practical relevance but, to the best of our knowledge, are not captured by any simple existing model. We can however get an idea of the behavior by instead defining $\tilde{h}$ as the voltage gain from transmitter to receiver, i.e. $v_R = \tilde{h}v_T$ plus noise, which is given by $\tilde{h} = \frac{f}{f_{res}} \frac{2R_R}{Z_R} \frac{2R_T}{Z_T} h_{res}$ and does indeed show transmit-side mismatch in the term $\frac{2R_T}{Z_T}$. Thereby $Z_T = 2R_T + j2\pi f L_T + \frac{1}{j2\pi f C_T}$ is the serial impedance of the transmit-side circuit. With the same high-Q approximation approach as earlier we obtain

$$\frac{|h_{\text{res}}|^2}{|\tilde{h}(f)|^2} \approx \left(1 + \left(\frac{f - f_{\text{res}}}{B_{\text{R}}}\right)^2\right)\left(1 + \left(\frac{f - f_{\text{res}}}{B_{\text{T}}}\right)^2\right) \tag{D.14}$$

which now exhibits signal attenuation due to resonance mismatch at the receiver and also the transmitter. Repeating all the calculations for this case (the details are omitted) yields the fifth-order equation

$$\frac{B_{\text{pass}}}{6}\left(\frac{B}{B_{\text{pass}}}\right)^3 + \frac{B_{\text{stop}}}{20}\left(\frac{B}{B_{\text{stop}}}\right)^5 = \frac{|h(f_{\text{res}})|^2 P_{\text{T}}}{N_0} \tag{D.15}$$

for the determination of the effectively used bandwidth B from all technical parameters. The equation uses the system bandwidth $B_{\text{pass}} = 1/\sqrt{1/B_{\text{T}}^2 + 1/B_{\text{R}}^2}$ and $B_{\text{stop}} = \sqrt{B_{\text{T}} B_{\text{R}}}$. They compare according to $B_{\text{pass}} \leq \min\{B_{\text{T}}, B_{\text{R}}\} \leq B_{\text{stop}} \leq \max\{B_{\text{T}}, B_{\text{R}}\}$ and their meaning is that $\tilde{h}(f)$ is hardly attenuated for $|f - f_{\text{res}}| < \frac{1}{2} B_{\text{pass}}$ but heavily attenuated for $|f - f_{\text{res}}| > \frac{1}{2} B_{\text{stop}}$.

In the power-limited regime the used bandwidth will be $B \ll B_{\text{pass}}$ and thus the cubic term dominates in (D.15). On the other hand, the fifth-order term dominates for $B \gg B_{\text{stop}}$. For these cases we can therefore approximate

$$B \approx B_{\text{pass}}\left(\frac{6\,|h(f_{\text{res}})|^2 P_{\text{T}}}{N_0 B_{\text{pass}}}\right)^{1/3} \qquad \text{for } B \ll B_{\text{pass}}, \tag{D.16}$$

$$B \approx B_{\text{stop}}\left(\frac{20\,|h(f_{\text{res}})|^2 P_{\text{T}}}{N_0 B_{\text{stop}}}\right)^{1/5} \qquad \text{for } B \gg B_{\text{stop}} \tag{D.17}$$

(both expressions are in fact upper bounds). We find that (D.16) is completely analogous to (D.2), it just uses the system bandwidth B_{pass} instead of the 3-dB bandwidth B_{R} of just the receive coil. The formula (D.17) shows the drastic effect of the effectively band-limited channel when a large $|h(f_{\text{res}})|^2$ and/or excess transmit power is available: one has to increase the transmit power by a factor of $2^5 = 32$ in order to warrant doubling the communication bandwidth.

Appendix E

The Role of Transients in Load Modulation Receive Processing

In this appendix we discuss the role of transients in near-field communication via load modulation. This modulation has not received much attention by the academic world despite being used by billions of devices for radio-frequency identification (RFID) and contact-less payment [63]. In particular, we will propose a receive processing that regards the signal transients after the load switching instants and, consequently, show possible improvements to the bit error probability when the transients are considered. Thereby we employ signal space concepts from basic communication theory.

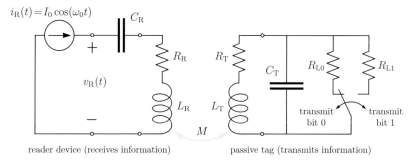

Figure E.1: Considered near-field communication system with load modulation. The reader device (R) drives a stationary current $i_R(t)$ through its coil which, by magnetic induction, powers the tag (T). The tag transmits bits to the reader by choosing either R_{L0} or R_{L1} as load resistance for $t \in [(n-1)T_b, nT_b]$ according to the value of the n-th bit. The load resistance affects the tag coil current and thus also $v_R(t)$ which is measured at the reader and allows to detect the transmitted bits.

Fig. E.1 shows a circuit description of the considered load modulation system, analogous to a description in [63]. An AC current source with frequency f_0 drives the reader coil which then also drives the tag coil due to their mutual inductance M. The tag chooses a load resistance $R_L \in \{R_{L0}, R_{L1}\}$ according to the value of the n-th bit which is transmitted during $t \in [(n-1)T_b, nT_b]$. The reader observes the signal $y(t) = v_R(t) - v_R^{\text{No Tag}}(t)$ in order to detect the bits. Thereby, $v_R^{\text{No Tag}}(t)$ is the voltage over the resonant reader coil for the case that no tag is present (we assume that this signal is known and can be canceled).[1]

The time-variant load resistance together with the reactive elements (inductances and capacitances) lead to transient signals and memory. Thus, strictly speaking, the signal $y(t)$ at any time t depends on the entire history of the load resistance (i.e. all previously transmitted bits) which results in inter-symbol interference. Due to this complication, the prevalent state-of-the-art approach for receive processing is to only regard the received signal at times where the transients died out for the most part [244, Fig. 27], [245, Fig. 8]. This approach seems suboptimal and negligent from the communication-theoretic point of view because vast portions of the received signal are being ignored. We want to understand the degree of this negligence and, if possible, propose a more suitable processing.

In order to study this system, we must describe the evolution of the electrical signals. We exploit the fact that R_L is constant for a bit duration. The corresponding circuit is depicted in Fig. E.2. It shall be noted that $i_T(t)$ is continuous because of the inductance L_T and $i_L(t)$ is continuous because of the parallel capacitance C_T. By

$$v_T(t) = M i_R'(t) = -\omega_0 M I_0 \sin(\omega_0 t)$$

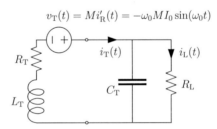

Figure E.2: Equivalent circuit description of the transmitting tag for the duration of a bit. Over this time interval the resistance $R_L \in \{R_{L0}, R_{L1}\}$ is constant which leads to a differential equation with constant coefficients. $v_T(t)$ is the voltage induced by the magnetic field from the reader device.

applying the Kirchhoff laws to the tag circuit we find the differential equation system

$$\begin{bmatrix} i_T'(t) \\ i_L'(t) \end{bmatrix} = \underbrace{\begin{bmatrix} -\frac{R_T}{L_T} & -\frac{R_L}{L_T} \\ \frac{1}{R_L C_T} & -\frac{1}{R_L C_T} \end{bmatrix}}_{=: \, \mathbf{A}} \begin{bmatrix} i_T(t) \\ i_L(t) \end{bmatrix} + \begin{bmatrix} -\omega_0 \frac{M}{L_T} I_0 \sin(\omega_0 t) \\ 0 \end{bmatrix}. \tag{E.1}$$

---

[1]In practical RFID systems (e.g. ISO/IEC 14443 standard) sideband modulation is used in order to facilitate the isolation of the information-bearing signal from the dominant f_0-oscillation at the reader. [63]

In particular, this is a linear inhomogeneous system of first-order ordinary differential equations with constant coefficients. It has the solution [246, Sec. 16.13.2]

$$
\begin{bmatrix} i_T(t) \\ i_L(t) \end{bmatrix} = \underbrace{D_1 e^{\lambda_1 t} \mathbf{v}_1 + D_2 e^{\lambda_2 t} \mathbf{v}_2}_{\text{homogeneous solution}} + \underbrace{\begin{bmatrix} g_{11} & g_{12} \\ g_{21} & g_{22} \end{bmatrix} \begin{bmatrix} \cos(\omega_0 t) \\ \sin(\omega_0 t) \end{bmatrix}}_{\text{particular solution}} \tag{E.2}
$$

where $\lambda_1, \lambda_2 \in \mathbb{C}$ are the eigenvalues and $\mathbf{v}_1, \mathbf{v}_2 \in \mathbb{C}^3$ the respective eigenvectors of $\mathbf{A}$. By requiring that the particular solution fulfills (E.1) and exploiting the orthogonality of sine and cosine, one can show that the related coefficients are given by

$$
\begin{bmatrix} g_{11} \\ g_{21} \\ g_{12} \\ g_{22} \end{bmatrix} = \begin{bmatrix} \mathbf{A} & -\omega_0 \mathbf{I}_2 \\ \omega_0 \mathbf{I}_2 & \mathbf{A} \end{bmatrix}^{-1} \begin{bmatrix} 0 \\ 0 \\ \omega_0 \frac{M}{L_T} I_0 \\ 0 \end{bmatrix} . \tag{E.3}
$$

The constants $D_1, D_2 \in \mathbb{C}$ follow from the initial value problem given the values of $i_T(t_{n-1})$ and $i_L(t_{n-1})$ at the switching instant $t_{n-1} = (n-1)T_b$. The resulting solutions $i_T(t)$ and $i_L(t)$ will of course be real valued. The received signal, which is the induced voltage at the reader coil due to $i_T(t)$, then follows from $y(t) = M i_T'(t)$. Fig. E.3 shows exemplary signals for the transmitted bit sequence 0110010. Significant transients can be observed after any switching instant (after a bit change).

Although the communication-theoretic implications of time-dispersive effects and inter-symbol interference are well-understood for linear modulation over linear time-invariant channels [37, 158], there seems to be no meaningful way to apply this theory to the load modulation system (which is governed by differential equations). This fact complicates the analysis of the communication performance. However, we can identify certain aspects that facilitate the analysis. A first simplification is the assumption of an ideal current source at the reader. Therewith, and without any further assumptions, the reactive elements at the reader do not get involved in the differential equations (a finite source resistance, e.g. 50 Ω, would lead to a system of four differential equations). The key simplification however is based on the observation that usually the time constant of the transients is smaller than a bit duration (i.e. $Q_T/f_0 < T_b$), which means that the transients effectively wear off before the next load-switching instant. Thus, only the previous bit interferes appreciably with the current bit because $y(t)$ approaches its stationary state near the end of each bit interval. As a result, the received signal can be described as a concatenation of waveforms that are drawn from only four prototype waveforms $y_{00}(t)$, $y_{01}(t)$, $y_{10}(t)$, and $y_{11}(t)$. These arise from the four possible bit transitions 0→0, 0→1, 1→0, and 1→1. For example, $y_{01}(t)$ is the waveform during a bit duration when the previous bit was 0 and the current bit is 1. This aspect can be observed clearly in Fig. E.3. For notational convenience we define that all four waveforms are supported for $t \in [0, T_b]$.

We are now going to expand orthonormal basis functions for these waveforms with the goal of later using them to transform received waveforms to signal space vectors.

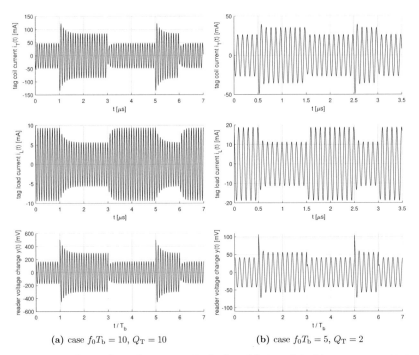

(a) case $f_0 T_b = 10$, $Q_T = 10$ (b) case $f_0 T_b = 5$, $Q_T = 2$

Figure E.3: Electrical signals over time for load modulation of the bit sequence 0110010. The currents at the tag and the received signal $y(t)$ (the tag-induced voltage difference at the reader) are shown. The quantities relate to the circuits in Fig. E.1 and E.2. The simulation uses a current source with $f_0 = 10\,\text{MHz}$ and peak amplitude I_0 such that $100\,\text{mW}$ are burnt at the resonant reader coil ($Q_R = 500$). Both coils have $5\,\Omega$ resistance. We consider two different tag coil qualities and bit rates (a) and (b). The coupling coefficient $\kappa = \frac{1}{100}$. The capacitances C_R and C_T are chosen such that both coils are resonant in uncoupled condition, i.e. C_T is chosen such that $\frac{1}{Z_T} = \frac{1}{R_T + j\omega_0 L_T} + j\omega_0 C_T$ is real valued. The two load resistances are set to $R_{L0} = \frac{1}{2} Z_T$ and $R_{L1} = \frac{3}{2} Z_T$.

Since $y_{00}(t)$ and $y_{11}(t)$ are very accurately described by stationary oscillations with constant amplitude and phase (depending on the validity of the assumption of the system having only a memory of one bit memory) we select the cosine (in-phase) and sine (quadrature-phase) as our first orthonormal basis functions

$$\phi_{\mathrm{I}}(t) = \sqrt{\frac{2}{T_{\mathrm{b}}}} \cos(\omega_0 t)\,, \tag{E.4}$$

$$\phi_{\mathrm{Q}}(t) = \sqrt{\frac{2}{T_{\mathrm{b}}}} \sin(\omega_0 t)\,. \tag{E.5}$$

We compute two more basis functions by applying the Gram-Schmidt process to $y_{01}(t)$ and $y_{10}(t)$, which are the waveforms with significant transients. We obtain

$$\phi_{\mathrm{rise}}(t) = \frac{z_{\mathrm{rise}}(t)}{\sqrt{\int_0^{T_{\mathrm{b}}} z_{\mathrm{rise}}^2(\tau)d\tau}}\,, \tag{E.6}$$

$$z_{\mathrm{rise}}(t) = y_{01}(t) - \phi_{\mathrm{I}}(t)\int_0^{T_{\mathrm{b}}} \phi_{\mathrm{I}}(\tau)y_{01}(\tau)d\tau - \phi_{\mathrm{Q}}(t)\int_0^{T_{\mathrm{b}}} \phi_{\mathrm{Q}}(\tau)y_{01}(\tau)d\tau\,,$$

$$\phi_{\mathrm{fall}}(t) = \frac{z_{\mathrm{fall}}(t)}{\sqrt{\int_0^{T_{\mathrm{b}}} z_{\mathrm{fall}}^2(\tau)dt}}\,, \tag{E.7}$$

$$z_{\mathrm{fall}}(t) = y_{10}(t) - \phi_{\mathrm{I}}(t)\int_0^{T_{\mathrm{b}}} \phi_{\mathrm{I}}(\tau)y_{10}(\tau)d\tau - \phi_{\mathrm{Q}}(t)\int_0^{T_{\mathrm{b}}} \phi_{\mathrm{Q}}(\tau)y_{10}(\tau)d\tau$$
$$- \phi_{\mathrm{rise}}(t)\int_0^{T_{\mathrm{b}}} \phi_{\mathrm{rise}}(\tau)y_{10}(\tau)d\tau\,.$$

As a result, $\{\phi_{\mathrm{I}}(t), \phi_{\mathrm{Q}}(t), \phi_{\mathrm{rise}}(t), \phi_{\mathrm{fall}}(t)\}$ is an orthonormal basis for the waveforms $y_{00}(t)$, $y_{01}(t)$, $y_{10}(t)$, and $y_{11}(t)$.

Following the proposal of basic communication theory we consider a receiver that correlates the received signal with each basis function and this way yields a vector in the signal space. In particular, the waveform $y(t)$ for $t \in [(n-1)T_{\mathrm{b}}, nT_{\mathrm{b}}]$ is transformed into a four-dimensional signal point via

$$\mathbf{y}_n = \begin{bmatrix} y_{\mathrm{I}} \\ y_{\mathrm{Q}} \\ y_{\mathrm{rise}} \\ y_{\mathrm{fall}} \end{bmatrix}_n = \int_0^{T_{\mathrm{b}}} \begin{bmatrix} \phi_{\mathrm{I}}(t) \\ \phi_{\mathrm{Q}}(t) \\ \phi_{\mathrm{rise}}(t) \\ \phi_{\mathrm{fall}}(t) \end{bmatrix} y\left(t + (n-1)T_{\mathrm{b}}\right) dt\,. \tag{E.8}$$

The associated signal flow diagram is shown in Fig. E.4. The correlator outputs in response to the four prototype waveforms yield four constellation points which we denote as $\mathbf{y}_{00}, \mathbf{y}_{01}, \mathbf{y}_{10}, \mathbf{y}_{11} \in \mathbb{R}^4$.

The basis functions that arise from the simulated waveforms in Fig. E.3 are shown in Fig. E.5. We observe that $\phi_{\text{rise}}(t)$ and $\phi_{\text{fall}}(t)$ naturally put their focus on the tran-

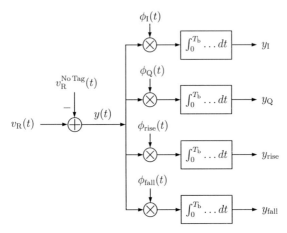

Figure E.4: Correlator bank for receive processing of load-modulated signals based on orthonormal basis functions. The output is a four-dimensional vector in the signal space.

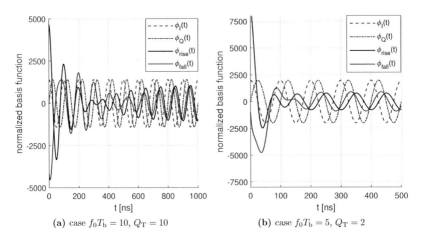

(a) case $f_0 T_b = 10$, $Q_T = 10$ (b) case $f_0 T_b = 5$, $Q_T = 2$

Figure E.5: Orthonormal basis functions that arise from the described approach, which involves the Gram-Schmidt process, to the prototype waveforms of the simulations in Fig. E.3. Their support is $t \in [0, T_b]$, i.e. one bit duration.

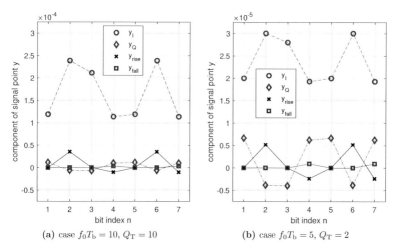

Figure E.6: The received signal vector, which is the output of the bank of correlators in Fig. E.4, is plotted versus the bit time index. In this example the transmitted bit sequence is 0110010 and the associated electrical signals are shown in Fig. E.3. The connecting lines are for visualization purposes only.

sients after the switching instant and (to some extend) frequency offsets from f_0. The evolution of the signal space vector over time, i.e. the result of subsequent application of Fig. E.4 to bit intervals for bits $n \in \mathbb{Z}$, is shown in Fig. E.6. Since the simulation is noiseless, any signal point $\mathbf{y}_n$ takes the value of one of the constellation points $\mathbf{y}_n \in \{\mathbf{y}_{00}, \mathbf{y}_{01}, \mathbf{y}_{10}, \mathbf{y}_{11}\}$ depending on which bit transition occurred (slight deviations occur because the assumption of only the previous bit interfering is not perfectly valid). We note that the occurring lie mostly in the I and Q dimensions, which is intuitive. However, the signal has a significant 'rise' component whenever the bit transition 0→1 occurs. The 'fall' dimension is always close to zero because the transients of the waveform $y_{10}(t)$ seem to get captured already by the other three dimensions (this is also caused by $y_{10}(t)$ being used last in the Gram-Schmidt process).

The signal space approach offers a rich set of analytic properties [37, 158] which are convenient for the evaluation of the communication performance of our approach. Foremost, when the received signal $y(t)$ is subject to additive white Gaussian noise with single-sided noise spectral density N_0 (a very common assumption), then $\mathbf{y}_n$ will be subject to Gaussian noise with covariance matrix $\frac{N_0}{2}\mathbf{I}_3$ (statistically independent between different time indices). Let us for the moment assume that the previous

223

bit was 0 and correctly detected as such. Then, the error probability of maximum-likelihood detection of the current bit is $Q(\sqrt{\text{SNR}})$ with SNR $= \frac{(d_{0x}/2)^2}{N_0/2}$ when both bits are equiprobable. Thereby, we use the Q-function and $d_{0x} = \|\mathbf{y}_{01} - \mathbf{y}_{00}\|$ is the Euclidean distance between the relevant constellation points. Likewise, if the previous bit was 1, then the same formulas hold with SNR $= \frac{(d_{1x}/2)^2}{N_0/2}$ and $d_{1x} = \|\mathbf{y}_{11} - \mathbf{y}_{10}\|$ instead.

In order to evaluate the merit of the proposed receive processing with the extra signal dimension, we want to compare the Euclidean distances d_{0x} and d_{1x} to those obtained with simpler receive processing. As a first step, we compare the four-dimensional approach to just I/Q processing (using only dimensions one and two). We can quantify the gain due to increased euclidean distance as either $G_{0x,\text{Full over I/Q}} = d_{0x}^2 / d_{0x,\text{I/Q}}^2$ or $G_{1x,\text{Full over I/Q}} = d_{1x}^2 / d_{1x,\text{I/Q}}^2$, depending on the previously transmitted bit. Thereby, $d_{0x,\text{I/Q}}^2 = (y_{I,01} - y_{I,00})^2 + (y_{Q,01} - y_{Q,00})^2$. The signal space approach with orthonormal bases ensures that this gain directly translates to an SNR gain in the case of additive white Gaussian noise. The values resulting from the conducted simulations are summarized in Table E.1. The table also states the gains of $d_{0x,\text{I/Q}}$ and $d_{1x,\text{I/Q}}$ over the Euclidean distances in the similar orthonormal basis

$$\phi_{I,\text{wait}}(t) = \begin{cases} \sqrt{5} \cdot \phi_I(t) & t \in [\frac{4}{5}T_b, T_b] \\ 0 & \text{Otherwise} \end{cases}, \tag{E.9}$$

$$\phi_{Q,\text{wait}}(t) = \begin{cases} \sqrt{5} \cdot \phi_Q(t) & t \in [\frac{4}{5}T_b, T_b] \\ 0 & \text{Otherwise} \end{cases} \tag{E.10}$$

which considers only the last 20% of a bit duration. This shall model the state-of-the-art approach of waiting for the transients to wear off (see above). For stationary oscillations with frequency f_0, e.g., the transient-free waveforms $y_{00}(t)$ and $y_{11}(t)$, this amounts to a 7 dB loss compared to I/Q processing over the entire bit interval since 80% of the signal power is discarded. The simulation results show that for bit intervals after a switch the difference can even exceed 7 dB as the transients are captured by I/Q processing which, surprisingly, increases the Euclidean distance.

Gains	case $f_0T_\mathrm{b} = 10$, $Q_\mathrm{T} = 10$		case $f_0T_\mathrm{b} = 5$, $Q_\mathrm{T} = 2$	
old bit	Full over I/Q	I/Q over Wait	Full over I/Q	I/Q over Wait
0	+0.357 dB	+9.038 dB	+0.530 dB	+7.737 dB
1	+0.052 dB	+7.444 dB	+0.158 dB	+7.067 dB

Table E.1: Symbol separation gains by projecting the received signal over a bit duration onto the four-dimensional basis (E.4),(E.5),(E.6),(E.7) versus only onto the two-dimensional I/Q basis (E.4),(E.5). Also shown is the gain of using I/Q basis (E.4),(E.5) over the full bit duration versus the I/Q basis (E.9),(E.10) that considers only the last 20% end of the bit duration to wait for the transients to die out.

Based on the observed results we draw the following conclusions:

- Most signal power is captured by I/Q processing at the operating frequency f_0.

- Processing the received waveform over the entire bit duration leads to vastly improved SNR compared to the state-of-the-art approach of waiting for the transients to wear off.

- Observing extra signal dimensions which relate to transients after a load switch yields an SNR gain which is appreciable but limited in the conducted experiments. The gains are more pronounced for low f_0T_b and Q_T and can exceed 1.5 dB for very small f_0T_b and Q_T (not shown). This could be interesting for high data rate transmission from micro-scale tags which necessarily have low Q-factor due to physical constraints [26]. In particular, a low Q_T causes stronger momentary signal disturbance, which could also be useful for synchronization.

Some interesting communication-theoretic questions about load modulation remain unanswered. Foremost, a study of the communication limits, e.g. in terms of the Shannon capacity of this channel, without any limited-memory assumption is a worthwhile research goal. An even higher-dimensional signal space expansion that pays regard to more than two neighboring bits could be a first step. Loads with more than two different states or even arbitrary evolutions $R_\mathrm{L}(t)$ over time give rise to further interesting opportunities. An interesting open research question is to identify all possible received signal waveforms that can be realized with an arbitrary $R_\mathrm{L}(t)$ within physical constraints, e.g., $R_\mathrm{L}(t) \geq 0$ $\forall t$, and the implications on communications performance.

Bibliography

[1] O. Bejarano, E. W. Knightly, and M. Park, "IEEE 802.11 ac: from channelization to multi-user MIMO," *IEEE Communications Magazine*, vol. 51, no. 10, pp. 84–90, 2013.

[2] A. Ghosh, R. Ratasuk, B. Mondal, N. Mangalvedhe, and T. Thomas, "LTE-Advanced: next-generation wireless broadband technology," *IEEE Wireless Communications*, vol. 17, no. 3, pp. 10–22, 2010.

[3] *Global Positioning System Standard Positioning Service Performance Standard*, 4th ed., DoD Positioning, Navigation, and Timing Executive Committee, 6000 Defense Pentagon, Washington, DC 20310-6000, Sep. 2008.

[4] "Sunday debate: Are phones getting too big?" GSMArena article. Online: https://www.gsmarena.com/are_phones_getting_too_big-news-33523.php, Sep. 2018.

[5] M. Koziol, "5G's waveform is a battery vampire," *IEEE Spectrum*, Jul. 2019.

[6] P. J. Mankowski, J. Kanevsky, P. Bakirtzian, and S. Cugno, "Cellular phone collateral damage: A review of burns associated with lithium battery powered mobile devices," *Burns*, vol. 42, no. 4, pp. 61–64, 2016.

[7] R. Du, A. Ozcelikkale, C. Fischione, and M. Xiao, "Optimal energy beamforming and data routing for immortal wireless sensor networks," in *IEEE International Conference on Communications (ICC)*, 2017.

[8] I. F. Akyildiz, W. Su, Y. Sankarasubramaniam, and E. Cayirci, "Wireless sensor networks: a survey," *Computer Networks*, vol. 38, no. 4, pp. 393–422, 2002.

[9] W. Pawgasame, "A survey in adaptive hybrid wireless sensor network for military operations," in *Asian Conference on Defence Technology (ACDT)*. IEEE, 2016, pp. 78–83.

[10] Z. Sheng, C. Mahapatra, C. Zhu, and V. C. Leung, "Recent advances in industrial wireless sensor networks toward efficient management in IoT," *IEEE Access*, vol. 3, pp. 622–637, 2015.

[11] C. Almhana, V. Choulakian, and J. Almhana, "An efficient approach for data transmission in power-constrained wireless sensor network," in *IEEE International Conference on Communications (ICC)*, 2017.

[12] "Cellular IoT evolution for industry digitalization," Ericsson white paper, Jan. 2019, online: https://www.ericsson.com/en/white-papers/cellular-iot-evolution-for-industry-digitalization.

[13] R. S. Sinha, Y. Wei, and S.-H. Hwang, "A survey on LPWA technology: LoRa and NB-IoT," *ICT Express*, vol. 3, no. 1, pp. 14–21, 2017.

[14] F. Benkhelifa, Z. Qin, and J. McCann, "Minimum throughput maximization in LoRa networks powered by ambient energy harvesting," in *IEEE International Conference on Communications (ICC)*, May 2019.

[15] W. Shi, J. Li, N. Cheng, F. Lyu, Y. Dai, H. Zhou, and X. S. Shen, "3D multi-drone-cell trajectory design for efficient IoT data collection," in *IEEE International Conference on Communications (ICC)*, 2019.

[16] M. Tortonesi, A. Morelli, M. Govoni, J. Michaelis, N. Suri, C. Stefanelli, and S. Russell, "Leveraging internet of things within the military network environment — challenges and solutions," in *World Forum on Internet of Things*. IEEE, 2016, pp. 111–116.

[17] V. C. Gungor, B. Lu, and G. P. Hancke, "Opportunities and challenges of wireless sensor networks in smart grid," *IEEE Transactions on Industrial Electronics*, vol. 57, no. 10, pp. 3557–3564, 2010.

[18] E. Slottke, M. Kuhn, A. Wittneben, H. Luecken, and C. Cartalemi, "UWB marine engine telemetry sensor networks: enabling reliable low-complexity communication," in *IEEE Vehicular Technology Conference (VTC Fall)*, 2015.

[19] W. Gao and J. Wang, "The environmental impact of micro/nanomachines: a review," *ACS Nano*, vol. 8, no. 4, pp. 3170–3180, 2014.

[20] H. Ceylan, I. C. Yasa, U. Kilic, W. Hu, and M. Sitti, "Translational prospects of untethered medical microrobots," *Progress in Biomedical Engineering*, vol. 1, no. 1, 2019.

[21] M. Sitti, "Miniature soft robots—road to the clinic," *Nature Reviews Materials*, vol. 3, no. 6, p. 74, 2018.

[22] M. Sitti, H. Ceylan, W. Hu, J. Giltinan, M. Turan, S. Yim, and E. Diller, "Biomedical applications of untethered mobile milli/microrobots," *Proceedings of the IEEE*, vol. 103, no. 2, pp. 205–224, 2015.

[23] B. J. Nelson, I. K. Kaliakatsos, and J. J. Abbott, "Microrobots for minimally invasive medicine," *Annual Review of Biomedical Engineering*, vol. 12, pp. 55–85, 2010.

[24] I. Akyildiz, M. Pierobon, S. Balasubramaniam, and Y. Koucheryavy, "The internet of bio-nano things," *IEEE Communications Magazine*, vol. 53, no. 3, pp. 32–40, 2015.

[25] C. Steiger, A. Abramson, P. Nadeau, A. P. Chandrakasan, R. Langer, and G. Traverso, "Ingestible electronics for diagnostics and therapy," *Nature Reviews Materials*, vol. 4, pp. 83—98, 2019.

[26] E. Slottke, "Inductively coupled microsensor networks," Ph.D. dissertation, ETH Zürich, 2016.

[27] S.-M. Oh and J. Shin, "An efficient small data transmission scheme in the 3GPP NB-IoT system," *IEEE Communications Letters*, vol. 21, no. 3, pp. 660–663, Mar. 2017.

[28] K. Alexandris, G. Sklivanitis, and A. Bletsas, "Reachback WSN connectivity: Non-coherent zero-feedback distributed beamforming or TDMA energy harvesting?" *IEEE Transactions on Wireless Communications*, vol. 13, no. 9, pp. 4923–4934, 2014.

[29] H. Ju and R. Zhang, "User cooperation in wireless powered communication networks," in *IEEE Global Communications Conference (GLOBECOM)*, 2014.

[30] J. Grosinger, W. Pachler, and W. Bosch, "Tag size matters: Miniaturized RFID tags to connect smart objects to the internet," *IEEE Microwave Magazine*, vol. 19, no. 6, pp. 101–111, Sep. 2018.

[31] E. Diller, M. Sitti *et al.*, "Micro-scale mobile robotics," *Foundations and Trends in Robotics*, vol. 2, no. 3, pp. 143–259, 2013.

[32] M. Beidaghi and Y. Gogotsi, "Capacitive energy storage in micro-scale devices: recent advances in design and fabrication of micro-supercapacitors," *Energy & Environmental Science*, vol. 7, no. 3, pp. 867–884, 2014.

[33] S. Bi, C. Ho, and R. Zhang, "Wireless powered communication: Opportunities and challenges," *IEEE Communications Magazine*, vol. 53, no. 4, pp. 117–125, 2015.

[34] B. Luo, P. L. Yeoh, R. Schober, and B. S. Krongold, "Optimal energy beamforming for distributed wireless power transfer over frequency-selective channels," in *IEEE International Conference on Communications (ICC)*, May 2019.

[35] T. D. Nguyen, J. Y. Khan, and D. T. Ngo, "An effective energy-harvesting-aware routing algorithm for WSN-based IoT applications," in *IEEE International Conference on Communications (ICC)*, 2017.

[36] A. F. Molisch, *Wireless Communications*. John Wiley & Sons, 2012.

[37] S. Haykin, *Communication Systems*. Wiley, 2001.

[38] M. Shafi, J. Zhang, H. Tataria, A. F. Molisch, S. Sun, T. S. Rappaport, F. Tufvesson, S. Wu, and K. Kitao, "Microwave vs. millimeter-wave propagation channels: key differences and impact on 5G cellular systems," *IEEE Communications Magazine*, vol. 56, no. 12, pp. 14–20, 2018.

[39] C. Andreu, C. Garcia-Pardo, A. Fomes-Leal, M. Cabedo-Fabrés, and N. Cardona, "UWB in-body channel performance by using a direct antenna designing procedure," in *European Conference on Antennas and Propagation (EUCAP)*. IEEE, 2017, pp. 278–282.

[40] C. Andreu, S. Castelló-Palacios, C. Garcia-Pardo, A. Fornes-Leal, A. Vallés-Lluch, and N. Cardona, "Spatial in-body channel characterization using an accurate UWB phantom," *IEEE Transactions on Microwave Theory and Techniques*, vol. 64, no. 11, pp. 3995–4002, 2016.

[41] Z. Sun and I. Akyildiz, "Magnetic induction communications for wireless underground sensor networks," *IEEE Transactions on Antennas and Propagation*, vol. 58, no. 7, pp. 2426–2435, 2010.

[42] A. Markham and N. Trigoni, "Magneto-inductive networked rescue system (miners): taking sensor networks underground," in *ACM International Conference on Information Processing in Sensor Networks*, 2012, pp. 317–328.

[43] S. Kisseleff, I. F. Akyildiz, and W. H. Gerstacker, "Survey on advances in magnetic induction based wireless underground sensor networks," *IEEE Internet of Things Journal*, 2018.

[44] I. F. Akyildiz, P. Wang, and Z. Sun, "Realizing underwater communication through magnetic induction," *IEEE Communications Magazine*, vol. 53, no. 11, 2015.

[45] A. A. Alshehri, S.-C. Lin, and I. F. Akyildiz, "Optimal energy planning for wireless self-contained sensor networks in oil reservoirs," in *IEEE International Conference on Communications (ICC)*, 2017.

[46] N. G. Franconi, A. P. Bunger, E. Sejdić, and M. H. Mickle, "Wireless communication in oil and gas wells," *Energy Technology*, vol. 2, no. 12, pp. 996–1005, 2014.

[47] Z. Sun, P. Wang, M. C. Vuran, M. A. Al-Rodhaan, A. M. Al-Dhelaan, and I. F. Akyildiz, "MISE-PIPE: Magnetic induction-based wireless sensor networks for underground pipeline monitoring," *Ad Hoc Networks*, vol. 9, no. 3, pp. 218–227, 2011.

[48] C. H. Martins, A. A. Alshehri, and I. F. Akyildiz, "Novel mi-based (fracbot) sensor hardware design for monitoring hydraulic fractures and oil reservoirs," in *Ubiquitous Computing, Electronics and Mobile Communication Conference (UEMCON)*, 2017, pp. 434–441.

[49] O. Renaudin, T. Zemen, and T. Burgess, "Ray-tracing based fingerprinting for indoor localization," in *IEEE International Workshop on Signal Processing Advances in Wireless Communications (SPAWC)*, 2018.

[50] D. Dardari, A. Conti, U. Ferner, A. Giorgetti, and M. Z. Win, "Ranging with ultrawide bandwidth signals in multipath environments," *Proceedings of the IEEE*, vol. 97, no. 2, pp. 404–426, 2009.

[51] H. Schulten, M. Kuhn, R. Heyn, G. Dumphart, A. Wittneben, and F. Trösch, "On the crucial impact of antennas and diversity on BLE RSSI-based indoor localization," in *IEEE Vehicular Technology Conference (VTC Spring)*, May 2019.

[52] D. Pozar, *Microwave Engineering*. Wiley, 2004.

[53] M. T. Ivrlac and J. A. Nossek, "Toward a circuit theory of communication," *IEEE Transactions on Circuits and Systems I: Regular Papers*, vol. 57, no. 7, pp. 1663–1683, 2010.

[54] A. K. Sharma, S. Yadav, S. N. Dandu, V. Kumar, J. Sengupta, S. B. Dhok, and S. Kumar, "Magnetic induction-based non-conventional media communications: A review," *IEEE Sensors Journal*, vol. 17, no. 4, pp. 926–940, 2016.

[55] R. K. Gulati, A. Pal, and K. Kant, "Experimental evaluation of a near-field magnetic induction based communication system," in *IEEE Wireless Communications and Networking Conference (WCNC)*, 2019.

[56] R. Barr, D. L. Jones, and C. Rodger, "ELF and VLF radio waves," *Journal of Atmospheric and Solar-Terrestrial Physics*, vol. 62, no. 17-18, pp. 1689–1718, 2000.

[57] A. Sheinker, B. Ginzburg, N. Salomonski, and A. Engel, "Localization of a mobile platform equipped with a rotating magnetic dipole source," *IEEE Transactions on Instrumentation and Measurement*, 2018.

[58] V. Pasku, A. De Angelis, M. Dionigi, G. De Angelis, A. Moschitta, and P. Carbone, "A positioning system based on low-frequency magnetic fields," *IEEE Transactions on Industrial Electronics*, vol. 63, no. 4, pp. 2457–2468, 2016.

[59] T. E. Abrudan, Z. Xiao, A. Markham, and N. Trigoni, "Distortion rejecting magneto-inductive three-dimensional localization (magloc)," *IEEE Journal on Selected Areas in Communications*, vol. 33, no. 11, pp. 2404–2417, 2015.

[60] D. D. Arumugam, J. D. Griffin, D. D. Stancil, and D. S. Ricketts, "Three-dimensional position and orientation measurements using magneto-quasistatic fields and complex image theory," *IEEE Antennas and Propagation Magazine*, vol. 56, pp. 160–173, 2014.

[61] O. Kypris, T. E. Abrudan, and A. Markham, "Magnetic induction-based positioning in distorted environments," *IEEE Transactions on Geoscience and Remote Sensing*, vol. 54, no. 8, pp. 4605–4612, 2016.

[62] C. A. Balanis, *Antenna Theory: Analysis and Design*. Wiley, 2005.

[63] K. Finkenzeller, *RFID Handbook*. Carl Hanser, 2015.

[64] E. Shamonina, V. Kalinin, K. Ringhofer, and L. Solymar, "Magneto-inductive waveguide," *Electronics letters*, vol. 38, no. 8, pp. 371–373, 2002.

[65] Z. Sun and I. F. Akyildiz, "On capacity of magnetic induction-based wireless underground sensor networks," in *INFOCOM*. IEEE, 2012, pp. 370–378.

[66] C. K. Lee, W. Zhong, and S. Hui, "Effects of magnetic coupling of nonadjacent resonators on wireless power domino-resonator systems," *IEEE Transactions on Power Electronics*, vol. 27, no. 4, 2012.

[67] S. Kisseleff, I. Akyildiz, and W. Gerstacker, "Throughput of the magnetic induction based wireless underground sensor networks: Key optimization techniques," *IEEE Transactions on Communications*, vol. 62, no. 12, pp. 4426–4439, 2014.

[68] C. W. Chan and C. J. Stevens, "Two-dimensional magneto-inductive wave data structures," in *European Conference on Antennas and Propagation (EUCAP)*. IEEE, 2011.

[69] A. Kurs, A. Karalis, R. Moffatt, J. D. Joannopoulos, P. Fisher, and M. Soljacic, "Wireless power transfer via strongly coupled magnetic resonances," *science*, vol. 317, no. 5834, pp. 83–86, 2007.

[70] A. K. RamRakhyani, S. Mirabbasi, and M. Chiao, "Design and optimization of resonance-based efficient wireless power delivery systems for biomedical implants," *IEEE Transactions on Biomedical Circuits and Systems*, vol. 5, no. 1, pp. 48–63, 2011.

[71] M. Kiani, U.-M. Jow, and M. Ghovanloo, "Design and optimization of a 3-coil inductive link for efficient wireless power transmission," *IEEE Transactions on Biomedical Circuits and Systems*, vol. 5, no. 6, pp. 579–591, 2011.

[72] A. RamRakhyani and G. Lazzi, "On the design of efficient multi-coil telemetry system for biomedical implants," *IEEE Transactions of Biomedical Circuits and Systems*, vol. 7, no. 1, pp. 11–23, 2013.

[73] O. C. Onar, S. L. Campbell, L. E. Seiber, C. P. White, and M. Chinthavali, "A high-power wireless charging system development and integration for a Toyota RAV4 electric vehicle," in *IEEE Transportation Electrification Conference and Expo (ITEC)*, 2016.

[74] D. Patil, M. K. McDonough, J. M. Miller, B. Fahimi, and P. T. Balsara, "Wireless power transfer for vehicular applications: Overview and challenges," *IEEE Transactions on Transportation Electrification*, vol. 4, no. 1, pp. 3–37, 2017.

[75] K. Agarwal, R. Jegadeesan, Y.-X. Guo, and N. V. Thakor, "Wireless power transfer strategies for implantable bioelectronics," *IEEE Reviews in Biomedical Engineering*, vol. 10, pp. 136–161, 2017.

[76] T. C. Chang, M. J. Weber, J. Charthad, S. Baltsavias, and A. Arbabian, "End-to-end design of efficient ultrasonic power links for scaling towards submillimeter implantable receivers," *IEEE Transactions on Biomedical Circuits and Systems*, vol. 12, no. 5, pp. 1100–1111, 2018.

[77] S. Wirdatmadja, P. Johari, A. Desai, Y. Bae, E. K. Stachowiak, M. K. Stachowiak, J. M. Jornet, and S. Balasubramaniam, "Analysis of light propagation on physiological properties of neurons for nanoscale optogenetics," *IEEE Transactions on Neural Systems and Rehabilitation Engineering*, vol. 27, no. 2, pp. 108–117, 2019.

[78] A. Bossavit, "Generalized finite differences in computational electromagnetics," *Progress In Electromagnetics Research*, vol. 32, pp. 45–64, 2001.

[79] G. Dumphart, B. I. Bitachon, and A. Wittneben, "Magneto-inductive powering and uplink of in-body microsensors: Feasibility and high-density effects," in *IEEE Wireless Communications and Networking Conference (WCNC)*, 2019.

[80] G. Dumphart, H. Schulten, B. Bhatia, C. Sulser, and A. Wittneben, "Practical accuracy limits of radiation-aware magneto-inductive 3D localization," in *IEEE International Conference on Communications (ICC) Workshops*, May 2019.

[81] W. H. Ko, S. P. Liang, and C. D. Fung, "Design of radio-frequency powered coils for implant instruments," *Medical and Biological Engineering and Computing*, vol. 15, no. 6, pp. 634–640, 1977.

[82] H. C. Jing and Y. E. Wang, "Capacity performance of an inductively coupled near field communication system," in *IEEE Antennas and Propagation Society International Symposium*, 2008.

[83] Z. Sun, I. F. Akyildiz, S. Kisseleff, and W. Gerstacker, "Increasing the capacity of magnetic induction communications in RF-challenged environments," *IEEE Transactions on Communications*, vol. 61, no. 9, pp. 3943–3952, 2013.

[84] K. Lee and D.-H. Cho, "Maximizing the capacity of magnetic induction communication for embedded sensor networks in strongly and loosely coupled regions," *IEEE Transactions on Magnetics*, vol. 49, no. 9, pp. 5055–5062, 2013.

[85] S. Kisseleff, W. Gerstacker, R. Schober, Z. Sun, and I. F. Akyildiz, "Channel capacity of magnetic induction based wireless underground sensor networks under practical constraints," in *IEEE Wireless Communications and Networking Conference (WCNC)*, 2013, pp. 2603–2608.

[86] S. Kisseleff, I. Akyildiz, and W. Gerstacker, "Digital signal transmission in magnetic induction based wireless underground sensor networks," *IEEE Transactions on Communications*, vol. 63, no. 6, 2015.

[87] S. Kisseleff, I. F. Akyildiz, and W. H. Gerstacker, "Magnetic induction based simultaneous wireless information and power transfer for single information and multiple power receivers," *IEEE Transactions on Communications*, 2016.

[88] M. Masihpour, "Cooperative communication in near field magnetic induction communication systems," Ph.D. dissertation, University of Technology, Sydney, 2012.

[89] A. Goldsmith, *Wireless communications.* Cambridge University Press, 2005.

[90] D. Tse and P. Viswanath, *Fundamentals of Wireless Communication.* Cambridge University Press, 2005.

[91] J. Jadidian and D. Katabi, "Magnetic MIMO: how to charge your phone in your pocket," in *ACM International Conference on Mobile Computing and Networking*, 2014, pp. 495–506.

[92] G. Yang, M. R. V. Moghadam, and R. Zhang, "Magnetic MIMO signal processing and optimization for wireless power transfer," *IEEE Transactions on Signal Processing*, vol. 65, no. 11, pp. 2860–2874, 2017.

[93] H. Sun, F. Zhu, H. Lin, and F. Gao, "Robust magnetic resonant beamforming for secured wireless power transfer," *IEEE Signal Processing Letters*, 2018.

[94] B. Lenaerts and R. Puers, "Inductive powering of a freely moving system," *Sensors and Actuators A: Physical*, vol. 123, pp. 522–530, 2005.

[95] F. H. Raab, E. B. Blood, T. O. Steiner, and H. R. Jones, "Magnetic position and orientation tracking system," *IEEE Transactions on Aerospace and Electronic systems*, no. 5, pp. 709–718, 1979.

[96] C. Hu, M. Q.-H. Meng, and M. Mandal, "A linear algorithm for tracing magnet position and orientation by using three-axis magnetic sensors," *IEEE Transactions on Magnetics*, vol. 43, no. 12, pp. 4096–4101, 2007.

[97] C. Hu, M. Li, S. Song, R. Zhang, M. Q.-H. Meng *et al.*, "A cubic 3-axis magnetic sensor array for wirelessly tracking magnet position and orientation," *Sensors Journal*, vol. 10, no. 5, pp. 903–913, 2010.

[98] H. Dai, S. Song, X. Zeng, S. Su, M. Lin, and M. Q.-H. Meng, "6-D electromagnetic tracking approach using uniaxial transmitting coil and tri-axial magneto-resistive sensor," *IEEE Sensors Journal*, vol. 18, no. 3, pp. 1178–1186, 2018.

[99] S. Kisseleff, I. Akyildiz, and W. Gerstacker, "Distributed beamforming for magnetic induction based body area sensor networks," in *IEEE Global Communications Conference (GLOBECOM)*, 2016.

[100] Y. Hassan, B. Gahr, and A. Wittneben, "Rate maximization in dense interference networks using non-cooperative passively loaded relays," in *Asilomar Conference on Signals, Systems, and Computers*, Nov. 2015.

[101] J. W. Wallace and M. A. Jensen, "Mutual coupling in MIMO wireless systems: A rigorous network theory analysis," *IEEE Transactions on Wireless Communications*, vol. 3, no. 4, pp. 1317–1325, 2004.

[102] B. K. Lau, J. B. Andersen, G. Kristensson, and A. F. Molisch, "Impact of matching network on bandwidth of compact antenna arrays," *IEEE Transactions on Antennas and Propagation*, vol. 54, no. 11, pp. 3225–3238, 2006.

[103] Y. Hassan, "Compact multi-antenna systems: Bridging circuits to communications theory," Ph.D. dissertation, ETH Zurich, 2018.

[104] Z. Zhang, E. Liu, X. Qu, R. Wang, H. Ma, and Z. Sun, "Connectivity of magnetic induction-based ad hoc networks," *IEEE Transactions on Wireless Communications*, vol. 16, no. 7, pp. 4181–4191, Jul 2017.

[105] F. Flack, E. James, and D. Schlapp, "Mutual inductance of air-cored coils: Effect on design of radio-frequency coupled implants," *Medical and biological engineering*, vol. 9, no. 2, pp. 79–85, 1971.

[106] E. S. Hochmair, "System optimization for improved accuracy in transcutaneous signal and power transmission," *IEEE Transactions on Biomedical Engineering*, no. 2, pp. 177–186, 1984.

[107] M. Soma, D. C. Galbraith, and R. L. White, "Radio-frequency coils in implantable devices: misalignment analysis and design procedure," *IEEE Transactions on Biomedical Engineering*, no. 4, pp. 276–282, 1987.

[108] K. Fotopoulou and B. W. Flynn, "Wireless power transfer in loosely coupled links: Coil misalignment model," *IEEE Transactions on Magnetics*, vol. 47, no. 2, pp. 416–430, 2011.

[109] B. Essaid and N. Batel, "Evaluation of RF power attenuation in biomedical implants: Effect of a combination of coils misalignment with a biological tissue absorption," in *IEEE International Conference on Modelling, Identification and Control (ICMIC)*, 2016, pp. 689–694.

[110] C. Angerer, R. Langwieser, and M. Rupp, "RFID reader receivers for physical layer collision recovery," *IEEE Transactions on Communications*, vol. 58, no. 12, pp. 3526–3537, 2010.

[111] G. Dumphart and A. Wittneben, "Stochastic misalignment model for magneto-inductive SISO and MIMO links," in *IEEE International Symposium on Personal, Indoor, and Mobile Radio Communications (PIMRC)*, Sep. 2016.

[112] R. Syms, I. Young, and L. Solymar, "Low-loss magneto-inductive waveguides," *Journal of Physics D: Applied Physics*, vol. 39, no. 18, 2006.

[113] R. R. Syms and T. Floume, "Parasitic coupling in magneto-inductive cable," *Journal of Physics D: Applied Physics*, vol. 50, no. 22, 2017.

[114] R. Syms and L. Solymar, "Noise in metamaterials," *Journal of Applied Physics*, vol. 109, no. 12, 2011.

[115] A. Servant, F. Qiu, M. Mazza, K. Kostarelos, and B. J. Nelson, "Controlled in vivo swimming of a swarm of bacteria-like microrobotic flagella," *Advanced Materials*, vol. 27, no. 19, pp. 2981–2988, 2015.

[116] G. Dumphart, E. Slottke, and A. Wittneben, "Magneto-inductive passive relaying in arbitrarily arranged networks," in *IEEE International Conference on Communications (ICC)*, May 2017.

[117] B. Gulbahar, "Theoretical analysis of magneto-inductive THz wireless communications and power transfer with multi-layer graphene nano-coils," *IEEE Transactions on Molecular, Biological and Multi-Scale Communications*, vol. 3, no. 1, pp. 60–70, 2017.

[118] T. Campi, S. Cruciani, V. De Santis, F. Maradei, and M. Feliziani, "Near field wireless powering of deep medical implants," *Energies*, vol. 12, no. 14, p. 2720, 2019.

[119] P. Feng, P. Yeon, Y. Cheng, M. Ghovanloo, and T. G. Constandinou, "Chip-scale coils for millimeter-sized bio-implants," *IEEE Transactions on Biomedical Circuits and Systems*, vol. 12, no. 5, pp. 1088–1099, 2018.

[120] R. M. Buehrer, H. Wymeersch, and R. M. Vaghefi, "Collaborative sensor network localization: Algorithms and practical issues," *Proceedings of the IEEE*, vol. 106, no. 6, pp. 1089–1114, 2018.

[121] S. Mazuelas, A. Conti, J. C. Allen, and M. Z. Win, "Soft range information for network localization," *IEEE Transactions on Signal Processing*, vol. 66, no. 12, pp. 3155–3168, 2018.

[122] V. Schlageter, P.-A. Besse, R. Popovic, and P. Kucera, "Tracking system with five degrees of freedom using a 2D-array of hall sensors and a permanent magnet," *Sensors and Actuators A: Physical*, vol. 92, no. 1, pp. 37–42, 2001.

[123] C. Hu, M. Q.-H. Meng, and M. Mandal, "Efficient magnetic localization and orientation technique for capsule endoscopy," *International Journal of Information Acquisition*, vol. 2, no. 01, pp. 23–36, 2005.

[124] S. Song, C. Hu, M. Li, W. Yang, and M. Q.-H. Meng, "Real time algorithm for magnet's localization in capsule endoscope," in *IEEE International Conference on Automation and Logistics*, 2009, pp. 2030–2035.

[125] H. Li, G. Yan, and G. Ma, "An active endoscopic robot based on wireless power transmission and electromagnetic localization," *The International Journal of*

Medical Robotics and Computer Assisted Surgery, vol. 4, no. 4, pp. 355–367, 2008.

[126] Y. Shen and M. Z. Win, "Fundamental limits of wideband localization - part i: A general framework," *IEEE Transactions on Information Theory*, vol. 56, no. 10, pp. 4956–4980, 2010.

[127] E. Leitinger, P. Meissner, C. Rüdisser, G. Dumphart, and K. Witrisal, "Evaluation of position-related information in multipath components for indoor positioning," *IEEE Journal on Selected Areas in Communications*, vol. 33, no. 11, pp. 2313–2328, 2015.

[128] G. Dumphart, E. Slottke, and A. Wittneben, "Robust near-field 3D localization of an unaligned single-coil agent using unobtrusive anchors," in *International Symposium on Personal, Indoor, and Mobile Radio Communications (PIMRC)*, Oct. 2017.

[129] S. Li, M. Hedley, and I. B. Collings, "New efficient indoor cooperative localization algorithm with empirical ranging error model," *IEEE Journal on Selected Areas in Communications*, vol. 33, no. 7, pp. 1407–1417, 2015.

[130] Y. Liu, Y. Shen, D. Guo, and M. Z. Win, "Network localization and synchronization using full-duplex radios," *IEEE Transactions on Signal Processing*, vol. 66, no. 3, pp. 714–728, 2018.

[131] R. Heyn, M. Kuhn, H. Schulten, G. Dumphart, J. Zwyssig, F. Trösch, and A. Wittneben, "User tracking for access control with bluetooth low energy," in *IEEE Vehicular Technology Conference (VTC Spring)*, May 2019.

[132] H. Wymeersch, J. Lien, and M. Z. Win, "Cooperative localization in wireless networks," *Proceedings of the IEEE*, vol. 97, no. 2, pp. 427–450, 2009.

[133] B. Alavi and K. Pahlavan, "Modeling of the TOA-based distance measurement error using UWB indoor radio measurements," *IEEE Communications Letters*, vol. 10, no. 4, pp. 275–277, 2006.

[134] D. B. Jourdan, D. Dardari, and M. Z. Win, "Position error bound for UWB localization in dense cluttered environments," *IEEE transactions on aerospace and electronic systems*, vol. 44, no. 2, 2008.

[135] K. Witrisal, P. Meissner, E. Leitinger, Y. Shen, C. Gustafson, F. Tufvesson, K. Haneda, D. Dardari, A. F. Molisch, A. Conti *et al.*, "High-accuracy localization for assisted living: 5G systems will turn multipath channels from foe to friend," *IEEE Signal Processing Magazine*, vol. 33, no. 2, pp. 59–70, 2016.

[136] G. Dumphart, M. Kuhn, A. Wittneben, and F. Trösch, "Inter-node distance estimation from multipath delay differences of channels to observer nodes," in *IEEE International Conference on Communications (ICC)*, May 2019.

[137] F. Trösch, A. Wittneben, G. Dumphart, and M. Kuhn, "System und Verfahren zur Positionsbestimmung in einem Gebäude," European Patent Office, Patent application No. EP18 200 011.7, Oct., 2018.

[138] ——, "Zugangskontrollsystem und Verfahren zum Betreiben eines Zugangskontrollsystems," European Patent Office, Patent application No. EP18 200 002.6, Oct., 2018.

[139] G. Dumphart, "System model for magnetoinductive communication," Communication Technology Laboratory, ETH Zurich, Tech. Rep., Jul. 2015.

[140] H. Schulten, "ISI-free load modulation with RRC pulses for passive near-field tags," semester thesis, ETH Zürich, 2016.

[141] B. Bitachon, "Magneto-inductive wireless networks: Active and passive cooperation for channel-aware transmission," Master's thesis, ETH Zürich, 2017.

[142] B. Bhatia, "Dimensionality reduction and phase recovery for magnetic near-field localization algorithms," semester thesis, ETH Zürich, 2017.

[143] M. De, "Development of a near-field 3D localization system prototype," semester thesis, ETH Zürich, 2018 (joint supervision with C. Sulser).

[144] B. Bhatia, "Calibration and evaluation of a magnetic-field-based 3D localization system," Master's thesis, ETH Zürich, 2019.

[145] M. Göller, "Inter-user distance estimation based on channel impulse responses for indoor localization," Master's thesis, ETH Zürich, 2017.

[146] G. Dumphart, E. Leitinger, P. Meissner, and K. Witrisal, "Monostatic indoor localization: Bounds and limits," in *IEEE International Conference on Communications (ICC) Workshops*, 2015, pp. 865–870.

[147] R. P. Feynman, R. B. Leighton, and M. Sands, *The Feynman Lectures on Physics (The New Millenium Edition) Volume II: Mainly Electromagnetism and Matter*. Basic Books, 2011, vol. 2, available at www.feynmanlectures.caltech.edu.

[148] J. C. Maxwell, "On physical lines of force," *Philosophical Magazine*, vol. 21, no. 139, pp. 161–175, 1861.

[149] O. Heaviside, *Electromagnetic Theory*. Electrician Printing and Publishing Company, 1893, vol. 1.

[150] S. J. Orfanidis, *Electromagnetic waves and antennas*. Rutgers University, 2002.

[151] S. Ramo, J. Whinnery, and T. Van Duzer, *Fields and Waves in Communication Electronics*. Wiley, 1994.

[152] M. Faraday, *Experimental Researches in Electricity*. Richard and John Edward Taylor, 1839, vol. 1 and 2.

[153] M. Stock, R. Davis, E. de Mirandés, and M. J. Milton, "The revision of the SI — the result of three decades of progress in metrology," *Metrologia*, vol. 56, no. 2, 2019.

[154] S. Ramo and J. R. Whinnery, *Fields and Waves in Modern Radio*, 2nd ed. Wiley, 1953.

[155] J. E. Storer, "Impedance of thin-wire loop antennas," *IEEE Transactions of the American Institute of Electrical Engineers, Part I: Communication and Electronics*, vol. 75, no. 5, pp. 606–619, 1956.

[156] M. S. Neiman, "The principle of reciprocity in antenna theory," *Proceedings of the IRE*, vol. 31, no. 12, pp. 666–671, 1943.

[157] F. E. Neumann, "Allgemeine Gesetze der inducirten elektrischen Ströme," *Annalen der Physik*, vol. 143, no. 1, pp. 31–44, 1846.

[158] R. G. Gallager, *Information theory and reliable communication*. Springer, 1968, vol. 2.

[159] E. Telatar, "Capacity of multi-antenna gaussian channels," *European transactions on telecommunications*, vol. 10, no. 6, pp. 585–595, 1999.

[160] S. Kisseleff, I. Akyildiz, and W. Gerstacker, "Interference polarization in magnetic induction based wireless underground sensor networks," in *IEEE International Symposium on Personal, Indoor and Mobile Radio Communications (PIMRC) Workshops*, Sep. 2013, pp. 71–75.

[161] *Propagation model and interference range calculation for inductive systems 10 kHz - 30 MHz (ERC Report 69)*, European Radiocommunications Committee (ERC) within the European Conference of Postal and Telecommunications Administrations (CEPT), Marbella, Feb 1999.

[162] K. Trela and K. M. Gawrylczyk, "Frequency response modeling of power transformer windings considering the attributes of ferromagnetic core," in *International Interdisciplinary PhD Workshop*. IEEE, 2018, pp. 71–73.

[163] C. Cuellar, W. Tan, X. Margueron, A. Benabou, and N. Idir, "Measurement method of the complex magnetic permeability of ferrites in high frequency," in *IEEE International Instrumentation and Measurement Technology Conference*, 2012, pp. 63–68.

[164] A. Tomitaka, A. Hirukawa, T. Yamada, S. Morishita, and Y. Takemura, "Biocompatibility of various ferrite nanoparticles evaluated by in vitro cytotoxicity assays using HeLa cells," *Journal of Magnetism and Magnetic Materials*, vol. 321, no. 10, pp. 1482–1484, 2009.

[165] D. W. Knight, "The self-resonance and self-capacitance of solenoid coils," 2016, DOI: 10.13140/RG.2.1.1472.0887. [Online]. Available: http://g3ynh.info/zdocs/magnetics/appendix/self_res/self-res.pdf

[166] G. Smith, "Radiation efficiency of electrically small multiturn loop antennas," *IEEE Transactions on Antennas and Propagation*, vol. 20, no. 5, pp. 656–657, 1972.

[167] H. C. Miller, "Inductance formula for a single-layer circular coil," *Proceedings of the IEEE*, vol. 75, no. 2, pp. 256–257, 1987.

[168] R. Medhurst, "HF resistance and self-capacitance of single-layer solenoids," *Wireless Engineer*, vol. 24, 1947.

[169] S. Kisseleff, X. Chen, I. F. Akyildiz, and W. Gerstacker, "Localization of a silent target node in magnetic induction based wireless underground sensor networks," in *IEEE International Conference on Communications (ICC)*, 2017.

[170] M. Ivrlac, "Network theory for communication," 2011, lecture notes, Technische Universität München.

[171] M. Catrysse, B. Hermans, and R. Puers, "An inductive power system with integrated bi-directional data-transmission," *Sensors and Actuators A: Physical*, vol. 115, no. 2–3, pp. 221–229, 2004.

[172] E. Slottke and A. Wittneben, "Single-anchor localization in inductively coupled sensor networks using passive relays and load switching," in *Asilomar Conference on Signals, Systems, and Computers*, Nov. 2015.

[173] M. T. Ivrlac and J. A. Nossek, "The multiport communication theory," *IEEE Circuits and Systems Magazine*, vol. 14, no. 3, pp. 27–44, 2014.

[174] *IEEE Std 1459-2010 (Revision of IEEE Std 1459-2000) Standard Definitions for the Measurement of Electric Power Quantities Under Sinusoidal, Nonsinusoidal, Balanced, or Unbalanced Conditions*, IEEE Power & Energy Society, IEEE, 3 Park Avenue, New York, USA, March 2010.

[175] Y. Hassan and A. Wittneben, "Adaptive uncoupled matching network design for compact MIMO systems with MMSE receiver," in *IEEE Wireless Communications and Networking Conference (WCNC)*, 2015.

[176] H. Nyquist, "Thermal agitation of electric charge in conductors," *Physical review*, vol. 32, no. 1, p. 110, 1928.

[177] R. Twiss, "Nyquist's and Thevenin's theorems generalized for nonreciprocal linear networks," *Journal of Applied Physics*, vol. 26, no. 5, pp. 599–602, 1955.

[178] *Recommendation ITU-R P.372-13 Radio noise*, ITU-R, Geneva, Sep 2016, available at www.itu.int/rec/R-REC-P.372.

[179] J. Dostal, *Operational Amplifiers*. Butterworth-Heinemann, 1993.

[180] T. M. Cover and J. A. Thomas, *Elements of Information Theory*. John Wiley & Sons, 2006.

[181] C. Berrou, A. Glavieux, and P. Thitimajshima, "Near shannon limit error-correcting coding and decoding: Turbo-codes. 1," in *IEEE International Conference on Communications (ICC)*, vol. 2, 1993, pp. 1064–1070.

[182] T. J. Richardson, M. A. Shokrollahi, and R. L. Urbanke, "Design of capacity-approaching irregular low-density parity-check codes," *IEEE Transactions on Information Theory*, vol. 47, no. 2, pp. 619–637, 2001.

[183] E. Arikan, "Channel polarization: A method for constructing capacity-achieving codes," in *IEEE International Symposium on Information Theory (ISIT)*, 2008, pp. 1173–1177.

[184] G. J. Foschini and M. J. Gans, "On limits of wireless communications in a fading environment when using multiple antennas," *Wireless personal communications*, vol. 6, no. 3, pp. 311–335, 1998.

[185] M. Vu, "MIMO capacity with per-antenna power constraint," in *IEEE Global Communications Conference (GLOBECOM)*, Dec 2011.

[186] S. Boyd and L. Vandenberghe, *Convex optimization*. Cambridge University Press, 2004.

[187] S. Goguri, R. Mudumbai, D. R. Brown, S. Dasgupta, and U. Madhow, "Capacity maximization for distributed broadband beamforming," in *IEEE International Conference on Acoustics, Speech and Signal Processing (ICASSP)*, 2016, pp. 3441–3445.

[188] D. Nie, B. M. Hochwald, and E. Stauffer, "Systematic design of large-scale multiport decoupling networks," *IEEE Transactions on Circuits and Systems I: Regular Papers*, vol. 61, no. 7, pp. 2172–2181, 2014.

[189] S. Shen and R. D. Murch, "Impedance matching for compact multiple antenna systems in random RF fields," *IEEE Transactions on Antennas and Propagation*, vol. 64, no. 2, pp. 820–825, 2016.

[190] R.-F. Xue, K.-W. Cheng, and M. Je, "High-efficiency wireless power transfer for biomedical implants by optimal resonant load transformation," *IEEE Transactions on Circuits and Systems I: Regular Papers*, vol. 60, no. 4, pp. 867–874, 2013.

[191] M. L. Morris and M. A. Jensen, "Network model for MIMO systems with coupled antennas and noisy amplifiers," *IEEE Transactions on Antennas and Propagation*, vol. 53, no. 1, pp. 545–552, 2005.

[192] S. Lu, H. Hui, and M. Bialkowski, "Optimizing MIMO channel capacities under the influence of antenna mutual coupling," *IEEE Antennas and Wireless Propagation Letters*, vol. 7, pp. 287–290, 2008.

[193] A. A. Abouda and S. Häggman, "Effect of mutual coupling on capacity of MIMO wireless channels in high SNR scenario," *Progress In Electromagnetics Research*, vol. 65, pp. 27–40, 2006.

[194] Y. Huang, A. Nehorai, and G. Friedman, "Mutual coupling of two collocated orthogonally oriented circular thin-wire loops," *IEEE Transactions on Antennas and Propagation*, vol. 51, no. 6, pp. 1307–1314, 2003.

[195] P.-D. Arapoglou, P. Burzigotti, M. Bertinelli, A. B. Alamanac, and R. De Gaudenzi, "To MIMO or not to MIMO in mobile satellite broadcasting systems," *IEEE Transactions on Wireless Communications*, vol. 10, no. 9, pp. 2807–2811, 2011.

[196] M. R. Andrews, P. P. Mitra *et al.*, "Tripling the capacity of wireless communications using electromagnetic polarization," *Nature*, vol. 409, no. 6818, pp. 316–318, 2001.

[197] L. Liu, W. Hong, H. Wang, G. Yang, N. Zhang, H. Zhao, J. Chang, C. Yu, X. Yu, H. Tang *et al.*, "Characterization of line-of-sight MIMO channel for fixed wireless communications," *IEEE Antennas and Wireless Propagation Letters*, vol. 6, pp. 36–39, 2007.

[198] J. J. Komo and L. L. Joiner, "Upper and lower bounds on the binary input AWGN channel capacity," *Communications in Applied Analysis*, vol. 10, no. 1, pp. 1–8, 2006.

[199] R. G. Vaughan, "Polarization diversity in mobile communications," *IEEE Transactions on Vehicular Technology*, vol. 39, no. 3, pp. 177–186, Aug 1990.

[200] R. U. Nabar, H. Bölcskei, V. Erceg, D. Gesbert, and A. J. Paulraj, "Performance of multiantenna signaling techniques in the presence of polarization diversity," *IEEE Transactions on Signal Processing*, vol. 50, no. 10, pp. 2553–2562, 2002.

[201] Archimedes, "On the sphere and cylinder," 225 BC.

[202] V. De Silva, "A generalisation of archimedes' hatbox theorem," *The Mathematical Gazette*, vol. 90, no. 517, pp. 132–134, 2006.

[203] F. Barthe, O. Guédon, S. Mendelson, A. Naor *et al.*, "A probabilistic approach to the geometry of the ℓ_p^n-ball," *The Annals of Probability*, vol. 33, no. 2, pp. 480–513, 2005.

[204] G. Merziger, G. Mühlbach, D. Wille, and T. Wirth, *Formeln + Hilfen zur höheren Mathematik*, 6th ed.   Binomi Verlag, 2010.

[205] H. A. Haus and W. Huang, "Coupled-mode theory," *Proceedings of the IEEE*, vol. 79, no. 10, pp. 1505–1518, 1991.

[206] M. Kiani and M. Ghovanloo, "The circuit theory behind coupled-mode magnetic resonance-based wireless power transmission," *IEEE Transactions on Circuits and Systems I: Regular Papers*, vol. 59, no. 9, pp. 2065–2074, 2012.

[207] A. Karalis, J. D. Joannopoulos, and M. Soljacic, "Efficient wireless non-radiative mid-range energy transfer," *Annals of Physics*, vol. 323, no. 1, pp. 34–48, 2008.

[208] L. Davis, *Handbook of genetic algorithms.*   Van Nostrand Reinhold, 1991.

[209] A. S. Poon, S. O'Driscoll, and T. H. Meng, "Optimal frequency for wireless power transmission into dispersive tissue," *IEEE Transactions on Antennas and Propagation*, vol. 58, no. 5, pp. 1739–1750, 2010.

[210] J. S. Ho, A. J. Yeh, E. Neofytou, S. Kim, Y. Tanabe, B. Patlolla, R. E. Beygui, and A. S. Poon, "Wireless power transfer to deep-tissue microimplants," *Proceedings of the National Academy of Sciences*, vol. 111, no. 22, pp. 7974–7979, 2014.

[211] S. Gabriel, R. Lau, and C. Gabriel, "The dielectric properties of biological tissues: III. parametric models for the dielectric spectrum of tissues," *Physics in Medicine & Biology*, vol. 41, no. 11, p. 2271, 1996.

[212] X. Zou, X. Xu, L. Yao, and Y. Lian, "A 1-V 450-nW fully integrated programmable biomedical sensor interface chip," *IEEE Journal of Solid-State Circuits*, vol. 44, no. 4, pp. 1067–1077, 2009.

[213] P. D. Teal, T. D. Abhayapala, and R. A. Kennedy, "Spatial correlation for general distributions of scatterers," *IEEE Signal Processing Letters*, vol. 9, no. 10, pp. 305–308, 2002.

[214] S. Gezici, Z. Tian, G. B. Giannakis, H. Kobayashi, A. F. Molisch, H. V. Poor, and Z. Sahinoglu, "Localization via ultra-wideband radios: a look at positioning aspects for future sensor networks," *IEEE Signal Processing Magazine*, vol. 22, no. 4, pp. 70–84, 2005.

[215] H. Xie, T. Gu, X. Tao, H. Ye, and J. Lu, "A reliability-augmented particle filter for magnetic fingerprinting based indoor localization on smartphone," *IEEE Transactions on Mobile Computing*, vol. 15, no. 8, pp. 1877–1892, 2015.

[216] A. Markham, N. Trigoni, D. W. Macdonald, and S. A. Ellwood, "Underground localization in 3-d using magneto-inductive tracking," *IEEE Sensors Journal*, vol. 12, no. 6, pp. 1809–1816, 2012.

[217] A. Sheinker, B. Ginzburg, N. Salomonski, L. Frumkis, and B.-Z. Kaplan, "Localization in 3-D using beacons of low frequency magnetic field," *IEEE Transactions on Instrumentation and Measurement*, vol. 62, no. 12, pp. 3194–3201, 2013.

[218] H. Godrich, A. M. Haimovich, and R. S. Blum, "Target localization accuracy gain in MIMO radar-based systems," *IEEE Transactions on Information Theory*, vol. 56, no. 6, pp. 2783–2803, 2010.

[219] D. B. Jourdan, D. Dardari, and M. Z. Win, "Position error bound for UWB localization in dense cluttered environments," *IEEE Transactions on Aerospace and Electronic Systems*, vol. 44, no. 2, pp. 613–628, 2008.

[220] H. G. Schantz, "A real-time location system using near-field electromagnetic ranging," in *2007 IEEE Antennas and Propagation Society International Symposium*. IEEE, 2007, pp. 3792–3795.

[221] A. Conti, S. Mazuelas, S. Bartoletti, W. C. Lindsey, and M. Z. Win, "Soft information for localization-of-things," *Proceedings of the IEEE*, 2019.

[222] D. Dardari, C.-C. Chong, and M. Z. Win, "Improved lower bounds on time-of-arrival estimation error in realistic UWB channels," in *2006 IEEE International Conference on Ultra-Wideband*. IEEE, 2006, pp. 531–537.

[223] S. Kay, *Fundamentals of Statistical Signal Processing: Estimation Theory.* Prentice Hall Signal Processing Series, 1993.

[224] P. Stoica and B. C. Ng, "On the Cramér-Rao bound under parametric constraints," *IEEE Signal Processing Letters*, vol. 5, no. 7, pp. 177–179, 1998.

[225] C. Hu, S. Song, X. Wang, M. Q.-H. Meng, and B. Li, "A novel positioning and orientation system based on three-axis magnetic coils," *IEEE Transactions on Magnetics*, vol. 48, no. 7, pp. 2211–2219, 2012.

[226] S. Song, B. Li, W. Qiao, C. Hu, H. Ren, H. Yu, Q. Zhang, M. Q.-H. Meng, and G. Xu, "6-d magnetic localization and orientation method for an annular magnet based on a closed-form analytical model," *IEEE Transactions on Magnetics*, vol. 50, no. 9, pp. 1–11, 2014.

[227] *Matlab Optimization Toolbox - Choosing the Algorithm*, MathWorks, MathWorks, 2019, online: https://www.mathworks.com/help/optim/ug/choosing-the-algorithm.html.

[228] T. F. Coleman and Y. Li, "An interior trust region approach for nonlinear minimization subject to bounds," *SIAM Journal on optimization*, vol. 6, no. 2, pp. 418–445, 1996.

[229] J. Dattorro, *Convex optimization & Euclidean distance geometry.* Lulu, 2010.

[230] W. Gander, "Least squares with a quadratic constraint," *Numerische Mathematik*, vol. 36, no. 3, pp. 291–307, 1980.

[231] I. Csiszár and G. Tusnády, "Information geometry and alternating minimization procedures," *Statistics and Decisions*, 1984.

[232] H. Schulten and A. Wittneben, "Magneto-inductive localization: Fundamentals of passive relaying and load switching," in *IEEE International Conference on Communications (ICC)*, Jun. 2020.

[233] A. F. Molisch, "Ultra-wide-band propagation channels," *Proceedings of the IEEE*, vol. 97, no. 2, pp. 353–371, 2009.

[234] M. O. Gani, G. M. T. Ahsan, D. Do, W. Drew, M. Balfas, S. I. Ahamed, M. Arif, and A. J. Kattan, "An approach to localization in crowded area," in *IEEE International Conference on e-Health Networking, Applications and Services (Healthcom)*, 2016.

[235] I. Dokmanić, R. Parhizkar, A. Walther, Y. M. Lu, and M. Vetterli, "Acoustic echoes reveal room shape," *Proceedings of the National Academy of Sciences*, vol. 110, no. 30, pp. 12 186–12 191, 2013.

[236] N. Ono, H. Kohno, N. Ito, and S. Sagayama, "Blind alignment of asynchronously recorded signals for distributed microphone array," in *Workshop on Applications of Signal Processing to Audio and Acoustics*, 2009, pp. 161–164.

[237] H. A. David and H. N. Nagaraja, *Order statistics.* Wiley, 1970.

[238] Y. Qi, "Wireless geolocation in a non-line-of-sight environment," Ph.D. dissertation, Princeton University, 2003.

[239] Y. Chen, "Evaluating off-the-shelf hardware for indoor positioning," Master's thesis, Lund University, 2017.

[240] C. Steiner and A. Wittneben, "Low complexity location fingerprinting with generalized UWB energy detection receivers," *IEEE Transactions on Signal Processing*, vol. 58, no. 3, pp. 1756–1767, 2010.

[241] E. W. Weisstein, "Spherical coordinates," From MathWorld – A Wolfram Web Resource. http://mathworld.wolfram.com/SphericalCoordinates.html.

[242] J. Rahola, "Power waves and conjugate matching," *IEEE Transactions on Circuits and Systems II: Express Briefs*, vol. 55, no. 1, 2008.

[243] P. Grover and A. Sahai, "Shannon meets Tesla: wireless information and power transfer," in *IEEE International Symposium on Information Theory (ISIT)*, 2010, pp. 2363–2367.

[244] *PN7120 Antenna Design and Matching Guide*, NXP Semiconductors, Apr. 2016, Application Note AN11564 Rev. 1.1, Online: https://www.nxp.com/docs/en/application-note/AN11564.pdf.

[245] *EMVco reference design*, austriamicrosystems AG, Apr. 2012, Application Note AS3911 Rev. 0V6, Online: https://www.all-electronics.de/wp-content/uploads/migrated/article-pdf/119201/ams-456ag1012-app9-pdf.pdf.

[246] G. Merziger and T. Wirth, *Repetitorium der höheren Mathematik*, 5th ed. Binomi Verlag, 2006.

Gregor Dumphart – Curriculum Vitae

Education

Nov 2014 – Jan 2020	**PhD Studies: Electrical Engineering** Communication Technology Laboratory, D-ITET, ETH Zürich
Feb 2011 – Apr 2014	**Master Studies in** **Information and Computer Engineering** Digital Signal Processing & System-on-Chip Design, Graz University of Technology
Oct 2007 – Jan 2011	**Bachelor Studies in** **Information and Computer Engineering** Graz University of Technology
Sep 2001 – Jun 2006	**Federal Technical College** Industrial Engineering, HTL BULME Graz

Professional Experience

Okt 2014 –	**ETH Zürich** Research Assistant, Postdoc
Okt 2009 – Jul 2013	**Graz University of Technology** Teaching assistant
Aug 2009 – Sep 2009	**NextSense Mess- und Prüfsysteme GmbH**
Apr 2007 – Jun 2007	**Joachim Maderer Web Information Systems**
Jul 2006 – Jan 2007	**Austrian Federal Armed Forces (Bundesheer)**
Jul 2004 – Aug 2004	**AVL List GmbH**
Jul 2003 – Aug 2003	**Thyssen Krupp Elevator**

Bisher erschienene Bände der Reihe
Series in Wireless Communications

ISSN 1611-2970

Alle erschienenen Bücher können unter der angegebenen ISBN-Nummer direkt online
(http://www.logos-verlag.de) oder per Fax (030 - 42 85 10 92) beim Logos Verlag
Berlin bestellt werden.